DRIFT
STATION

ALSO BY WILLIAM F. ALTHOFF

Sky Ships:
A History of the Airship in the United States Navy

Arctic Mission:
By Airship and Submarine to the Far North (2nd ed.)

USS Los Angeles:
The Navy's Venerable Airship and Aviation Technology

DRIFT STATION

Arctic Outposts of Superpower Science

William F. Althoff

Potomac Books, Inc.
Washington, D.C.

Copyright © 2007 by William F. Althoff

Published in the United States by Potomac Books, Inc. All rights reserved. No part of this book may be reproduced in any manner whatsoever without written permission from the publisher, except in the case of brief quotations embodied in critical articles and reviews.

Library of Congress Cataloging-in-Publication Data

Althoff, William F.
  Drift station : Arctic outposts of superpower science / William F. Althoff. – 1st ed.
     p. cm.
  Includes bibliographical references and index.
  ISBN 1-57488-771-8 (hardcover : alk. paper)
  1. Arctic regions–Discovery and exploration–Russian. 2. Arctic regions–Discovery and exploration–American. 3. Drifting ice stations–Arctic Ocean. I. Title.
  G630.R8A48 2006
  910.9163'2–dc22

                                                                      2006014143

ISBN-10: 1-57488-771-8
ISBN-13: 978-1-57488-771-6

Printed in the United States of America on acid-free paper that meets the American National Standards Institute Z39-48 Standard.

Potomac Books, Inc.
22841 Quicksilver Drive
Dulles, Virginia 20166

First Edition

10 9 8 7 6 5 4 3 2 1

CONTENTS

List of Illustrations — vii
Preface — ix
Acknowledgments — xi

CHAPTER ONE
  The White Desert — 1
CHAPTER TWO
  Red Star: "North Pole-1" — 33
CHAPTER THREE
  Fletcher's Ice Island, T-3 — 65
CHAPTER FOUR
  International Geophysical Year — 109
CHAPTER FIVE
  ARLIS, Acoustics, AIDJEX, and SPs — 153
CHAPTER SIX
  Global Change: Advancing the Case — 209

Epilogue — 255

APPENDIX 1
  Arctic Ocean Drifting Stations Since the Ice Ship *Fram* Expedition of 1893–96, 1937–2004 — 265

APPENDIX 2
    Civilian Scientific Investigators, U.S. IGY Drifting
    Station ALPHA, June 1957–November 1958   269

APPENDIX 3
    Primarily Acoustics U.S. Drifting Stations, Arctic
    Ocean, 1962–96   271

APPENDIX 4
    Research Projects Conducted on ARLIS II, May
    1961–May 1965   275

APPENDIX 5
    Polyarnik Personnel at *Severnyy Polyus-22*,
    Fourth Shift of the Drifting Station, 15 April
    1976 to 16 April 1977   277

APPENDIX 6
    Ice-Island Regulations, Polar Continental Shelf
    Project, Government of Canada, 1990   279

APPENDIX 7
    Arctic Research Budgets, Individual U.S. Federal
    Agencies, 2000–2004 (In Millions of Dollars)   283

APPENDIX 8
    Plotted Tracks of Selected Drift Stations   285

Notes   289
Glossary   335
Selected Bibliography   337
Index   349
About the Author   357

ILLUSTRATIONS

Overview map of the Arctic Ocean — Frontispiece

PHOTOGRAPHS
The Greenland Sea in April — 2
The ice ship *Fram* during its epic three-year drift — 9
The ice-strengthened freighter SS *Chelyuskin* founders in the Chukchi Sea, off northeastern Siberia, 13 February 1934 — 22
A twin-engine ANT-4 reaches Shmidt Camp, 5 March 1934 — 30
The four-plane squadron of ski-equipped ANT-6s rendezvous on sea ice in the central Arctic, May 1937 — 44
Dr. Otto Shmidt and Ivan Papanin watch the descent of an ANT-6 transport — 45
In 1950, *Severnyy Polyus-2* was deployed off eastern Siberia — 72
A navy party augers a "hydrohole" during SKI-JUMP II — 78
Sea ice offers a natural yet dangerous runway — 79
The Ward Hunt Ice Shelf, Canadian High Arctic — 81
Tents and Jamesway hut on T-3 — 88
Spring replacing winter at SP-2 and SP-3, 1954 — 100
Severnyy Polyus-4 — 101
A *polyarnik* braves the dark season at SP-6 — 113
A C-124 on the runway at ALPHA — 115
Geophysics at ALPHA–part of USL's study of under-ice phenomena — 118
Biologists prepare to explore at ALPHA, July 1958 — 119
T-3 camp trailers at summer's end, 1958 — 123

vii

Skate (SSN 578) at ALPHA camp, 14 August 1958, following her run out from the pole	136
Arctic Research Laboratory Cessna 180s en route to ice selected for a seismic refraction shot, spring 1960	150
The nuclear icebreaker *Lenin* standing by station SP-10, 14–17 October 1961	163
A pioneer team coring to confirm floe thickness prior to setting camp	164
Ski landing gear of a Lisunov Li-2, SP-10	165
The T-3 encampment, summer 1960	166
A typical hut of the ARLIS series	178
Nascent camp for AIDJEX, its first C-130 on the ice off North Alaska	194
Fracturing threatens the main station floe, AIDJEX, 1975.	198
The APLIS 86 ice camp, March–May 1986, off North Alaska	201
Hydrographic station airlifted from *Fram I*, April 1979	206
Geologists examine a bottom core, Canada's LOREX, April 1979	211
Divers and photographic equipment beneath LOREX ice, April 1979	212
Welcome to *Severnyy Polyus-25*, May 1983	216
The ice island occupied by Canada, 13 August 1989	238
Ice Station SHEBA, April 1998	250
View from ship's bridge at SHEBA, April and August 1998	254

TABLES

Food, Fuel, and Equipment for the Soviet "North Pole" Expedition of 1937–38	46
Airlift Support for IGY Drifting Station "A," or ALPHA, 1957–1958	121
Air Temperature Data for the Anomalous June–September 1958 Ablation Season, IGY Drifting Station BRAVO (Ice Island T-3)	122
Rate of Annual Expenditure of U.S. Federal Funds for Arctic Activities, 1968	183
Charter Aircraft Rates for Logistical Support, Arctic Canada (March 1986)	225
Occupancy at Canada's Ice Island Drifting Station, 1987 Field Season	229
Operations Overview for the CEAREX Drift Experiment, 1988, Eastern Arctic Ocean–Eurasian Basin	234

PREFACE

THIS IS A HISTORY OF ICE STATIONS from which scientific research has been conducted in the Arctic Ocean—a punishing work environment a world apart.

What one needs to penetrate an ocean are *platforms*. In a normal open sea, an oceanographic research vessel serves well enough. Until the 1970s, however, surface ships could not reach the central arctic for survey or basic research. Though the arctic submarine was demonstrable fact with *Nautilus* (SSN 571) and *Skate* (SSN 578), the navy's under-ice missions were infrequent, short in duration, and classified—prime data denied the open literature. An alternative was exploitation of the natural canopy itself, installing ice-rafted camps by airplane and then supporting them by air.

Explorer-scientist Fridtjof Nansen (1861–1930) all but invented arctic oceanography. Still, modern drifting stations are a Soviet logistic invention. In May 1937, the progenitor camp was air-deployed onto pack ice mere miles off the geographic pole—a feat of maritime daring. The *Severnyy Polyus* ("North Pole") camp survived a nine-month, 1,273-mile (2,050-kilometer) transit via the East Greenland Current. Its floe disintegrating, the party was retrieved by ship in the ice-rich waters of the Greenland Sea.

The first aircraft-supported ice camp for the United States was deployed by the air force in March 1952 to Fletcher's Ice Island, T-3. As well, two pack-ice stations—ALPHA and CHARLIE—were established in support of the International Geophysical Year of 1957–58 and follow-on field research (1959). "Those were *rich* sources of information," an American geophysicist recalls, "because almost anything you discovered up there was new at the time." A succession of ice-borne bivouacs ensued, deployed by both superpowers and by Canada, to help advance Cold War science and research in the basin.

Climate-model outputs are strongly dependent on how sea ice is incorporated within them, so twenty-first-century science still looks poleward—from instrumented outposts ashore and afloat, from the air, from space. Ships and aircraft sortie, space-tracked buoys drift, remote-sensing satellites orbit and image. U.S. Arctic policy is concerned particularly with understanding the region's role in global environmental processes. Multinational programs assail the obstinate issues attending global ecological trends and the implications of climate changes for humankind.

Danger is companion to ice-based missions and programs: intense cold, months-long night, the hazards of airlift resupply. Heavy equipment, bears, firearms, and explosives boost the probability of emergencies. As for fracture and breakup, the threat of losing one's foundation is unceasing.

This is a story without a full written history—until this book.

ACKNOWLEDGMENTS

FEW, IF ANY, HISTORIES ARE RESEARCHED in isolation; certainly, this project was not. I am indebted to a host of agencies, institutions, and individuals from Russia, Canada, and the United States. I am deeply grateful to Sergey Priamikov. His courtesies helped me gain entrée into the renowned Arctic and Antarctic Research Institute, St. Petersburg (formerly Leningrad), and interviews with its *polyarniks* (polar researchers). Thanks to Nickolai Yagodnitsin, director of Leningrad's Arktika Musee, and Senior Scientist Ilya P. Romanov, additional materials were located–and a delightful tour granted. Throughout my stay, Tatiana Kalliazina served ably–and charmingly–as translator. The 1992 invitation to the former Soviet Union was of inestimable value.

The National Archives of Canada and the National Library of Canada are rich sources of information on the circumpolar Arctic and on the Canadian experience in that extreme realm. In Ottawa, Brig. Gen. Keith R. Greenaway, Royal Canadian Air Force (Ret.) and his wife, Hazel, hosted this researcher and granted fruitful hours in the general's own library. Selected files at the Polar Continental Shelf Project, an agency within Energy, Mines, and Resources, Canada, were made available through former director George D. Hobson as well as by director Bonnie Hrycyk.

Tom B. Kilpatrick, chief, Ice Reconnaissance Branch, Environment Canada, approved my accompanying, in 1989, three surveillance missions over Eastern Arctic Canada. Thanks to George Hobson, director, at Polar Shelf in 1990, the writer visited Ottawa's ice-island science base in the eastern Arctic Ocean. And it was courtesy of Imants J. Versnieks, Arctic Systems Undersea Warfare Systems

xi

Command, and the 109th Airlift Wing, New York Air National Guard, that I was flown via C-130 to Thule Air Base in northwest Greenland and from thence offshore to an ice-borne U.S. Navy camp (Ruby) in the Greenland Sea.

As for participants in the events described, I am exceedingly fortunate: the roster is a long one. Beaumont M. "Beau" Buck and Robert E. Francois provided an indispensable appreciation of the U.S. Navy's research programs in the Arctic Ocean relative to underwater acoustics. The private papers and reminiscences of these seasoned researchers were particularly useful, as were those made available by Dr. Gerald "Gerry" H. Cabaniss, George D. Hobson, Dr. Kenneth L. Hunkins, and Cdr. Edward M. Ward, U.S. Navy (Ret.).

One-on-one audio interviews were recorded with Dr. Maxwell L. Britton, Dr. Arthur Collin, Mr. Norman Goldstein, Dr. Zalman Gudkovich, Brig. Gen. Keith R. Greenaway, Royal Canadian Air Force (Ret.), George D. Hobson, Dr. Kenneth L. Hunkins, Valery S. Ippolitov (notes), Sergey V. Karpekin, Thomas Krochak, Dr. Waldo K. Lyon, David A. Maloney, Cdr. Ronald McGregor, U.S. Navy (Ret.), Wayne Parton, Dr. Albert A. Romanov, Dr. Ilya P. Romanov, Vadimir Sokolov (notes), Igor Tsigelnitsky, Nikolai D. Vinogradov, Imants J. Versnieks, and Nikolay Yaganitsin. (Institutional affiliations are included in the selected bibliography.)

Correspondence contributed much. Among those assisting were Dr. Terence Armstrong, Hazard C. Benedict, Beau Buck, Ms. Vivian C. Bushnell, Gerry Cabaniss, Robert D. Cotell, Cdr. Ronald Denk, U.S. Air Force (Ret.), Bob Francois, Andreas "Andy" Heiberg, Tatiana Kaliazina, Thomas B. Kilpatrick, Vice Adm. John H. Nicholson, U.S. Navy (Ret.), Christopher Pala, Dr. Donald Plouff, Dr. Charles Plummer, Juan G. Roederer, Michael Schmidt, Dr. Boris I. Silken, Capt. Brian Shoemaker, U.S. Navy (Ret.), Dr. Norbert Untersteiner, Dr. James A. Van Allen, Imants J. Versnieks, and Cdr. Edward M. Ward, U.S. Navy (Ret.).

For the book's images, Beau Buck, Gerry Cabaniss, R. E. G. "Ron" Davies (by Boris Vdovienko), George Hobson, John L. Schindler, and Michael Schmidt were especially helpful. Thanks to Hazard C. Benedict for images and information from the files of the Air Force's Cambridge Research Center became available.

James Bitterman (now with ABC News Paris) recovered videotape of his 1977 drop-in and Canadian Broadcasting Corporation (CBC) interview at "North Pole-22." The Honorable Barnett J. Danson, Canadian defense minister in 1977, offered reminiscences concerning that long-lived Soviet station. Dr. D. K. Perovich sent along both images and technical publications concerning the SHEBA drift experiment (1997–98). Thanks to Dr. Von Hardesty, I received a short-term visitor grant at the National Air and Space Museum, Smithsonian Institution. There I reviewed selected materials held by the museum's Aeronautics Division.

Portions of the draft manuscript were kindly reviewed by Beau Buck, Drs. Gerry Cabaniss and Von Hardesty, George Hobson, Adm. John Nicholson, and

John Schindler. Each offered helpful suggestions and corrections. Librarians Dorothy McLaughlan and Mary Kearns-Kaplan were simply indispensable.

My oral-history research was supported in part by generous grants from the New Jersey Historical Commission.

Penny—again—was understanding and supportive.

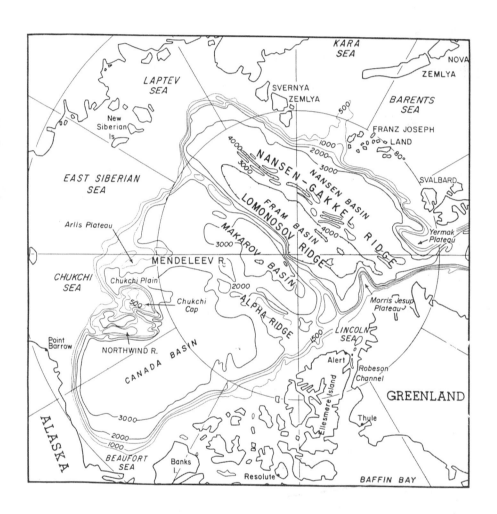

Chapter One

THE WHITE DESERT

We cannot command nature except by obeying her.
—Francis Bacon (1561–1626)

Man belongs wherever he wants to go—and he'll
do plenty well when he gets there.
—Wernher von Braun (1912–1977)

The world draws together at the poles. Cruel, vast, faraway, polar wastes both estrange and inspire. Geoscientists are shameless enthusiasts for high latitudes; certain phenomena, they know, are centered upon the circumpolar zones.

Earth is largely a marine habitat. The extreme north has a unique environment—an icy, maritime cap to the blue planet.[1] Because of its inaccessibility, the Arctic Ocean is relatively unknown. It fills an immense crustal basin. One of its singular features is a seemingly uninhabitable, impassable canopy of pack ice. It is the only ocean that at present supports a permanent sea-ice cover. Perennial ice cloaks all the central Arctic Ocean in winter, with open water developing off the surrounding landmasses during summer.[2] Bewildering and trackless, the interior pack is a shifting barrier centered (roughly) on the geographic pole. At that place, a dozen or so feet of ice float upon two and a half miles of blue-black sea.

An ocean mirrors atmospheric energy. For Arctic waters, two liquids—air and water—are coupled by a solid yet deformable plate. Remote and complex, veneered and climate-altering,

the Arctic Ocean is the last frontier in oceanography. Important issues are

Unlike Antarctica, a continent, the central Arctic is maritime—morphologically, an oceanic basin. This is the Greenland Sea in April. Unlike a normal open ocean, here floats a near-continuous canopy of snow-draped floes, broken by occasional leads. A coupled system of air, ice, and seawater, the Arctic Ocean canopy is a sensitive expression of environmental change during recent geologic time. (Author)

related to the global carbon cycle, freshwater balance, circulation, heating, transport of sediments and pollutants, and spreading of the sea floor—as well as the volume, flow, and properties of sea ice.[3]

—•—•—•—•—•—

Arctic research has always been data-limited. The tropics collect most solar radiation and are central to Earth's heat and water-vapor budgets. As well, the cryosphere (sea ice, permafrost, snow cover, glaciers) has impacts on global systems of climate, weather, and oceanic circulation. Here ecosystems are dominated by sea ice. The oceans store as well as convey heat and (plainly) are reservoirs of moisture. Because of their high albedo (reflectivity), snow and ice influence the global heat budget. Understanding climate has granted great cachet to polar research programs. However, we are changing the earth faster than we are learning about it. No ecosystem is free of human influence. Global change, predictive (computer) models show, is a strong function of sea-ice change in the Arctic. Accordingly, paleocirculation and paleoclimate (fossil climate) records along with long-term observations of coverage and ice thickness are vital to assessing human-induced (or greenhouse) warming.

The Arctic Ocean, like the Mediterranean, is largely landlocked; it has but one deepwater connection to subpolar seas. (Shallow subsidiary export occurs through the interisland channels of the Canadian Arctic Archipelago.) Basin hydrodynamics are fundamental. Why? The main deepwater masses in the global ocean originate in high-latitude waters, namely, the high North Atlantic and in the ocean basins surrounding Antarctica. Cold, dense and oxygen-rich, these ventilate the abyssal world ocean and provide a mechanism for planetary-scale heat transfer as well as the exchange of salt, water, nutrients, and carbon.[4] Salinity is an operative variable because the density of the upper ocean—thus its stability—depends upon temperature and salinity. On the North Pacific side of the Arctic Ocean, inflow sweeps through the Bering Strait where, in summer, warm southern winds and currents push north. Rendered passable early in June, the slot is free of ice until mid-October.

Certain forces are immutable. The borders of the sea-ice disc pulse in seasonal rhythm; in the average cycle of freezing and advance, coverage approximately doubles by close of winter.[5] The configuration of the fringing coasts—Alaska, Greenland, Spitsbergen (Svalbard), northern Scandinavia, Russia—greatly influences the distension. Riverine inflow also affects the canopy as well as hydrology and sedimentation. At its maximum, in February–March, sea ice covers about 10 percent of the total sea surface. Extensive perimeter sheets form as well, accreting into regional basins.[6] When this frozen world begins to melt, the ice edge retreats, decaying to its smallest area in September.

Seen from its surface, the dynamic nature of sea ice is hard to appreciate. The floating white is restive, a field of contending forces. Moving in more or less predictable directions, the ice is driven mainly by the prevailing winds assisted by tides and currents. These keep the canopy in almost continual motion, causing floes to diverge, converge, shear, and crush, thus maintaining the canopy's broken character. Shoved in distinct directions at different rates, floes crack, ridge, and rubble—a slow-motion chaos that can be appalling. "If you happen to be camped on the weaker one," explorer Vilhjalmur Stefansson wrote of a ridging event, "it behooves you to move quickly. Pieces of your floe the size of a city lot will rise on edge, tower up and crumble toward you. The ice around camp and under it will begin to groan and buckle and bend. Where it bends down, little rivers of sea water come rushing in; where it buckles up, small pressure ridges form."[7]

Surface travel here is horrendous—a challenge to sledge parties and to naval operations, formidable even for ice-capable shipping. Pack ice, in short, dictates where man can and cannot venture in the interior ocean. The first to sledge into its permanent canopy was Sir Edward Parry, who gained a record northing of 82°

45′ in 1827. Parry's chief difficulty was a perverse drift that effectively revoked his northing. Large segments of this realm retained the label "Unexplored"—most notably the blank between Alaska and the pole. The northernmost Arctic stood inaccessible, guarded by ice too thick for ships and—until Nansen (in 1895), then Peary (1909)—too distant for sledging.

A howling, implacable world, this is a habitat not to be taken lightly—in the words of one nineteenth-century explorer, a "kingdom of death." It demands humility.

Native hunter-gatherers, their survival skills superb, have peopled the northlands for countless generations. In the busier cosmos of western Europe, the geography of that rim world was but vaguely known. Indeed, save for the wide-ranging seagoings of the Norse, less than half a millennium embraces European probings into that drear, uncolored realm. An epic tale of greed, glory, and exploration, it is at once enthralling and disturbing.

—·—·—·—·—·—

The Vikings were the first polar voyagers. Traversing the open, unknown seas, small fleets or single boats navigated through the world as it was known between AD 800 and 1200, reaching (and settling) Iceland, then coastal Greenland, before crossing Davis Strait to Arctic Canada until, at the last, Viking craft gained the sea fringe of Newfoundland. In a more recent era of exploration, in ships of wood, men again penetrated the far world to fight the floes—a shifting curse that might enclose abruptly and entrap absolutely. Trusting to skill, courage, and fate, the first (sixteenth-century) Europeans pushed in. The savagings they endured are legend. The pack follows its own course: beset one day, vessels might be proffered passage the next. For the incautious, drift ice was fatal. The object of these ventures was the Northwest Passage, a navigable seaway from a covetous Europe to the "spice islands." But by the mid-nineteenth century, the essential motive had changed. The new grail was prestige—to be first to gain the pole. A prodigious appetite for adventure and glory pulled men poleward. At different times during this competition, the record "farthest north" was claimed by British, American, Norwegian, and Italian explorers. Not until 1909 did men tread all the way to 90° north latitude: Robert Peary and his party effectively ended this high-latitude steeplechase.[8]

Eight nations share the maritime Arctic—among the most punishing waters on the planet. Canada is bordered by three oceans and has the longest coastline in the world. Opposite North America, arcing as one landmass across the top of the Old World, is Arctic Russia—by far, the dominant geographic entity. A country with strong historical and strategic interests in the north, the Russian republic is unquestionably a maritime nation with an enormous stake in the seas, particularly its north-facing littoral. As a people, Russians love untamed nature. Incubated by geography, they are obsessed with security and sovereignty. And long has this people's gaze been northward and eastward. *Sibir*, the Silent Land, is embedded deep in the "soul" of Russia. During the eleventh century, fisherman and peasants moved into the White Sea region and the Pechora Basin. During the sixteenth and seventeenth centuries, fur traders and cossacks made themselves heard and seen in the faraway district east of the Urals. Thereafter, between 1580 and 1640, motivated by the search for fur, Russians made a frenetic advance eastward, seizing territory from the Urals across the great arc of Asiatic Russia–Siberia–to the far Pacific.

Under the imperial czars, the interior of north Eurasia was explored almost exclusively by Russians or by persons in service to the czars. Peter the Great (1672–1725) was the first to realize the significance of Russia's north-facing coastline. Before his time, no one knew what lay east of the Kara Sea, off European Russia. Rumors of Siberian natives engaging in sea trade with Alaska had suggested a waterway separating Asia from America. A man of singular will, intelligent as well as curious, Peter was determined to modernize his realm. The "Great Northern Expedition" (in fact, a series of campaigns) was organized. It was launched in February 1725. The objectives were bold: to determine the relation of Asia to North America; to gather information concerning Russia's northern periphery; and to prepare a map of that terrain. These were the first properly equipped Russian expeditions into those desert wastes. Driven by Vitus Bering, an indomitable Dane in the service of the czar, seven detachments–about one thousand men over ten years–charted the whole of the Siberian littoral. The first comprehensive map of the Northern Asia territories resulted.

Bering had "discovered" America–from the west. Commencing in the 1740s, these explorations ranged as far as present-day Alaska and the arc of the Aleutians. The Russians, in short, were a persistent, pathfinding force. "And they did not merely travel along the coasts," Nansen writes, "but crossed the drift-ice itself to the New Siberian Islands, and even north of them."[9] Russia claimed Alaska in 1784, establishing the first European settlement and inaugurating the sea-otter fur trade. Fifteen years thereafter, in 1799, Russian emperor Paul I chartered the Russian–American Company to manage the empire's American holdings.

Whaling ships from the American colonies visited the eastern Arctic in the 1730s. But more than a century passed before an American-flag whaler, *Superior*, transected the strait named for Vitus Bering. Two years thereafter, in 1850, the United States logged its first attempt at serious polar exploration by contributing two vessels to the Franklin Sea Search, the Admiralty's hunt for the missing explorer Sir John Franklin and his naval expedition. In 1867, with the purchase of Alaska, the United States assumed the status of an Arctic sovereignty.

In 1878, preparations concluded for the American North Polar Expedition, commanded by Lt. George W. DeLong, a naval officer. The intent was to sail as far as possible up the coast of Wrangel Island, off northeastern Siberia, then sledge all the way to the geographic pole. (Wrangel was then thought to be a continent-sized landmass.) Newspaper baron James Gordon Bennett Jr. understood his times as well as the value of melodrama. As owner-heir of the *New York Herald*, the flamboyant Bennett had funded this iffy venture. (Four years earlier, he had dispatched Henry Stanley in search of the "lost" Livingstone.) DeLong's avenue of approach was unusual: the unpromising Bering Strait. His ship would exploit a current DeLong believed arced across the top of the world, a kind of arctic Gulf Stream that proffered, he maintained, a clear-water passage through. The plan's fantastical aspect notwithstanding, DeLong was a superb arctic commander, and no fool.

A protracted search for an appropriate vessel produced a three-master. Equipped with auxiliary steam engines, refitted, and rechristened *Jeanette* (for Bennett's sister), the vessel stood out from San Francisco in June 1879. Steaming boldly through the slot, *Jeanette* was steered northwest—the first American scientific vessel to work Siberian waters.

The expedition was beset by ice in the neighborhood of Wrangel Island—the very lip of the sea DeLong hoped to transit. On 6 September, a drift to the west-northwest commenced—a slow, unremitting agony that consumed twenty-one months. *Jeanette* foundered on 12 June 1881, near 77°15′ N, 154°59′ E—north of the New Siberian Islands. Half a thousand miles of desolation interposed to the mainland.[10]

DeLong had been conveyed less than six degrees nearer the pole. Marooned and unreachable, the castaways began sledging south toward unbroken water and, beyond that, the coastal belt of Asia. Theirs proved a trail to break the heart. Manhauling three boats, the men endured a grotesque sleepwalk through a broken, torturous icescape.[11] It is difficult to imagine their moil or to appreciate the strict regard for discipline that drove the march. Following weeks of unspeakable suffering, and free of the pack, DeLong and his crew set sail for shore. One boat was lost with all hands, but two managed to reach land. Stupid with exhaustion, desperate for calories, the survivors—DeLong with thirteen others—waded ashore onto the alluvial shallows of the Lena River delta. Stranded unknown miles from an unknown place, one by one his party succumbed. But the second group was saved. Seeking its (dead) commander and comrades, a search party found the pitiful camp. DeLong's journal (found with his body) ends on 30 October 1881, 140 days after the foundering. The ill-starred party was interred in a cairn above the frozen delta that had taken its every life. Twenty-three had perished—a dreadful failure, it has been said, though a magnificent one.

Human achievement is rife with serendipity. In 1884, flotsam unquestionably belonging to *Jeanette* was found on Greenland's southwest coast—more than 3000 miles (4830 kilometers) from the point of foundering. That autumn, Dr. Fridtjof Nansen—the first man to cross the Greenland ice cap, zoologist, artist, Nobel Peace Prize winner—happened upon an article by Professor Henrik Mohn, director of the Norwegian Meteorological Office. It appeared that the newfound oddments had been couriered from Siberian waters. If this were true, it meant that a westward-flowing transpolar current swept the top of the world, just as the luckless DeLong had believed. "It immediately occurred to me," the Norwegian recounts in *Farthest North*, his seminal work, "that here lay the route ready to hand. If a floe could drift right across this unknown region, that region might also be enlisted in the service of exploration—and my plan was laid."[12]

A man of insight and imagination, Nansen had long reflected on arctic oceanography. To his mind, a consistent pattern of circulation was plain. Among the clues was driftwood of Siberian origin found along Greenland's treeless coast and thick offshore ice that, evidently, had survived several melt seasons. The ice off Siberia, in contrast, was young and relatively thin. *Jeanette*'s relics tendered confirmation. "Putting all this together," Nansen observes, "we seem driven to the conclusion that a current flows at some point between the Pole and Franz Josef Land from the Siberian Arctic Sea to the east coast of Greenland."

An utterly original plan was propounded—a drifting research station. No glory grab, his would be the first scientific expedition of its era and arguably the most consequential of modern times. An unprecedented amount of information would be collected about the deep ocean of the Eurasian Basin. In an elegant performance, Nansen would find his concourse through the heart of the maritime Arctic.[13]

—•—•—•—•—•—

An idea is not enough; one needs to get it into the minds of others. "I propose," Nansen declared, reconsidering the prevailing dogma, "to have a ship built as small and as strong as possible—just big enough to contain supplies of coals and provisions for twelve men for five years. A ship of about 170 tons (gross) will probably suffice."[14] The man had ignited the scientific imagination of his time and was granted support and funding (notably from Norway). "I think I may say," one luminary remarked, "this is the most adventurous programme ever brought under the notice of the Royal Geographical Society." Doomsayers abounded. That imposing figure, Adolphus Greely, U.S. Army, a figure intimate with northern travail, dismissed Nansen's audacity as folly. "It strikes me," he wrote in 1891, "as almost incredible that the plan . . . by Dr. Nansen should receive encouragement or support."[15] In addition to suffering and death, the expedition, in Greely's view, was doomed to barren results. The officer-explorer was hardly alone in this; any vessel reckless enough to tempt the floes of the central basin, prophets insisted, would purchase sure destruction.

Scrupulously careful, Nansen had calculated his risks.

To succeed, the platform would have to be purpose-built. Nansen turned to Norwegian boatbuilder Colin Archer. The design had to satisfy twin objectives: a hull shape as smooth as possible (to frustrate embrace) yet solidly built to withstand the pressures of the ice. Further, the ship had to be small enough to maneuver readily among the floes, yet sufficiently large to equip and support a dozen men for up to five *years*. A wholly new design, Archer's hull had "a plump and rounded form." With its bow, stern, and keel rounded off, a grip for the ice would be denied. (This, in fact, proved to be the forerunner of present-day icebreaker hull forms.) Constructed of wood—as were all polar vessels of the time—the resulting platform was exceptionally strong, perhaps the strongest wooden ship ever fashioned. Her frames—of the finest Italian oak, well seasoned—were twenty-one inches square and spaced one inch apart. These spaces were filled with a sawdust-and-tar mixture to preserve watertightness even if the planking were penetrated. The planking itself consisted of two layers of oak—the inner three inches and the outer one four inches thick—plus an outer skin of greenheart six inches thick at the waterline. Bow and stern received extra protection in the form of iron chafing strips. The whole interior was shored with extra beams and knees, and the hull divided into three watertight compartments.

Wood is elastic. If distorted, Archer's hull would regain its shape after the battering subsided. Further, timber is an insulator. Without proper insulation, Nansen knew, moisture in the living spaces would condense and then drip from overhead and freeze on bulkheads—a threat to morale. To combat this effect, the ceiling was covered with felt, cork linoleum, and paneling to a thickness of sixteen inches.

In 1892, in Archer's shipyard at Rekkevik, the first vessel designed for science-centered sailing was christened *Fram* (forward). In June of the following year, expedition leader Nansen and his team stood out from Norway. All hands—thirteen men—had prepared themselves to spend three, four, or even five years in the ice.

Fram pushed east, across the rim of Russia. Off the New Siberian Islands, the ship was driven in. Ironically, her crew was at pains to accomplish what sailors before had been powerless to resist. The ice pack failed clutch. On 22 September 1893, with autumn well advanced, the men secured to a floe and the hull allowed to freeze in. Their point of seizure was not far from where *Jeanette* had succumbed. No release was expected until the other side of the pole. Accordingly, the rudder was hauled up and the engine dismantled, oiled, and laid away. Three years were to pass before *Fram*, thoroughly disabled now, would again be under power. On deck, a windmill (to drive a "dynamo") was put up and various workshops—joiner's, mechanical, tinsmith's, shoemaker's, sailmaker's—laid out for repairs and the making of parts. Of cardinal import was the syllabus of work—the program of tenacious, direct observation into the theories, facts, and mysteries of the maritime Arctic.

An ocean lay undiscovered before them; working the frontiers of knowledge, ship's company—a highly organized island population in an unstudied waste—soon was rapt with discovery. From stations atop the pack, regular weather and magnetic observations commenced. Soundings embraced temperature as well as salinity at varying depths; indeed, the expedition was the first to record the layered nature of the water column. As well, observations of the formation of the ice and its growth and thickness were noted, along with currents. Spot soundings and dredgings also would be logged and data on astronomical, aurorae, biology, and other phenomena recorded.

Soundings promptly established abyssal ocean—more than 13,000 feet (nearly 4000 meters). Nansen was not prepared for such depths. A constricted, shallow basin had been assumed whose flow, he had calculated, drove a strong transpolar current. Yet the numbers couldn't have been clearer. The party was traversing a

BEFORE EARTH-ORBITING SATELLITES, ONE HAD TO GO TO SEA TO STUDY THE OCEAN. THE PREDECESSORS OF MODERN DRIFTING STATIONS WERE THE STRENGTHENED SHIPS OF EARLIER EXPLORERS, FROZEN IN AND ALLOWED TO DRIFT. THIS IS THE ICE SHIP *FRAM* DURING ITS EPIC THREE-YEAR DRIFT (1893–96)—PERHAPS THE MOST PRODUCTIVE ARCTIC EXPEDITION EVER MOUNTED. (GEORGE D. HOBSON)

covered sea averaging two miles in depth. Wind, not current, was controlling slow-motion drift. This was data of the rarest kind—the first oceanographic measurements for the eastern Arctic (Eurasian Basin)—an enormous, unpeopled blank untouched by science. First to measure the deep-ocean Arctic, the expedition also would detect discrete layering (of temperature, salinity) with depth. Imagine being the very first to plumb an ocean. "You could say," oceanographer Ned Ostenso observed, "that Nansen had discovered the Arctic Ocean."[16]

When near-surface stresses climb to a failure limit, large-scale cracking, lead formation, rafting, and ridging occur. On 9 October, the pack closed in. "All at once in the afternoon," Nansen said of the assault, "as we were sitting idly chatting, a deafening noise began, and the whole ship shook. This was the first ice pressure. Everyone rushed on deck to look. One pushed the ice with steady pressure, but down under us it had to go, and we were slowly lifted up. These squeezings continued off and on all afternoon and were sometimes so strong that the *Fram* was lifted several feet; but then the ice could no longer bear her, and she broke it below her."

Archer's design proved a masterwork. By virtue of the vessel's rounded form, ice pressure forced the craft to rise, eel-like, away from the nipping pack. Again, on 25 October: "We had horrible pressure last night. I awoke and felt the *Fram* being lifted, shaken, and tossed about, and heard the loud cracking of the ice breaking against her sides. After listening for a little while I fell asleep again, with a snug feeling that it was good to be on board the *Fram*." Instead of being caught and crushed, the platform would respond thus for every assault, a spectacle at once impressive and reassuring, Otto Sverdrup, ship's master and second in command, was to marvel.[17]

Though confident of their "little oasis," they could not relax their vigilance, and they kept their emergency plans. Sverdrup explains:

> We did not believe . . . that *Fram* ran the slightest risk of being crushed in any ice-pressure; but it was obviously possible, or at least *conceivable*, so that it was our duty to be prepared for all contingencies. Accordingly we devoted much labor and care to securing ourselves against being taken by surprise.[18]

The men believed in their work; all had accepted the risks and hardships and sacrifices of this uniquely isolated work. Motivation and morale, Nansen knew, were crucial. Along with a certain austerity, their quarantine tendered certain comforts: indeed, the crew was to savor a near-indolent, collegial ease. To dispel the awful desolation, Nansen had insisted on a comfortable salon within their sanctuary; there, together, ship's company would take its meals. And secluded niches had been provided throughout the vessel where, alone, a man could own his own time. In contrast to modern expeditions, the therapy of radio communications was denied them.

For the course of this odyssey, one day varied little from another. The crew turned out at eight, to breakfast on hard bread, cheeses, corned beef or corn mutton, luncheon ham or Chicago tinned tongue or bacon, cod-caviar, anchovy roe—in short, everything that could be reasonably desired. "Three times a week we had fresh-baked bread as well, and often cake of some kind. As for our beverages, we began by having coffee and chocolate [each] day . . . ; but afterwards had coffee only two days a week, tea two, and chocolate three." Obligations followed the meal, with men attending to the dogs or whatever task needed doing. All hands took a turn in the galley helping the cook to wash up, set table, and table-wait. "Some of us would take a turn on the floe to get some fresh air, and to examine the state of the ice, its pressure, etc. At one o'clock, we were assembled for dinner, which generally consisted of three courses. . . . I think we all agreed that the fare was good; it would hardly have been better at home; for some of us it would perhaps have been worse." After dinner, smokers gathered in the galley and then some took a short siesta. Each man then returned to his work until summoned to supper at six, when the regulation workday was done. The evening meal was almost the same as breakfast. And, "Afterwards there was again smoking in the galley, while the saloon was transformed into a silent reading room."

Minute by latitude-minute, *Fram* edged on—a meandering motion without apparent movement. Early in February 1894, months adrift, a plot of position touched the 80th parallel. (By aiming at the sun, moon, planets, and stars, via sextant, and knowing the azimuth and the exact time, one can reduce sightings to a calculated position line.) "Hurrah! Well sailed!" Nansen exulted. "I had offered to bet heavily that we had passed 80, but no one would take the bet. Dinner menue: Ox-tail soup, fish-pudding, potatoes, rissoles, green peas, haricot beans, cloudberries with milk, and a whole bottle of beer to each man. Coffee and a cigarette after dinner. Could one wish for more?"[19]

Predictably, wind and weather and the state of the ice, especially as they pertained to drift, were favored topics of conversation. According to the plots, their northing was not encouraging: a fitful and meandering track with many backsets. But the household's great enemy was a pernicious boredom, which could invade the most enthusiastic spirit. "A dismal, dispiriting landscape—nothing but white and gray," Nansen recorded that June—nine months adrift and a year since home. "No shadows—merely half-obliterated forms melting into the fog and slush. Everything is in a state of disintegration, and one's foothold gives way at every step."[20]

The men were tiring of grayish whiteness.

By the second winter—early 1895, it was clear that the platform, then at 84° (in the longitude of Franz Josef Land) would fall short of the pole by six or seven degrees. With his characteristic energy and restless impatience, Nansen elected to uphold the assault by other means. Equipped with three sledges, twenty-eight dogs, provisions for one hundred days, and two bamboo-and-canvas kayaks, Nansen and Lt. F. H. Johansen parted company with the rest of the expedition, taking their departure from *Fram* on 14 March.

Cut completely from *Fram*, the pair mushed a void of white and water and light to a high northing of 86°13′ (on 8 April)–a feat of colossal assurance. But famine and the drift were against them and compelled them to retreat. Pressing south, they killed off the weaker dogs one by one to feed the others. Patches of open water interposed. In mid-August 1895, concluding 400 miles (644 kilometers) of travel over mixed terrain atop a shifting labyrinth, Nansen and Johansen reached unknown shores of the Franz Josef group–land under their feet after two years. Supremely fatigued, the pair wintered miserably in a crude hut, hibernating in hermetic isolation until May 1896. For food, fuel, and clothing, they relied on polar bear and walrus. Then again, over water, snow, and ice, the two slogged south–chancing almost unbelievably to meet an English expedition under Frederick Jackson that was then working the islands. Though famous, the adventurers were barely recognizable–"wild men, clad in dirty rags, black with oil and soot, with long uncombed hair and shaggy beards, black with smoke."

Nansen was returned to Norway aboard Jackson's ship. Neither those aboard *Fram* nor Nansen yet knew the fate of the other.

The ice ship had caromed on. By summer 1895, all hands, shuttered away, were tantalized by escape. Any weather that spurred progress westward was hailed. "It must not be understood," Sverdrup, Nansen's closest advisor, reminds his readers, "that we dreaded another winter in the ice before getting home. We had provisions enough, and everything else needful to get over some two or three polar winters, if necessary, and we had a ship in which we placed the fullest confidence, in view of the many tests she had been put to." In large part, their mission was accomplished. Little prospect remained of further northward drift; whatever could be done to explore the regions to the north would be done by sledge. "It was our object, therefore, in compliance with the instructions of Dr. Nansen, to make for open water and home by the shortest way and in the safest manner, doing, however, everything within our power to carry home with us the best possible scientific results."[21]

The 22nd of September 1895 found a celebratory mood on board. "We had reason to be satisfied with the second year's drift," Sverdrup continues, "since we had advanced nearly double as far as during the first year, and, if this continued, there could scarcely be any doubt that we should get clear of the ice in the autumn of 1896."[22] That October *Fram* gained a farthest north: 85°57′. Through the dark boreal stage (her second "night"), the ice ship traced a ragged run south and west.

By spring 1896, with heavy thoughts of home, preparations for getting under way were effected. The engine was reassembled and reactivated. As crewmen crowded by, steam pressure was allowed to build, to test the apparatus. "Our gallant ship," Sverdrup wrote (20 May), "had awakened to renewed life after her long winter sleep, and we rejoiced to feel the first pulsations of her strongly beating heart. It seemed as if the *Fram* understood us, and wanted to say: 'Onward! Southward! Homeward!'"[23]

Weeks passed before the pack opened close aboard. Following a night of chopping ice and blasting wind, *Fram* gave a lurch–and floated free. The date: 3 June 1896, their third summer. Not until the 12th did the pack itself open. On the horizon, a dark sky signaled open water. Fighting for every mile, repeated rammings allowed only a faltering progress. On 13 August, after a month in close pack, *Fram* was steaming on various southerly headings, the ice giving way, ever slacker. A dark expanse was sighted. In less than an hour, the plucky ship steered through the final floes.

The adventurers–with a staggering hoard of observations–were back in Oslo on the 20th. Total sequestration: thirty-five months.

Nansen had all but invented modern-day arctic oceanography, a benchmark in exploration.[24] For one thing, Nansen's ideas on the relationship between wind and ice motion formed the basis of early mathematical treatment of oceanography. (Indeed, the response of ice to external loads is still today not fully understood.) Besides providing fundamental data like oceanographic measurements, the drift had proved fairly conclusively that the pole was occupied by moving ice. (Not until the 1940s was it established beyond reasonable doubt that no more land remained to be discovered.) A feat of grit, ingenuity, and imagination, the saga of *Fram* stands as the greatest of all drifts into the pack.

Thus was parted the curtain of mystery for that far sea. It was a hard act to follow, but Nansen urged others to continue his efforts. Marshaling personal resources, they gained magnificent northings. At first a symbol for the closing age of discovery, the North Pole became an end in itself. Most authorities agree that this international sweepstakes for one of the great geographical goals ended in 1909 with Peary's sledging to the place.[25] Abruptly, the prestige attending high-latitude expeditions dimmed. "Peary's splendid achievement in 1909 had destroyed the value of that prize for all later explorers," Roald Amundsen, a Norwegian inspired by Nansen's triumph, writes. The engine of northern exploration slowed. A romantic chapter–yelping dogs, marching men, wooden ships–had ended, consigned to the museums. And to historians.

—•—•—•—•—•—

Transportation and communications offer the primary difficulties in polar spaces. Russia sprawls over 150 degrees of longitude–eleven time zones. And most of the river flow entering the Arctic Basin is Russian. Although traffic on Siberian streams is as old as settlement, such "roads" had yet to figure into a coherent system of transport. A key impediment was river and sea ice—the perennial physical barrier. The Asiatic sector is endowed with many rivers that drain the subarctic interior, some of them—the Ob, the Yenisei, and the Lena—among the world's largest in volume. Most empty northward, their mouths thus linked by the Arctic Ocean. Unhappily for human commerce, the North Asia seaboard is iced in for at least five months, its debouching rivers for about eight.

The continental shelves of northern Asia underlie immense sweeps of bottom. These shoals (650 feet [200 meters] or less) are seasonally ice-free, leaving traversible lanes for shipping. Along the Eurasian margin lie a string of offshore basins from Murmansk to the Bering Sea: the Barents near European Russia, the Kara and Laptev off central Asia, the Chukchi and East Siberian in the east.[26] Here navigable seaways are necessarily coastwise. In the eighteenth, nineteenth, and early twentieth centuries, the czar's officials had appreciated the strategic importance of a sea-lane across the top of Russia; a viable route would help link the Atlantic and Pacific flanks of the empire. The Soviets inherited considerable knowledge of that shore in the form of the charts made during various expeditions put forth during the eighteenth, nineteenth, and early twentieth centuries. Still, overcoming the treachery and power of the ice had to await the age of technology—and the Bolshevik successors to Russian rule.

Commitment to icebreaking is long-standing in Russia. Vice Adm. Stan Osipovich Makarov (1848–1904) conceived the idea of working the pack with ice-capable ships. Makarov advocated settlement of the Russian North. "If Russia is compared with a building," he remarked, "then it must be recognized that its façade extends to the Arctic Ocean." To that "façade" flowed the great rivers draining interior Siberia, arteries that might help establish and then sustain a population in a continental space—Russia's raw-materials annex. In short, if the empire was to prosper, the north-facing coast had to be made a Russian lakeshore.

The world's first icebreaker was completed in 1899, in a Newcastle shipyard, though the plans were Russian. *Yermak* was large and powerful at 8500 tons and 3000 horsepower—for its time, a near optimum combination of strength, weight, and power. Its primary mission: convoy duty in Baltic waters and along the north Asian coast, the shortest route between European and Asiatic Russia.

To an empire exhausted by war and deprivation came revolution. And imperial collapse. In November 1917, in the Great October Socialist Revolution, Vladimir Lenin led the Bolsheviks to power. Support from Britain and other governments for the opposing White Army left the infant state trustful of no foreign government. Safely entrenched by 1921 and craving strategic self-sufficiency, the regime deemed the far northern reaches property to be exploited. Eager both to shorten transportation routes *and* to lengthen the shipping season, the Soviets threw abiding, systematic attention—time, energy, funds—into a large-scale marine transportation system. Theirs would be a through route, an arctic highway linking the great north-flowing rivers of Asiatic Russia—roads offering access far into the interior.[27]

As in all things Bolshevik, brutality attended these cold-water trails. And there would be no chance for appeal in the event of failure.

> The Soviet pioneers' efforts in the Arctic regions are marked by courage, proficiency, and indescribable suffering. The pioneers knew full well that if they were not successful on their expeditions, there would be no return trip. Leadership in the Communist regime, as well as under the Czars, had been incredibly severe, throwing Arctic expeditioners into the slough of disfavor if they were not able to bring home positive and vital results. Tricky weather and ice conditions have never been accepted as reasonable alibis for failure.[28]

Still, the Soviets brought forward-looking planning plus coordination to their push. Study of the north, indeed, was made a centralized national enterprise.[29]

Trading from ports of western Europe to the Ob and Yenisei rivers had been logged in prerevolutionary days. Early in the Soviet period (1921), airplanes helped open the eastern sector of the route for systematic shipping to both rivers—the annual so-called Kara Sea Expeditions. The Sea Route to the Russian Arctic developed steadily in terms of tonnage, seasonal gains, ice, and weather reporting.

The Northern Ocean washes nearly 18,000 kilometers (11,000 miles) of Eurasian coast, from the Barents Sea to the Bering Strait—the full reach of Russia. It is the shortest sea-lane joining the northern European and the far Eastern ports. Imperfectly known, this crenellated shore was just appearing on the Soviet horizon. Hydrographic work (anointed now as urgent) was organized, a network of shore-based stations conceived. Established in 1923 by special resolution, the earliest bead for this faraway polar thread went operational at Matochkin Shar, a fiord-like strait that slices the island of Novaya Zemyla in two. Its waters are difficult for shipping; hence the need for systematic observation of the state of the ice. Initially, services were limited to local needs: radio communication with ships at sea, reporting on ice and the weather ahead.[30] Longer-term, the goal was to establish a scientific basis for predicting the opening and closing of the short navigation season.

Radio operator Ernst Krenkel, a member of the second wintering party at Matochkin Shar, recalled the great hopes placed upon that station. "Its primary importance," he writes, "lay not in scientific research (although scientific work, naturally, was planned for the station), but in the practical assistance it could give to Soviet polar navigation." Before his arrival, Krenkel often turned to the map of that huge island mass, "but reality, naturally, had nothing in common with the precise delineations of the cartographers."[31]

—·—·—·—·—·—·—

In April 1926, unfettered by treaty law, the Presidium of the USSR Central Executive Committee announced itself in favor of the sector theory, decreeing all land in the Arctic Ocean between the north-facing coast and the pole to be Soviet, including that yet undiscovered. In one bold action, it laid claim to a triangle of land, ice, and islands arcing from the extreme eastern and western meridians of the USSR (32° E and 169° W), its base the mainland shore of European and Asian Russia, its apex the North Pole.[32] Notes dispatched to the northern littoral countries elicited a diplomatic silence, or, instead, a formal reservation of opinion as to the newly declared Soviet beachhead.[33]

The seasonal Norwegian settlements on Franz Josef Land lay beyond this claim. Still, the Kremlin rumbled. Under the command of Dr. Otto Julievitch Shmidt, a geographer-crusader of growing celebrity, an expedition was dispatched, and in 1928, the Red flag flown over Hooker Island and ceremonies attendant to taking possession conspicuously conducted. Within months, the Soviet government further decided to erect a radio station and weather observatory.

A northern seaway was now fanatically promoted. The riches of the eastern sector—the minerals, oil, timber, and furs of Asiatic Russia—lay tantalizingly beyond practicable access. What was to be done? As various schemes were put forth, their merits and defects received endless scrutiny and debate. Eighteen different routes (for instance) were studied for a Great Northern Railway alone. "The question," Krenkel said of that time, "was not whether a northern route was needed, but what sort of route it should be—a sea passage or a railway? Or perhaps an air route?" Among the options, the potential of a fleet of dirigibles was examined.[34]

Planners knew that the system of internal trade and transportation was eroding gains in Soviet national output. To help vitalize its arctic sectors, a new administration to replace the Committee for the Northern Sea Route was created in 1920–the *Komitet Severnogo morskogo puti* or KSMP. The old committee was elevated in status by government decree, becoming the Central Administration of the Northern Sea Route–*Glavnoe upravlenie Severnogo morskogo puti* or GSMP, "Glavsevmorput," as the Russians called it. Under its sovereignty was placed "final development" of a seaway "from the White Sea to the Bering Strait, full equipment of this route, maintenance of it in proper condition, and procurement of means of safety of navigation over the same." Meteorological and radio stations on the Eurasian Arctic coast and islands were thus transferred to GSMP jurisdiction.[35]

All of the territory north of 62° N was assigned to Glavsevmorput for exploitation. The northern coast of the Soviet mainland was lined with weather stations. Air routes were established, ice reconnaissance was instituted, and icebreakers were stationed along arctic shoreline at difficult locations. By coordinating efforts, the Northeast Passage [Sea Route] was open to sea traffic three months of the year. Polar towns grew into important seaports and export of Russian produce to European countries was carried out by way of the northward flowing Siberian rivers and the Northern Sea Route.[36]

Appointed as head of this new administration was Professor Otto Shmidt. A vivid, energetic personality, the "Ice Commissar" dominated Soviet Arctic enterprise until the decade's end. Shmidt held sweeping responsibilities, among them supervision of the activities of Leningrad's Arctic Institute, the center for the study of all natural phenomena in the Arctic.

The institute traces its bureaucratic genealogy to 1920. The work of the Soviet Union in its frigid zones increased to such a degree that, by 1925, a shuffle brought about the Institute for the Study of the North. The first five-year plan (1928–33) sponsored a further mutation–a radical reorganization of the Academy of Sciences. So as to link scientific research more closely with the practical problems of the Soviet economy, a number of new technical institutes were organized (and many transferred). And so an Arctic Institute was fashioned.[37]

The old institute customized its students for circumpolar work. A vital (if now diminished) force for the purely scientific study of polar regions, the Arctic and Antarctic Research Institute (AARI) of today has sponsored hundreds of official expeditions, fielding a well-selected, well-educated, well-disciplined corps of men and women. In the 1930s, the institute was the scientific center of the new Sea Route administration. Spreading the gospel of development, this central-government organ trained many cadres of *polyarniks* (polar researchers). The still-lovely-if-fraying St. Petersburg remains the soul of Russian polar enterprise.

Early in Soviet times, Arctic Russia comprised one of the earth's least-known expanses. More than 20 percent of the motherland lies north of the Arctic Circle (66°32′ N), its coast tracing a line almost entirely between 71° and 77°. (Cape Chelyuskin, at latitude 77°43′, is its most northerly point.) The north-facing outline had yet to be established, particularly its eastern Asiatic sector. Accordingly, each hydrographic expedition added to the official charts and maps.

Applied science *and* the Sea Route were emphasized after 1929. Polar stations became multipurpose–"new citadels of science" in Shmidt's label. When 1933 closed, twenty-two stations dotted Russia's north-facing coast; two years later, the network boasted three dozen. And by mid-1956, a total of 107 were under GSMP jurisdiction. These stations produced extensive records on polar weather and communications—data assets indispensable to long-range forecasting. To build qualified staffs, the institute trained cadres of biologists, hydrologists, hydrographic experts, magnetologists, meteorologists, and geologists. For communications (essential to any such network), especially radio, facilities acquired quantum improvements.[38]

Use of icebreakers changed, as well. At positions where ice usually caused difficulties, *Yermak, Lenin, Krassin,* and other units stood by, steam up, ready for the brief navigation season.[39] Each served double duty, with research staffs conducting hydrologic and related work as it steamed. At far-flung stations and in the small, isolated settlements, ships' comings were understandably special. If a layover was needed, say, for taking on coal or to allow the following caravans to catch up, a wireless message was tossed ahead. Waiting in boats, shoreside personnel scrambled aboard the visitor-ship, anticipating such treats as fresh fruit or flowers or invitations to attend its cinema. "The ice-breakers, these Arctic flotilla leaders, are not only technical but also cultural pioneers of the Far North."[40]

Each northern country had participated in the International Polar Year (IPY) of 1882–83, the first coordinated, multidisciplinary study of physical phenomena of the earth. Despite jealousies and glory grabbing, eleven nations had dispatched fourteen separate expeditions to locations in the arctic and subantarctic.

> The contribution of the IPY to world baseline data in a number of subjects, and to the concepts and understanding of planetary geophysical processes is impressive; but what is perhaps more important . . . is the influence that this one coordinated activity had in bringing about a change in the character of science. From being an exclusive activity, pursued by a privileged few and often jealously guarded for seasons of national prestige world science developed into an open activity, where everyone who was qualified could take part, and where the results belong to the whole world.[41]

The Second Polar Year began in 1932. Researchers from twenty-eight nations recorded observations at thirty-two stations. Russia supported nine. As Krenkel recorded, polyarniks "received every support." And why not? Moscow's self-assigned goals were titanic. A cadre of specialists was absolutely vital simply to man the polar stations, to say nothing of achieving the overall Soviet ambitions for the northlands. In a thin line through the Soviet north, afloat and ashore, Russia was observing, measuring, testing, analyzing, recording. "Life was roaring," one participant enthused, "along the entire coastline." "In general everything was happening in the North," another recalled of the 1930s.

A further project was inspired: a scientific transit of the north waterway (actually a series of sea-lanes confined to Barents, Kara, Laptev, East Siberian, and

Chukchi seas of the Arctic Ocean) *without* wintering over. The SS *Sibiryakov* expedition was the result. Leadership for this venture fell to Shmidt, with the institute undertaking to equip the vessel. "It is no light matter," Krenkel explains, "to dispatch an expedition on an independent cruise, especially to little known Arctic regions. Moreover, no one could guarantee how long the voyage would last. If we were lucky it would be measured in months; if not, it would require two years to complete. We had to be prepared for all contingencies, most of which would not be very pleasant."[42]

On 28 July, scientists embarked and *Sibiryakov* weighed and proceeded from Archangel.

The Northern Sea Route lies largely within sight of the mainland or islands. The Eurasian sector was traversed handily; Matochkin Shar stood astern within three days. Next came a transit of the Kara Sea. Early August saw the expedition moored at Dickson, on the Tymyr Peninsula, the first port to be constructed on the Siberian coast and host to its largest polar station. Favored by exceptionally easy ice, *Sibiryakov* next rounded the northern tip of the Severnaya Zemlya (North Land) archipelago, into the Laptev Sea. But the sector beyond longitude 167° E held heavy ice. "The East Siberian Sea gave us a less friendly reception," Krenkel records, making particular mention of the seaway beyond North Cape. In the Chukchi, the pack is particularly vexing: the navigation season is brief, the coverage severe, its pack ice perennial.

By 16 September, *Sibiryakov* had steamed 5635 kilometers (3500 miles) on its outward passage. But it was late in the hour. Ice is a danger to screws and rudders. With 161 kilometers (100 miles) to steam, *Sibiryakov* trembled from three blows. Chunks had taken one propeller blade; three more had been gnawed to near-stumps. Thoroughly disabled, she lay helpless before wind, current, and the shifting white.[43] A solution was improvised. Weight would be shifted forward, the hull put hard-down by the bow, thus exposing the shaft and screw so as to effect repairs. (No diver was aboard.) For six days and nights all hands—from stoker to scientist—strove mightily, hauling and tugging coal, cargo, movable gear. Repairs made, headway resumed. But the shaft sheared free two days later, taking its propeller to the bottom.

Sibiryakov again lay powerless—the world's best-equipped buoy for checking polar currents (a few joked). "The situation," Krenkel writes, "was infuriating. We had enough fuel and the engine was in full working order, but we were unable to move." Yet the net drift was *east*ward. So as to enhance this gift and spur progress, sails were jury-rigged; the final miles were logged under canvas. On 1 October 1932, two months and four days out of Archangel—the Bering Strait lay beneath the keel. A full west-to-east transit had been made in one navigation season—a feat never before logged.

The transit signaled a glowing chapter for the Soviet Arctic. Still, deficiencies had been highlighted. Meteorological services stood wanting: the network of stations required augmentation. And measurements taken well to the northward—in the icebound central Arctic—would be basic to any improvements in prognostications that would facilitate the proposed scheme of coastal freighting. As one polyarnik observed: it had become "absolutely clear that the conditions in the central Arctic influence to a great extent the coastal climate, ice and navigation conditions along the Route."[44] Thus, that December, GSMP was decreed. Its charge: the final development of a working seaway.

Stalin's second five-year plan commenced in 1933—a difficult year of numerous at-sea operations. For the first time in the Soviet North, freight vessels (guided by airplane, escorted by icebreaker) pushed to the mouth of the Lena, which debouches into the Laptev Sea. Exploiting the enthusiasm attending *Sibiryakov*, Moscow decided to repeat the full west-to-east passage. The freighter SS *Chelyuskin* was assigned to traverse the route in one navigation season. The ship put Leningrad astern in July, a blaze of publicity attending her wake. Wanting seasoned hands (in ice, experienced shiphandlers are vital), Shmidt had recruited a group of Sibiryakovians. The scientific staff were from the Arctic Institute, among them a soon-to-be famous hydrobiologist, Peter Shirshov. In all, 121 men stood out that day plus six women and one child. The family groups and cargo were bound for the colony of Soviets on Wrangel Island, which lies off the far-eastern corner of Siberia.[45] The leader of the expedition was the omniscient, omnipresent Shmidt.

—•—•—•—•—•—

Chelyuskin's bow cut the Kara Sea by mid-August. Although the ship had been ice-strengthened, the designers' calculations for *Chelyuskin* were manifestly in error. Bulling its first real pack, the ship's hull sustained damage. "This made it quite clear," Shmidt records, "that we must not expect too much from the ship." Despite leakage, the ship's wake foamed eastward (aircraft assisting). She navigated Cape Chelyuskin to cross the Laptev Sea en route to the next basin, the East Siberian thence the Chukchi. Maneuvering to avoid ice, she attended the business of science as she steamed: parties were dispatched ashore, radiosondes lofted, soundings and meteorological observations recorded.

In nature as in human affairs, power prevails. Sea-ice concentrations (expressed in tenths) are usually heavier to the eastward. In the Chukchi sector the freighter met serious ice. Despite the need for haste—it was late in a stubborn melt season—the advance slowed as *Chelyuskin* forced its way through consolidating pack (ten-tenths concentration), further wounding an infirm hull. Off the Gulf of Kolyouchin, forward momentum halted altogether. Icebound. The delay proved fatal. September is winter—not the most favorable period for ship operations in the Far East. A labyrinthine drift commenced. Unexpectedly mushed out from the mainland, dog teams returned with eight evacuees. Aboard a hapless *Chelyuskin*, no one could predict the ordeal to come.

Release came five days into October. Steaming resumed. Early in November, yearning toward its objective, the freighter dispatched the hopeful signal, "We are in sight of Bering Straits." It was not to be. The pack again took hold, and a rudderless drift resumed. Direct assistance proved unsummonable: in the course of a difficult ice season, the nearest icebreakers were either damaged or themselves beset.

Inscribing fitful, erratic loops but granted favorable drift, the expedition found itself conveyed into the strait. *Chelyuskin* entered on 3 November—it was on the Pacific doorstep. Two days more saw her halfway through. "From the bridge," Krenkel writes, "the edge of the field could be seen with the naked eye, lying to the south. Beyond, open sea stretched to the horizon." Alongside, a marathon two-day assault with tools and explosives failed to gain release; the hull seemed welded in. Disappointment soured to alarm when southward drift faltered then died altogether. The ice field then reversed its track, bearing its captive back into the Chukchi.

So as to conserve human strength, efforts to break the ship free were abandoned and, instead, preparations effected for that most grievous of eventualities: emergency abandonment. Leningrad lay eight months astern. Now, to seaward of the Siberian Far East, senior radio operator Krenkel tapped out the words that were to echo worldwide: "February 13, 15.30 hours, 155 miles from North Cape and 144 miles from Cape Wellen the *Chelyuskin* sank, crushed by the pressure of the ice . . ."

The end had been swift, dramatic, final. A huge crack had appeared in the ice close aboard. Lying normal to the ship, it pointed, swordlike, directly at her. Then forces beyond human power had slid countless tons along the line of fracture. Krenkel was witness to the unstoppable: the steel hull bulged outward, he said, "like a piece of cardboard," followed by a crackle like machine-gun fire—failing rivets. The ship trembled and groaned "like a living thing." Fatally mauled, *Chelyuskin* began to settle. As she succumbed, Krenkel's urgent transmissions were accompanied, topside, by frantic unloading.

The derelict took one man; the others suffered no physical harm. In a dark, biting cold, "Shmidt Camp" was urgently put together. And so the world learned of the end. "*Chelyuskin* sunk. Passengers marooned on drift ice." Instantly, the offshore drama seized an audience, centering attention worldwide on activity in the northern Soviet Union.

The ice-strengthened freighter SS *Chelyuskin* founders in the Chukchi Sea, off northeastern Siberia, 13 February 1934. Rescue of the ice-stranded ship's company (104 persons) — the first large-scale Arctic airlift — was to profoundly impact Soviet aeronautics and Soviet science in the maritime North. (R. E. G. Davies)

Unwilling colonists had disembarked onto a ghastly, blasted habitat. Pack ice is dangerous and unwelcoming. To the drifters–their world now ice, it threatened both privation and disaster, the potential for which was excellent. The physical plight must have been terrifying. To survive, panic would have to be mastered, order and a ruthless discipline–on which everything now hinged–imposed. Such was the force of his personality that, from the first, Shmidt was held to be head taskmaster.

Within days of tent-camp, the carpenters had erected a barracks and kitchen using lumber and salvaged remnants intended for the colony on Wrangel Island. Airstrips, though, were crucial. These would have to be sculpted from the enclosing chaos and replaced as the ice took them. Accordingly, the ice camp's full attention became focused on leveling suitable sites in the near neighborhood. Intruding hummocks were attacked by hand. As the chopping proceeded, ice scraps were hauled off and surface crevasses "caulked" with fragments. As this brutish labor proceeded, it was recorded by the assorted journalists, artists, cameramen, and photographers on hand. Of paramount importance was the radio– the camp's umbilical to the outside. "We had only one task," Krenkel records, "to prevent the invisible thread linking us to the mainland from snapping." The batteries were crucial; their precious spark had to be husbanded. Therefore, no private messages were tolerated—an edict which infuriated the pressmen who, by order, were obliged to sit on irresistible copy.

Krenkel's transmissions had stunned officialdom. Finding its nerve, reaction proved swift and, at the last, masterly. *Pravda* and *Izvestiya* headlined the high-stakes opera unfolding off northeastern Siberia, as did a startled Western press.

Rescue was the responsibility of a (compelled) governmental committee, hastily organized, which reported to Stalin and the Central Committee. (In the purges to come, most of its members would lose their status—or worse.) On 15 February, orders were issued to retrieve the stranded party. The commission's first communiqué briefed Shmidt of developments on his behalf: every polar station, he was advised, would maintain a constant vigil for his radiograms, which would be relayed instantly. To assist the rescue, stations along the eastern sector would issue four reports daily on weather, ice conditions, transport preparations, and the organization of provisions and depots between the stations and the ice camp of Soviets.

Inventive schemes and means were floated about how (not whether) to effect rescue. Opinions differed wildly, yet only two means seemed practicable: dogs or aircraft. Travel by either dog team or foot for so large a party was ruled out; rescue would be by air campaign. Spring—and breakup—were imminent. As the world watched, furious plans were organized and presented for official imprimatur. These would center on the new and powerful Wellen station. It lay in northeastern Siberia amid an impossible waste, the farthest corner of Soviet hinterlands. No pilot had ever surveilled its winter taiga. Yet, as Mikhail Vodopyanov exulted, "The decisive role in these operations was assigned to aviation."

Largely unacknowledged in the West, Soviet aeronautics has a glorious history. Initially, flying in the Soviet North had been concerned with development. Most missions were brief and were for reconnaissance, mapping, or ice patrol. Efforts peaked with a series of remarkable operations, not least a North Pole expedition in 1937. And why shouldn't they have? Soviet airmen had devised the world's most advanced techniques for cold-weather, long-range flying. "All that is really important in the world," one boasted, grandly, "lies in the Northern Hemisphere. Russia, Europe, America. Quite naturally, therefore, the shortest air routes are over the Arctic and not east-west round the waist of the earth."[46] Prodded by need, hubris, and Comrade Stalin, the USSR was helping pioneer—and sustain—transpolar commercial air transport.[47]

A 1913 hydrographic expedition made by pre-Bolshevik Russia was probably the first deployment of airplanes in the Arctic. This expedition took a Maurice Farman plane into the Laptev Sea. But no flights were made. The next year, Ivan Josifovich Nagursky, a twenty-five-year-old naval lieutenant, accompanied the Vilkitskiy expedition to Novaya Zemlya. His mission was to assist the search for the missing Russian polar explorer Georgy Sedov. A Farmon hydroplane flew two trial sorties—the first on 8 August—and five long searches skirting Novaya Zemlya, mechanic Kuznetsov in the second seat. Although he found none of the missing men, Nagursky in his sweeps was able to make numerous corrections to the charts and, further, to tender reconnaissance to the search ships—by helping to select the weakest sections of the ice and which leads to follow. Thus had he demonstrated the practicability of flying in the Arctic. This paragraph from his report, submitted in October 1914, bears quoting.

> Although flying in the arctic regions is difficult it is entirely feasible and in the future aviation can make a major contribution to hydrography. . . . Past expeditions attempting to reach the North Pole have all been unsuccessful since they badly overestimated man's endurance and energy when faced with the thousands of versts [a Russian measure of distance, 1.06 kilometer] which must be covered, full of obstacles and very difficult conditions. . . . As being an enormously rapid means of transport aviation is the only way to resolve this problem.[48]

Nagursky's flights were the last to be conducted by Imperial Russia.

The new Soviet State proved air-minded. The first sorties north of the mainland were ordered in 1924, in support of a hydrographic expedition.[49] The pilot was B. G. Chukhnovski, a colleague of Nagursky. Flying a two-seater Junkers hydroplane, he made a dozen probes to reconnoiter ice—perhaps the first such use of aircraft—and map coastline in the Kara-Barents Seas region, including the approaches to Matochkin Shar. (A scientist occupied the second seat.) Pleased with its results, the regime ordered Chukhnovski back the next shipping season. Accompanied by a second plane, the fliers departed Leningrad; in twenty-five days, the squadron of two reached Matochkin Shar. In the Kara Sea, the planes did cartographic work and reconnoitered ice, assisting shipping in choosing the best routes, thereby saving the regime both time and fuel. Again in 1926, several flights were made for cartographic purposes, ice reconnaissance, and (for the first time) seal hunting in the White Sea. During these years the focus was primarily on development and on access—via the Sea Route—to ports and rivers and settlements dotting the coast. In terms of Soviet military and industrial development, aviation played but a minor role.

Arctic Russia is a bewilderment, a stupendous high-northern geography of tundra, forest, and endless steppe. Siberia's longitudinal extent—eleven time zones—confounds systems of transportation and communications. Sprawling across the roof of the Old World, it spans more than 8050 kilometers (5000 miles) west to east and enfolds 15.6 million square kilometers (6 million square miles). As late as

1933, only sixty-five large cities splotched this vacancy. Today, population centers are dispersed and divided still—isles of humanity in a gray ocean, quarantined by an all but roadless taiga and tundra. Even rail networks are limited. In the early Soviet years, land travel across forest and swamp was restricted to reindeer and dog driving, and even that was restricted to winter.

These difficulties in transportation were impeding the whole of the Soviet economy. (Infrastructure suffers still: currently, much of the communications lines and railroads stand in need of repair.) With vast spaces to vault and a hinterland to serve, civil aviation assumed huge national import. Soon almost every town had its own airport.

> You can imagine [airman Vodopyanov writes] what it is like when a plane appears over such a locality. The inhabitants have never seen one before, but they know it brings them letters, newspapers, ammunition for their guns, medicine and everything else they need. Besides, the arrival of an aeroplane signifies regular communication with the outside world. That is why they are so overjoyed at sighting a visitor from the sky, why they make so much of the pilot.[50]

Its peoples largely peasants, czarist Russia had languished—a backward preindustrial giant. Through huge exertions, the Soviets had shed the fetters of underdevelopment and yet striven for independence from foreign influences. Their aviation industry, for example, relied on foreign equipment, designs, and ideas. (In 1928, 70 percent of all aircraft engines in use were imported.) Experienced designers were lacking, so design became largely a matter of copying foreign models. Stalin's command "counterrevolution" would be guided by five-year plans of development, the first of which was announced in 1928. In the absence of electoral pressures, Stalin could subordinate social necessity to economic necessity. Ruthlessly forcing technological superiority, this crash industrialization in fact insulated the Soviet Union from worldwide depression.[51] Results proved stunning. Under the first plan, about half the total funds invested in the national economy were for industry. More than fifteen-hundred new plants were erected, among them the world's largest. In the period 1927–32, Soviet research and development spending sextupled and industrial production doubled.

A shrewd strategist, Stalin had instincts second to none. He knew that aviation enjoyed the glamor of modernity. A prestigious and powerful symbol and political tool, aviation would proselytize for the regime—the space technology of the time. The first five-year plan represented the takeoff phase. To end dependence on airframes, engines, and spare parts manufactured under capitalist systems, officially disdained, a native industry was nurtured by means of massive state assistance. Factories were enlarged and improved, new engine plants erected. Goals proved overambitious; nonetheless, output soared. By 1933, the Soviet tyrant could boast, "We had no aviation industry. Now we have one." Between 1930 and 1937 and the second five-year plan, fighter production rose tenfold, bomber units well over fivefold.[52]

The civil air-transport system blossomed as well. The Soviet Union's internal network of airlines nearly tripled. As well, passenger, mail, and freight miles exploded.[53] For the distant precincts of Siberia and the Far East, Central Asia and Kazakstan, airlines constituted the sole means of mechanical transport and mail communication. To stimulate interest, clubs and glider schools–operating on government funds–sprouted like seeds, as did official training schools. Flying became a countrywide obsession. The Soviet press, a visitor observed, devoted "a disproportionate share" to aviation news of all kinds. "This country is not air-minded," he decided, "but air-crazy!" Still, foreign pilots on flying visas found the civil system an exasperation of permits and delay worsened by ancient maps, meager food, and dysfunctional hotels. And fear of foreigners. Everything was regimented but nothing quite organized. "I began to realize," a private pilot wrote of his air journey, "that in Russia civil aviation of the sporting kind . . . is quite unknown. They regard the aeroplane purely as a war engine, and the foreign aviator as a spy."[54]

Hungry for glory, Soviet fliers entered domestic as well as international competitions. Throughout the USSR, designers and fliers strained to realize the popular national slogan, "We shall fly faster and higher and further than anyone else." By 1937, only the United States led the Soviet Union in civil air traffic. The airmen themselves were young and fit, hence readily portrayed as strong, fearless, attractive Bolsheviks–perfect examples of the new Soviet man.[55] Esteemed, privileged stars, "Stalin's falcons" received lavish rewards–money and, for the very best, honor and special favor. The blandishments of celebrity attracted the ambitious, the skillful, the courageous.

Aerial reconnaissance was an integral element of Central Administration function; indeed, aviation was key. Without it, navigating the route and maintaining reliable commerce for a three-month season would have been unthinkable. (Surviving) pilots quickly earned a superb knowledge of northern operations obtainable only through experience. In a report to Stalin, Shmidt referred to the work done for 1936, "The most important role in the success of Arctic navigation was played by our polar aviation." Soviet aircraft logged twelve thousand hours of polar flying and covered nearly 1.4 million miles.[56]

The Soviets saw themselves, rightly, as leaders in ice navigation. Witnessing a demonstration aboard *Yermak*, a British journalist marveled at the symbiosis of ice-breaking ship and aircraft:

As soon as we set out to fight for a passage we sent a radio message for the dispatch of an aeroplane from the shore base. Within half an hour the ice watch pilot was circling over our bridge. He left us and surveyed hundreds of square miles, rising high above the sea. His observer watched the slowly drifting fields and floes below. He drew a map of their constellation. His first advice reached us by wireless. After two hours the ice watch returned and dropped a parcel with a small parachute. It contained a map of our route ahead. The captain and his staff laid out their plans of attack. A strategic line of approach was fixed; the course decided by the thickness of the ice on the one hand and by the time factor on the other.[57]

Northern flying—Soviet-style—peaked with a series of astounding operations. In 1935, a trio of winter flights sortied from the capital. And that August, Sigismund Levanevsky, an exceptional airman (and Stalin favorite) attempted to leap from Moscow to San Francisco via the pole, but an oil leak forced him to turn back. The next year, four outstanding operations were logged. The third (and most spectacular) was piloted by Chkalov, at the controls of an ANT-25, a huge wide-winged, all-metal, single-engine monoplane.[58] Chkalov and his crew—a copilot and navigator—flew from a Moscow aerodrome to the Far East. Flying over Franz Josef Land, the machine pressed coastwise past Cape Chelyuskin to Tiksi. It then steered overland, southeast, landing at Nikolayevsk on Amur in southeast Russia, almost at the Japanese border—9338 nonstop kilometers (5800 miles) in fifty-six hours, twenty minutes.

In 1937 came long-distance operations of universal repute. Despite strong evidence and widespread speculation of a transpolar sortie, the government kept silent. Flying the same ANT-25 as before, the same crew—Georgiy Baidukov, copilot; Alexander Belyakov, navigator—Chkalov lifted off from an airfield near Moscow early on 18 June. After twenty-seven hours, from an altitude of 4148 kilometers (13,600 feet), the aircrew claimed the pole. Chkalov himself, asleep, did not mark the moment.

As the airplane neared North American airspace, listeners attended their radios, thrilled by this first arrival via the geographic pole. As thousands waited, the airmen were forced down—by near-zero visibility and weariness—at the Pearson Army airfield, near Portland. "Bewhiskered, red-eyed and tottery," the trio tumbled from their machine—thus becoming the first to chart the polar air route from Moscow to the American Pacific. Their airtime had been sixty-three hours, seventeen minutes.

Declared "world heroes" by the Soviet ambassador, the aircrew met a frenzied reception.[59] "When Russia established a base at the North Pole last month," (next chapter) *Time* reported, "many were skeptical about the Soviet's announced intention of inaugurating a Moscow–San Francisco airline. Last week, skeptics were astounded when a Russian plane nonchalantly flew non-stop from Moscow to the U.S. via the North Pole." Granted a whirlwind tour, the trio was received by President Franklin Roosevelt. For the homecoming, Stalin himself was at the field. "Flowers showered down from the windows" on the Moscow cortege, a journalist recorded, "streets were beflagged, Russia went wild over them as America over Lindy."[60]

Riding this burst of publicity, a second transpolar trip to America was launched in July. Departing Moscow with three aboard, an ANT-25 piloted by Mikhail Gromov leapt to San Diego in 62.3 hours, a distance of 10,000 kilometers (6262 miles)–a nonstop distance world record. In the twentieth year of Bolshevism, Soviet aeronautics had announced itself to the world–and cut a strong international figure. Propaganda organs trumpeted these leaps. The aeronautical world was impressed, even incredulous: few observers had supposed that the Soviets were informed of the advances in airplane design.

That August, Shmidt was discussing flying air passengers over the pole. On the 12th, overreaching, Moscow went for three. Levanevsky, with a crew of five, lifted off for Fairbanks, Alaska, in a four-engine ANT-6. The fliers crossed Ninety North–and disappeared. By the 14th, the entire north was listening for the missing airmen. Despite months of searching by both the Russians and the Americans, no sign of plane or crew was found.[61]

—•—•—•—•—•—

Roughly 161 kilometers (100 miles) to sea, at Shmidt Camp, life on ice had assumed a sustaining rhythm. The first man up lit the stove and put on a bucket of ice for water. (During the night, shelter temperatures would descend to near ambient.) The radioman–usually Krenkel–was awakened for the 6 a.m. chat with a Wellen operator. (Any delay raised concern ashore.) Everyone soon awakened–and the latest information poured onto the radiomen. Reports on visibility and the ice were made to Shmidt, weather conditions and news of the "aerodrome" delivered, along with their coordinates (position). To help sustain morale, the residents were kept fit by regular, compulsory exercise. Hydrobiologic and other research observations were dutifully recorded. Dr. Shmidt gave uplifting disquisitions on science, history, mathematics, and modern theories of psychology. Unspirituality was the canon. "Being atheists," a Western journalist sneered, "they were denied the consolation of prayers." Illiterate carpenters reportedly were taught reading, writing, elementary arithmetic.

On 27 February, a radiogram from the leadership was read to the drifters, duly assembled. "We are following your heroic struggle against the elements with admiration," it said, "and are taking all measures to assist you. We are confident of the favorable outcome of your glorious expedition and are confident, too, that you will write new and shining chapters in the history of the struggle for the Arctic."

The first plane, a twin-engine ANT-4, was heard on 5 March. Ambient air temperature was −38°F. Pilot Anatoly Lyapidevsky circled the camp—a clot of color amid the jumble—to evaluate the rough, broken surface before daring to set down. Though faint, the late-winter light assisted, conferring good contrast. Atop the white, the women were ordered to the landing site. Deftly avoiding a crevasse, Lyapidevsky set down on uneven ice and slid to a stop. Smiling, cheering faces surrounded the airman, the greetings ardent and undisguised. Wasting no time, ten evacuees were bundled in, after which the machine retired bearing eight women and *two* children to shore. (A birth had occurred during the Kara Sea transit.) The soon-to-be famous *Chelyuskin* airlift was under way.

Determined on success, Moscow had dispatched three airborne groups, along with hordes of men and equipment, to effect a full evacuation. But weeks passed before all the Chelyuskinites stood safely ashore. Blowing snow and low ceilings interposed, to say nothing of a rash of accidents and needed repair and false starts, all of which delayed the campaign.

Not until 7 April did a second aircraft appear over the shifting bivouac. Devouring its hard-won "aerodrome," the Shavrov Sh-2 crashed—no injuries, the damage repairable. A crate of American beer, chocolate, and cigarettes had gained camp along with eight dogs. An hour or so later, two planes more dropped down and took on evacuees.

Two days later, ice pressure delivered a severe trembling—the worst since the foundering. As the day lengthened, icequakes grew progressively more violent. Hummocking took the galley and a boat. Then one of the airstrips succumbed. "The ice buckled," Krenkel recounts, "and the ice-floe on which the barracks stood suddenly moved towards the floe on which the tents stood. Fortunately, the ice billow stopped at the last moment." Incited by gale-force winds, the tremors persisted; cracking nearly took the radio mast. Because of the conditions, including snowfall, the oncoming aviators were warned off. The camp struggled to resurrect itself. On the 10th, a plane came for the next evacuees, snatching away three. The airlift accelerated. Peeling in, another dot appeared out of the south; this plane took six. A third plucked away fourteen. Exploiting the weather, seven sorties more retrieved thirty-five the following day.

Late on the 12th only six remained afloat. The final shuttle came on the 13th—exactly two months after the foundering. A three-plane squadron pressed offshore from Vankarem, a ninety-minute flight away, piloted by Vodopyanov, Vassily Molokov, and Nikolai Kamanin. Among the last to leave, Krenkel shut down his precious station, closing out transmissions with "Three aircraft arrived.

First to arrive! Exploiting clear and calm, a twin-engine ANT-4 reaches Shmidt Camp, 5 March 1934. The pilot is Anatoly Lyapidevsky. Air temperature −38°F. Ten evacuees were plucked away this day. The famed *Chelyuskin* airlift was under way. (R. E. G. Davies)

Landed safely. Removing radio. Now leaving Shmidt Camp." Then he repeated slowly three times "RAEM!"–*Chelyuskin*'s call sign.

A complete evacuation–104 individuals–had been effected from the merest splash of habitation in a shifting, pallid blank. Even the dogs were spared a frozen end. The airmen were duly acclaimed. The title Hero of the Soviet Union–created by Stalin for the rescuers–was bestowed upon seven of the pilots directly involved, Vodopyanov among them. Along with the mechanics, each man also received the Order of Lenin.

Throughout Russia, the *Chelyuskin* adventure commands an exalted regard. Moreover, it proved to be a creative spur, a rehearsal. A practical equivalent of Shmidt Camp was soon to burst onto the circumpolar stage.

Chapter Two
RED STAR: "NORTH POLE-1"

> Almost in the beginning was curiosity.
> – Isaac Asimov

THE NOTION OF A NORTH POLE SCIENCE STATION was first floated by Fridtjof Nansen, who understood the potential of ice.[1] Untouched by land or sea roads, the sea desert posed an unexplored blank well into the twentieth century.

The Arctic Ocean offers an appalling logistical environment. For anyone afloat or afoot, polar ice is adversary; its reputation for malevolence is deserved. Denied direct access, Nansen saw air routes as the only practicable means. The strategy was simple but revolutionary. A scientific party (the thinking went) would penetrate by dirigible to the heart of the central Arctic, where, sustained by modern equipment, men would establish a semipermanent ice base and settle in to observe—a colony for science in an otherwise unassailable blank.

The potential rewards were incalculable. The north lacked an observation network comparable to that in more congenial latitudes. Yet vastly complex synergies—air-ice-ocean—drive circumpolar systems. Science had yet to settle (in some cases, even address) the phenomena that occurred at the poles. Atmospheric processes, for example, were virtually unrecorded. Explorer G. Hubert Wilkins and his pilot, Carl Eielson, had been first to land on, and sound through, the ice. "The [ice] base," Sir Hubert argued as late as 1931, "would be particularly valuable in connection with a plan of meteorological investigation with relation to seasonal forecasting and for the collection of data of value to airplanes and airships flying in high northern latitudes." A host of disciplines would profit, the geosciences particularly.

The tectonic and sedimentary processes that over geologic time had shaped the continental margins and Arctic Ocean basin lay undeciphered. Arctic Eurasia in 1928, a Russian geologist admitted, "is one of the least known parts of the earth's surface whether we consider its geology or its topography. Large sections of it have not been surveyed at all or in sketchy reconnaissance only. Even the Arctic shore line cannot be considered well established, especially in the Asiatic sector, where every new expedition makes important corrections upon the maps."[2] To seaward, the basin's sedimentary history and architecture stood as obscure as Martian mare. Further, the paucity of data prevented full explanation of the fringing massifs. (The time and manner in which the basin had formed determines the age and type of the continental margins.) Accordingly, certain keys to tectonic evolution of the adjacent land geology would be found beneath the maritime canopy of white.

Circulation of the lower atmosphere and the surface ocean are, it now is known, tightly linked. As traced by the drift tracks of ships, mean ice circulation has long been known. In the second half of the last century, patterns of flow became plain, courtesy of the movements of drifting shelf-ice fragments, ice camps, and buoys. During the 1930s, the mechanisms were scarcely imagined, deepwater circulation especially.[3] Systematic probing awaited the instruments of the explorer-scientist. "As to the nature of the Central Arctic," writes Yevgeny Fedorov, a geophysicist, "it remained the object of often contradictory hypotheses based chiefly on indirect data. To redress these deficiencies, "a new method of work had to be found, one providing for a long-term, planned and comprehensive study of the Central part of the Arctic Ocean."[4]

—•—•—•—•—•—

Passionate about everything arctic, Soviet scientists upheld Nansen's concept of an ice station—a vision intriguing polar circles everywhere.[5] Inside the USSR, prodigious initiatives were under way and in planning. An air-deployed camp in the interior ocean was not beyond possibility. "The drone of the airplane," marveled the *New York Times*, "is heard all day over Soviet Russia. . . . In the far north, all year round, planes are kept busy exploring, doing photography, experimenting, keeping hundreds of far-flung bases and camps in communication."[6] But Moscow discouraged curiosity; aircraft production statistics were unobtainable— even through official channels. Doubtless the Red Air Force was huge, Western reports conceded, if largely obsolete.

Ernst Krenkel had learned of drift bases in 1930; later, he had heard them discussed aboard *Sibiryakov*. The *Chelyuskin* affair energized the debate. As for Shmidt, the explorer had gone to the mountaintop, so to speak. As part of Soviet plans for its Arctic properties, not least to improve the ice and weather service, in 1933 a pole expedition was announced that would include the landing of scientists in the central polar basin. After all (as Stefansson had remarked), the Arctic is the only sea one can walk upon. The projected date of takeoff was 1937–the end of Stalin's second five-year plan. In a venture made practicable by aircraft, Moscow would pioneer ice-borne bivouacs for survey and basic research.

Following Peary's trek, little had been done to enlarge upon his handful of measurements. Airplanes and airships had flown over 90° north latitude; as yet, though, no one had dared to set down. For Dr. Otto Shmidt, the idea of merely gaining the place or flying across held scant appeal. But to land and work there! "It had become clear," he would write following his triumph, "that a different technique was needed; that it was not sufficient just to fly over the Pole, but that what was required was to land [there] and to stay on the ice long enough for extensive scientific observations. That was the task we had set for ourselves."[7]

What drove this concern for science at and near the north geographic pole?

[T]here were no observations from the great expanse between Siberia and North America that is filled with floating ice. The absence of any observations complicated the analysis of processes in the atmosphere and in the sea and ice. That is why in the 1930's there arose among Soviet polar explorers the idea of organizing the long-term research stations on the floating ice of the central Arctic.[8]

In the Soviet press, a preformulated debate played on. What was the best method for deploying men at the pole? The cardinal unknown was precise data on ice conditions: Were there suitable floes? Could ski-equipped airplanes alight? In the central Arctic, some maintained, ice fields were hummocky and young, broken by cracks, leads and pocked with melt pools–an intolerable surface for heavy aircraft. Some thought its broken, buckled aspect–the chaos of drift ice– doomed any landing. Others contended that old floes offered flat, unbroken zones.

Sir Hubert Wilkins, the Australian-born veteran of Stefansson's 1913–18 penetrations, had reported "splendid landing fields" upon which planes might drop, both along the coast (where pressure is harshest) and offshore–wide smooth patches affording, he opined, "perfect runways." In his own diary, the great Nansen had remarked upon "smooth ice sheet." And Peary had noted floes "large and old, hard and level." Wilkins had logged the first land-plane descent onto drift ice. In March 1927, exploring off North Alaska, he'd selected points of touchdown in his field of view upon which his pilot, Ben Carl Eielson, had then landed. Five hundred miles (805 kilometers) to sea, the first sounding had been taken–16,000 feet (4850 meters) of water, the deepest yet recorded. "From the writer's experience," Wilkins recounts, "there seems to be a possibility of finding many safe landing fields on the Arctic pack ice, both for airplanes and for men from an airship."[9]

Which view was correct? Perhaps neither. Ice texture is an artifact of ceaseless churn and change.

Mikhail Vasilievitch Vodopyanov, a peasant boy, had seen his first motorized transport in 1917. A qualified pilot by the mid-twenties, he logged hundreds of hours along Soviet Arctic air routes. In 1929, he flew across Siberia. The first airman to fly to Franz Josef Land from Moscow, Vodopyanov had helped execute the famed *Chelyuskin* rescue. ("Certainly the brightest page in his biography," Leningrad told the author.) As for the geographic pole, he was a partisan of the optimistic school. For Vodopyanov, an aerial approach had become an *idée fixe*. "Is a landing at the Pole possible?" he wrote in an apparent run of intuition. "Of course, it is. I have done a lot of flying in the Arctic. . . . Unlike the Chukchi Sea where the ice is a mass of tiny fragments as if it had been put through a giant mincer, the ice on the Barents Sea . . . is fairly smooth. Several expeditions have confirmed that flat ice-fields are frequently to be found there."[10]

Vodopyanov would realize his "pilot's dream." In the mid-thirties, the Polar Professor (a Shmidt moniker) was head of Central Administration of the Northern Sea Route, a professor of mathematics, a member of the Council of People's Commissars—in short, a gigantic figure, one of Russia's most popular. Decisive, strong-minded, energetic, he had become absorbed in a fresh enterprise. He had resolved to challenge the ice using aircraft.

Mid-February 1936. Along with such airmen-luminaries as Valeriy Chkalov and Sigismund Levanevsky, Shmidt was summoned to the Kremlin for a most sacrosanct of briefings. (The *Chelyuskin* loss and rescue was two years past.) The subject was the possibility of transpolar flights originating inside the Soviet Union. Stalin had taken note of the political dividends attending a leadership reputation in aeronautics. And Comrade Leader had been much impressed by the gripping effects of record flights, including round-the-world jaunts over Soviet territory. The campaign to transform Soviet aviation was about to soar.

Sensing an opportunity, Shmidt presented his case. He would marry aircraft to the drifting laboratory technique validated by Nansen. Stalin was interested: a landing would affirm the strengths of socialism, as would the transpolar flights the dictator now wanted.[11] Further, an outpost in the central basin could serve as a weather station for lights across the geographic pole.

So the Ice Commissar was instructed to get on with it. The pole camp would be no impromptu loiter. To support the requirements of a scientific station, transport aircraft would be basic. In this respect, Dr. Shmidt's expedition was to be well served. The TB-3 heavy bomber was a four-engine, all-metal monoplane—at twenty-two tons, the largest land-plane anywhere. Built for endurance and known to be reliable, the ANT-6 (the civil version) could hang aloft for sixty-five hours. Further, a low landing speed was adaptable to imperfect airfields. Vodopyanov joined the circle of decision. Summoned before Shmidt, the startled pilot was told, "Your dream may come true." "The Government has directed us to begin the organization of a scientific station on a drifting ice floe in the North Pole area. We shall have to fly the men, cargo and scientific equipment to the Pole." Ordered to draft the overall plan, Vodopyanov also would lead the air group bearing the assault.[12]

Particular preparation was called for, covering every aspect of operations from the grand to the most banal. The aircraft, engines, equipment, instruments, clothing, provisions—all had to be selected, designed, tested, made ready. And the ice-camp party itself had to be chosen. Pre-field planning is always a busy time. "Traditional wisdom," one member wrote, "has it that thorough preparation is three-quarters of success. We had not only to live on drifting ice with a basic minimum of comfort—we also had to work there. It was hard to say where most of the attention should be concentrated, for while we had to be well-fed and clothed in order to work, our astronomical instruments and the radio station had to function perfectly if a steady flow of information was to be maintained."[13]

Bureaucracies have an in-built tendency not to cooperate or coordinate. A skilled autocrat, Shmidt knew the intricacies of Soviet statecraft: armed with Stalin's fiat, he roused the central government—already a tangle of inefficiency, incompetence, and red tape.[14] The People's Commissariat for Heavy Industry was commissioned to recondition the necessary aircraft. Another agency was set the task of designing engines for low-temperature operations, a third special radio gear. In Leningrad, Moscow, and elsewhere, "The whole country—factories, laboratories and institutes—were hard at work to fit us out for the long journey," Vodopyanov records.[15] Various ministries and other key machinery would provide and test the myriad of items to be designed (or redesigned), procured, and delivered: tailored clothing, provisions, shelters, scientific instruments, radio and other equipment and supplies. The working list of phone numbers for Shmidt's expedition—eighteen or so months of preparations—ranged from the USSR State Planning Committee to a solitary workshop creating felt boots.

Andrei Nikollayevich Tupolev and his team became the most renowned among Soviet designers for multiengine aircraft, called ANT—Tupolev's own initials. Four ANT-6s would sortie for the assault on 90° north latitude, each specially modified (ANT-6A). Extra wing and fuselage spaces were annexed for cargo, for example, and the pilot cabin widened. The magnetic compass would be prejudiced by the magnetic pole, so celestial navigation had to be substituted. The navigation compartments were retailored into near solariums, enabling the navigators to deploy sextant and sun compass—the latter a virtual twenty-four-hour clock used as a sundial. Here also were fitted the last word in instrumentation, plus a set of flight controls. Radio power was increased, ski undercarriages designed, an additional fuel tank installed within each wing. The M34 engines—

twelve cylinders/970 h.p.—were modified as well; each was fitted with special cowlings and warming lamps and equipped for cold-weather fueling. The expedition's four machines were painted in orange-red and blue, to enhance visibility in the event of forced landing.

No less than 818 of Tupolev's ANT-6s were built. Give or take, the empty weight of most versions was about 48,500 pounds fully equipped. "The British and French industries had nothing in the same league," historian Ron Davies writes, "and the U.S.A. had not yet thought of the B-17." To equip the North Pole party, each of the four allotted transports had to carry almost two tons extra—total expedition weight, nine tons. In Leningrad, the expedition's stockpile fattened. The prime concern now was to limit weight. Although each machine could deliver a reasonable load in a single sortie, the aggregate of men, equipment, and supplies restricted the camp-to-be to nine tons. Crucial items had therefore to be redesigned, including the ice camp's radio, living tent, instruments, and provisions. As for the ANT-6A, each was bare metal—all nonessential gear stripped from the interiors.[16]

Normal range of the big plane: 1350 kilometers (840 miles). Still, no aircraft had the range—fully fueled and loaded—for a single leap into the central Arctic. Shmidt required a forward air base. In March 1936, Vodopyanov and fellow airman V. M. Makhotkin set off in a two-plane reconnaissance, to scout an air route to Ostrov Rudolph. Located just shy of latitude 82°, Prince Rudolf Island is the northernmost of the Franz Josef group. When a weather station was established there (for the second IPY), it had been the northernmost scientific outpost anywhere. The most northerly soil in the Soviet Arctic, located on the very lip of the deepwater basin, this speck plotted within striking distance of the geographic pole—902 air kilometers (560 miles), well within operating parameters.

Two icebreakers were dispatched. Transiting the Barents Sea into the island group, the pair bulled their way close in, to a landfall at Teplitz Bay. Trails were cleaved through hummocky shorefast ice to a base camp, supplies hauled ashore. A working party set about transforming the chosen site, on the glacier cap above the bivouac, into a landing strip. About four kilometers (2.5 miles) off, a settlement arose boasting living quarters, a radio station, workshops, tractors, storehouses—and stockpiles for wintering.

For any bold stroke, men are the indispensable ingredient. The aircrew members were all seasoned comrades. The prospective colonists were no less carefully vetted. Ivan Papanin would lead the on-ice party. Chief of the polar-station network, long-experienced, Papanin at forty-two was a fast-track civil servant and a proven *polyarnik*. (His northern biography dates from 1929.) "He was short, stout and thick-set, walked quickly, had a rather commonplace appearance, but was extremely attentive and quick-witted," a colleague opined. "It was not by chance that he had been chosen from among the many candidates to head [in 1930] the most difficult and important polar observatory of that time—on Franz Josef Land."[17] No scientist, Papanin was a well-known specialist and administrator possessed of a formidable organizational skill. A born organizer and expedition leader, he had well-proven gifts. And his sunny disposition offered a distinct asset; in isolated camps, uncongenial temperaments can be disastrous.

Ice-leader Papanin helped pick his comrades. Pyotr Shirshov, thirty-two, a hydrologist and hydrobiologist, had been aboard *Chelyuskin*. A dedicated researcher, Shirshov had exploited the now-fabled drift to record observations. The two-month sequestration, indeed, had convinced him that it was "quite feasible" to winter on the floating canopy of the central Arctic. Yevgeny Fedorov, twenty-seven, would be the youngest member of the North Pole party of four, a relative newcomer—having graduated in 1932. A specialist in geophysics and terrestrial magnetism, Fedorov shunned laboratories, preferring the outdoors and expeditions. The renowned Ernst Krenkel would be radioman. The four had been cross-trained; all major field responsibilities would be duplicated. Krenkel could make the astronomical observations, if need be, he and Papanin reading weather. Fedorov would stand in at the radio. And biologist Shirshov, designated as camp physician, logged months in a Soviet clinic, honing his skills.

Self-possessed, hard-edged, finely disciplined, resourceful—each man was exceptional. On the ice, this particular blend of skills and personality would triumph. The four were politically correct as well: from the people, the *narod*. "They were," the author was told in St. Petersburg, "all sons of workers, of peasants; that was very good."

On one level, it was all theater—a great lunge and heartfelt, but theater nonetheless. Much was being dared. So the Great Leader was kept informed throughout the planning. When the pole party and its cadre of airmen were presented, patron Stalin pronounced his satisfaction. "Certainly [Stalin] was very much interested in this station," the Arctic and Antarctic Research Institute offered, "because he always wanted to show all the world that we are first in the Arctic."[18] Not for science alone were these Soviets being flung to sea. The pole camp was, as well, a powerful political act.

As the winter of 1936–37 ebbed, heavy supplies and a party, including the North Pole team and its major equipment were readied for Rudolf Island and its aerodrome (the world's most northerly). Lytoff awaited April or May, when the sun was high yet the air amply cold, thus granting a skiable surface.

It would be no timid overture. Leningrad was posed for a historic leap—a venture that was to cast a long shadow, shifting the course of northern science and observation.

Departure day—22 March 1937, Moscow. Chaperoned by Shmidt and attended by cameras, commotion and bustle, the Shmidt-Papanin expedition prepared to take its leave. Takeoff was delayed by rebellious engines and weather along the planned route that barely met the demands of safety. Then snow thawed to slush, obliging some off-loading. (The gear would be sent on by rail.) At last, the men strode to their respective machines. Setting forth were the N-169 (registered SSSR-N169), N-170 (the flagship), N-171, and N-172. One ANT-7 would accompany this squadron of four. Twin-engined, the smaller craft would fly long-range reconnaissance, probing ahead of the big transports, seeking weather favorable for the airlift assault on the pole.

Not until 18 April did the first machine, repeatedly delayed, plop heavily onto the glacial ice draping Rudolf Island. The intensity of polar weather governs all human exertion. As the full expedition rested, a watch on the skies commenced (operating requirements were strict). Days added to weeks. Storms are frequent in the Franz Josef group; man has to abide his chances. One worry at this phase was that the men would have to analyze conditions at the pole indirectly, from half a thousand miles away. Another concern nagged: the improbability of congenial atmospherics at both sites simultaneously.

In May, a series of probes were flung poleward. A preliminary reconnoiter was logged on the 5th by the ANT-7, Pavel Golovin at the left-seat controls. As its crew sped on, "Golovin gaily reeled off the parallels his aeroplane had crossed and told us that the weather was clear and the ice good," Krenkel writes. Though solid overcast was met at 88°, Golovin, irresistibly tempted, dashed further against orders—the first Soviet overflight of the north geographic pole. He returned, tanks empty, with good news—"there were plenty of smooth ice fields in the area of the Pole."[19] The next reconnaissance was logged on the 11th; yet another four days thereafter. On the afternoon of 21 May, the resident "weather god," Boris Dzerdzeyevsky, approved the main event.

Pockets stuffed with last-minute "contraband," the lead group shuttled to the flagship. Set on a dome-shaped plateau above the base camp, the runway angled downslope, thus granting departing aircraft a boost in speed and, hence, lift. The ice-camp party would sortie first in Vodopyanov's N-170; the sister transports would go later. Throttling to gain all possible speed, the overloaded machine galumphed free, Vodopyanov at the left-seat controls, copilot Mikhail Babushkin to his right. In his own cabin, Ivan Spirin—one of the Red Air Fleet's most able navigator's—was assisted by Fedorov. The aircrew this mission consisted of nine, accompanied by Expedition Leader Shmidt and the would-be pole-party. Navigator Spirin set his course.

In less than an hour, light clouds materialized. Climbing, Vodopyanov soon was skimming a thick, seemingly endless cloud field that, at intervals, rose up to enfold his machine. Maintaining bearing and position, Spirin "shot" an insipid, milky sun. A radiator leak was discovered in the inboard port engine; tension increased. Vanishing into the wing tunnel, the engineers proceeded to soak up the dripping, scalding coolant with rags. Having been squeezed into a bucket, the lifeblood was returned to the threatened engine. Consigned (until landing) to the status of mere passengers, the pole-party sat idly amidships, intent on the windows or their shuttling, stone-faced comrades.

Transecting 89°–deep in the white void of the central ice pack–Spirin and Fedorov nervously reiterated their calculations. They must not miss the place. Shortly after 11 a.m. (six hours' flying), Spirin and Fedorov "suddenly ran up in great excitement," Krenkel recounts. "Beaming, Fedorov shouted in my ear, 'The Pole!'" Vodopyanov, elated, resisted the urge to shout "for the whole country to hear me!" "And here was I–once a simple village lad, now an aviator, trained by the Communist Party–over the Pole! In a few minutes I would land at a place where no plane had ever alighted."[20] Through scudding gaps in the undercast, the surface was spied. The pilots pressed a bit further–"to make sure," Spirin explained. Radioman Ivanov began tapping out a message–whereupon the radio–abruptly temperamental–expired, severing all communication. The converter coil had burned out. Banking steeply, the pilots reversed their course.

Everything rested now upon chief pilot Mikhail Vodopyanov. Throttled to half-power, the transport spiraled down. At 2100 feet (640 meters) ice flashed into view, then disappeared. At 1800 feet N-170 broke through. "Before our eyes lay a panorama of the roof of the world," the airman recalled. "Boundless sparkling ice-fields were seamed with blue lanes of open water, and the vast ocean looked as if it had been paved with slabs of varying shape and size, like irregular geometric figures drawn by a child's unsteady hand."

Two circling flybys were sent to study the ice. Several suitable floes were apparent. A smoke box was dropped, to confirm wind direction, after which Vodopyanov banked to use it. Another virtue of Tupolev's genius was about to be exploited: low landing speed. The pilots chose. Maneuvering deftly, "I gently eased the engines, flattening out to make the landing; the tail dropped and for a second or two the machine traveled about three feet above the ice. Then I pulled the control lever and the skis softly touched the virgin snow." Jolting over uneven ice, its parachute air brake deployed, the bird slid to an abrupt halt after only 750 feet. It was 11:35 a.m. The expedition was sixty-five days out of Moscow.[21]

By an act of singular audacity, the North Pole was theirs–the first ever landing at the fabled extremity.[22] Everywhere, in all directions, on all sides, in a seemingly endless white, there was only "south." A cheer arose. "The happiness we felt," Papanin writes, "is hard to convey." All aboard were seized by a powerful emotion. "We rejoiced, exchanged impressions, poured congratulations on one another."[23] Beyond the windows, the icescape awaited. A hatch was opened, the ladder shipped.

A flag was set, brandy broken out (a gulp for everyone), images snapped. Beneath thin spring sunshine, the pole-party set to work straight away. It had traveled heavy. Radio and scientific equipment, food, fuel, clothing, assorted supplies and paraphernalia were spirited off, into scattered piles. Spirin and Fedorov took astronomical sightings while the mechanics, quite manic, fussed over their charges. Ivanov and Krenkel arranged the radio gear as the rest helped to erect the mast.[24] The need to send a transmission was urgent; standing by, the other planes would not come until word of their landing. In short order, a tent for the wireless was erected. Then a sleeping tent, into which collapsible mattresses and sleeping bags were unrolled. From the radiomen came word of further delay: the batteries were discharged. The penalty would be several hours more severed from the outer world. Hands behind his back, Shmidt began to pace.

After ten and a half hours, the pole camp found its voice: dots and dashes flashed to a waiting world. "We are all alive and the aeroplane is safe," Krenkel began. As Shmidt crafted a telegram, Krenkel chatted with Rudolf Island, each side offering details of the hours empty of contact. "The ice is in excellent condition . . ." Telegram number one, addressed to the Chief Directorate for the Northern Sea Route, outlined the landing, their approximate position. The floe, Shmidt continued, was "entirely suitable" for a scientific station. And once an airstrip was fashioned, the waiting machines and supplies could move.

Congratulations sped poleward—followed closely by a demand for a weather report. At 9 a.m., 22 May, the camp wired its 6 a.m. readings: the first such report ever from the interior Arctic: "Atmosphere pressure 761 comma temperature minus 12 comma wind westerly from Greenwich gusty stop fog stop sun hardly showing through stop visibility one kilometer stop slight snowfall."

The first transmission-day concluded, the party turned in, exhausted. Excitement, fuss, and anxiety had drained each man. As light snow powdered the overcast, a gusty wind swept the outpost. "Amid the boundless snowy expanses," pilot Vodopyanov records, "on a floe fringed by ice blocks, stood a big orange plane. Near it were silk tents, also orange-coloured, where thirteen Soviet citizens were calmly sleeping."[25] Maybe. Their rest shallow and troubled, two of the four slated to remain—Papanin and Krenkel—lay awake.

The public relations dimension is seldom absent from any venture of this magnitude. Repressing an instinct for secrecy, Moscow would nourish all possible publicity. This was a high-voltage story. Fussing over its "ice-comrades," the official Soviet media lavishly chronicled the ice-borne bivouac, including copy telegraphed directly from camp. And three of the four drifters published accounts. Sensing the eye of history, Papanin wrote a rendering that is ebullient yet self-restrained. Unfettered by self-doubt, his is a bold paean in the official line: iron resolve, cheerful stoicism, taut discipline—the regulation bonhomie of that square-jawed icon of impossible virtue, the New Soviet Man. The expedition was a community of Soviets, in capsule form.

Impatient to be up and doing that first morning, Papanin awakened the men with tea and biscuits—and hardworking good cheer. "In about ten minutes I'll have an omelette for you," the *polyarnik* smiled. A debate ensued: Was it forenoon or night? Without radio or chronometer, no answer was possible. As for the ice under their feet, a test hole revealed ten feet—a suitable raft. Hereafter, all would depend on its whims and vagaries. The four slated to remain would now either make history . . . or be history.

On 23 May, Rudolf Island relayed a radiogram. Shmidt called his men together—their first conference—and read greetings from Comrade Stalin ("our dear teacher," in Papanin's phrase) and the high *nomenklatura*. "The Party and government," the message began, "warmly greet the glorious members of the polar expedition to the North Pole and congratulate them on carrying out their mission—the conquest of the North Pole. . . . This victory for Soviet aviation and science sums up a brilliant period in developing the Arctic and the northern routes which are so necessary for the Soviet Union." Before closing with Bolshevik greetings, Moscow voiced its conviction that the party would conduct itself "splendidly."

On the third day of occupation an article appeared in *Pravda*, the mouthpiece of the Communist Party. Over the byline "O.Y. Shmidt," a headline proclaimed, "We Fulfilled the Tasks Given Us by Comrade Stalin." "The work of putting up the camp goes on," diarist Papanin scribbled. "We are pitching a tent for the kitchen and making a storage dump for rucksacks, apparatus, and various minor articles essential in our household. . . . The weather is bad, with a blustering wind. But the sun came out from behind the clouds for a short while, and Fedorov was able to make astronomical observations. Shmidt and Babushkin cut huge blocks of snow and used them for building walls around the tent, to keep the wind out. We know from experience that unless one puts ice or snow walls around this kind of tent, the wind blows right through the silk fabric."[26]

Rudolf Island, for its part, was suffering blizzard conditions. Urgency hung heavy: most of its gear had yet to reach the ice camp. Not until late on the 25th did meteorologist Fedorov forecast agreeable skies, whereupon the advance base was instructed to make ready. At the pole, the arch of sky cleared.

Three orange-red painted transports strained aloft, after which the squadron of three dispersed. Proceeding independently, the flight crews pressed campward.

N-171—piloted by Vasily Molokov, navigated by Alexei Ritsly—arrived exactly at the place, turned onto its meridian—and came heavily to rest in camp. It was 25 May. The airmen were accorded a hero's welcome, and the ANT-6A ("a regular warehouseful") unloaded. Camp mess now boasted twenty-two personnel. Proceeding independently, pilots Anatoly Alexeyev and Ilya Mazuruk carried out a search without result. Unable to promptly locate the Papanin camp–the merest speck in a pallid blank–the pilots elected to land on open ice in the suburbs, so as to conserve fuel and check position (by sextant). Nearer the pole, concern mounted. Key equipment–a windmill generator, the main living tent, the winch for hydrologic equipment–had yet to reach the station. All would arrive only with the absent machines now likewise adrift somewhere beyond the horizon-circle of white.

Alexeyev gained the nascent base a full week after the pole touchdown, on the 28th. His machine was unloaded "at once." (The camp had now to feed forty-two.) Brought out with his N-172 were food and petrol–and the main living tent. Its erection began.[27] As for pilot Mazuruk, on the ice twenty-five miles off, transport N-169 had lost its voice: failed radio. And a takeoff strip had to be cleared. Last to arrive, Mazuruk completed his delivery on 5 June. The airmen found an improvised aerodrome, a bustling, ramshackle bivouac: thirteen tents, including the main living quarters and radio tent, a galley, stores, and a weather station. Over the tiny settlement, a windmill generator spun dutifully.

PROCEEDING INDEPENDENTLY, THE FOUR-PLANE SQUADRON OF SKI-EQUIPPED ANT-6S RENDEZVOUS ON SEA ICE IN THE CENTRAL ARCTIC, MAY 1937. MISSION: TO ESTABLISH A SCIENCE CAMP AT THE POLE. PILOT IN CHARGE OF THIS AUDACIOUS, BRILLIANTLY EXECUTED EXPEDITION: MIKHAIL VODOPYANOV. LANDING SITE: 89°41' N, 87°41' W—19 KILOMETERS (12 MILES) FROM THE NORTH GEOGRAPHIC POLE. (ARKTIKA MUSEE)

Expedition leader and academician Dr. Otto Shmidt (left) and *polyarnik* Ivan Papanin (team leader) watch as one of the ANT-6 transports, delayed by weather, gains the nascent *Severnyy Polyus* encampment. A four-man team would persevere for 274 ascetic days on the ice. Among its discoveries, the "North Pole" camp confirmed the presence of a "reverse current"—inflowing Atlantic water beneath the cold, relatively low-salinity surface layer. (*Arktika Musee*)

The last of the belongings had reached camp—six ice landings. Among other supplies, a typewriter, a chess set, books, pots and pans, guns, collapsible boats, sledges, and chairs were manhandled off N-169. Everything out, someone calculated that leader Papanin—entitled officially to nine tons of food, fuel, and equipment—had somehow managed to squeeze in over ten. (See Table 1.)

The aircrews—transients—were consigned to tents, pitched beneath the wings of their machines. Fifteen days found the planes empty and quarters for the ice-camp party ready in all essentials.

A pole camp: a place of risk-taking, innovation, and enterprise. "Everything was new," Fedorov records, "a first-time thrill, discoveries all around." Singularly positioned, he and Shirshov got on with the work at hand. A hydrochemical lab was set up. On 4 June, Fedorov took gravitation observations and Shirshov managed his first lowering—to 1000 meters (nearly 3300 feet). The hydrological winch was still en route, courtesy of Mazuruk, so an improvised windlass was rigged up. Hands trembling with impatience, Shirshov recorded his inaugural deepwater temperature—premiere data for this ocean. The reading proved startling. The Soviets had detected the relatively warm (above 0°C) Atlantic water. This lies below the surface layer and is derived from inflow into the Norwegian Sea from

TABLE 1.
FOOD, FUEL, AND EQUIPMENT FOR THE SOVIET "NORTH POLE" EXPEDITION OF 1937–38.

Fuel, Food, Equipment	Tons
Food for 2800 man-days (770 days, ice party of four)	3.5
Fuel (primus stoves, lamps, engines)	2.5
Scientific instruments	0.7
Radio receiver-transmitters	0.5
Power apparatus	0.5
Clothing/personal items, tents, rafts, maps, domestic needs	1.3
Total (divided among four ANT-6A aircraft)	9.0

Source: Adapted from R. E. G. Davies, *Aeroflot: An Airline and its Aircraft* (Rockville, MD: Paladwr Press, 1992), 30–31.

the North Atlantic Current. (The underlying bottom water has temperatures below 0°C). Water warmed by the Florida sun, one Papanite wrote, presciently, finds its way to the North Pole.

This Atlantic water overspreads much of the eastern Arctic, not mixing with the surface layer. Oceanic heat conveyed northward by this mass, as well as absorption of solar radiation through leads (passages through pack ice) and *polynyas* (natural runways of flat ice) helps regulate global climate. If brought to the surface, this Atlantic layer would melt the ice cover. In physical oceanography, the layer of rapid salinity increase is termed a "halocline." At the water temperatures of this ocean, the halocline between the surface and Atlantic layers suppresses vertical mixing with the warmer water, allowing sea ice to form readily in winter and inhibiting melt. In 1937, however, the mean surface circulation was poorly understood, the deep return-flow of cold water unfathomable. The main features of world ocean circulation had yet to be charted—let alone simulated with even the simplest numerical computer models. (Such attempts would not begin until the late 1960s.) Nor, yet, was the fundamental role of the ocean in climate change at all understood.[28]

Following Shirshov's discovery, the first sounding was recorded—a deep look. Nearly three *hours* were needed to gain bottom—4265 meters (14,075 feet). Yanking against gravity and drag, six hours *more* were expended in raising the weight.

As the drift days added, this data trickle would swell. Meantime, the station was declared open, a flag put to the winds. Rudolf was reporting satisfactory conditions for return of the planes—"We can't miss this weather!" Shmidt announced. Just before takeoff, on 6 June, Papanin scrambled up each ladder, claiming last-minute prizes: pots and pans, canned food, emergency reserves. All were granted unbegrudgingly. Accompanied by shouts and waves of benediction, the larger expedition then trundled off, relegated once more to a supporting role. "It was a very unusual silence around us," Papanin observed, the stack-up of transients gone from camp, "and we realized that we were absolutely alone and only the ice and snow around us. . . . We went into the tent and drank a cup of wine and wished each other a successful wintering."[29]

This wee camp was the ultimate prison: emptiness enclosed as inescapably as walls. Harnessing phenomenal will, its party would persevere—engaged, disciplined—for nine months. What that meant is hard to imagine. Isolation stood total. In dirt and discomfort and danger, through days of routine interactions and months-long dark, the men would work and record and endure. Gone were the days of surfeit. Fuel, food, their very energy and rest—all now would be dear. Following a sorting out, equipment was distributed (by sledge) among three depots—a hedge against ice pressure and fracture. With their lodgings at its center, food, fuel, spare clothes, and rubber boats were cached at the corners of a huge triangle, to preserve assets and resources. For care and comfort, the living tent now was home—a physical space and social place.

The world's northernmost observatory was operational. Working in a scientific blank, its every measurement—meteorological, hydrographic, biologic, magnetic, gravity—tendered a contribution to human knowledge. The "North Pole" camp would host a staggering regimen of grinding routine plus simple, patient observation. On the mainland, as the weeks unfolded, waves of prepackaged image—uncritical dollops of Stalinist-style prose, a melange of bombast, information, and rhetorical excess—would recount its every element. The "Polar Heroes" were to inhabit a near-public stage. The equipment sustaining them (for instance) was perused fastidiously by news gathering inquisitors—clothing, food, beds, even the dishes. The darker realities of life on ice—cold, damp, and wet; illness, tedium, and drudge; the threat of breakup—were glossed, either expurgated or not much reported save as a test for socialism. As discoveries were realized, however, each was rhapsodized. For the ideology factories, the outpost was an instrument of policy—and a useful poultice for the grim everyday scarcity of planned economic life, the ugly domestic reality of "building socialism." And an escape from the growing terror.

Thirsty for tales of the Soviet "experiment," the West found the venture irresistible. Brash, hence newsworthy, and universally lauded (rightly) as a tour de force, the ice camp garnered international coverage. Shmidt and Papanin burst into the headlines (Moscow obliging), then refused to depart.[30] Who could deny the heroic quality of these Russians? Their acts were not mere showmanship for foreigners but, instead, real expertise driven by need, nationalism, and autocracy. Richard E. Byrd lent his praise but predicted that drift would force a shifting of the new bivouac. Stefansson had comment, as did the U.S. Weather Bureau–citing the value of a station in those wastes. In addition to bylined news from Moscow, Western readers thrilled to reports relayed directly from the "Soviet North Pole Camp." Late in the expedition, pressmen would not have to struggle to report true drama.

Workaday routines for these deliberate exiles would prove obsessive, peculiar, extreme. Supplies and manpower (plainly) were finite. Through the months to come, the hidden seams of character would be taxed in unforeseeable ways, under the daily rub of conditions that can only be imagined. The white world is no place for those averse to discomfort. And the hazard is unceasing. As in native cultures, teamwork, cooperation, and sharing–to the point of compulsion–would be key to survival and to the success of the ice party.

Regular meteorological readings held special urgency–the core of the at-sea program. (Because of sparse coverage, the interior pack is particularly important as a location for regular synoptic observations.) "Four times a day, with unfailing regularity," as Papanin tells it, "Krenkel's small radio station transmitted weather bulletins to the mainland where they were of great service in composing synoptic weather charts." To a large degree, weather in the western sector of the former Soviet Union depends on conditions in the central polar basin. Hence, combined with data from shoreside stations, the accuracy of North European forecasts was much enhanced by Krenkel's data pearls.

To assist its airborne dashes across the pole, Moscow demanded further of its "heroic wintering party." On 10 June, informed of preparations, Papanin was instructed to broadcast weather reports and provide wireless service for an aerial sortie to America. Eight days later, Chkalov lifted away. All night Krenkel was at the wireless. The Pananites yearned for a drop. Amid great excitement, the machine was heard. Then it faded. Low-level conditions were abysmal. The ablation of snow in the central Arctic begins in early June, finishing by month's end. Large meltwater ponds form, covering at least half the surface. Ice melt begins early in July. Wet snow from a weepy dead-gray fog wreathed a dreary camp. Disappointment stung deep. "Wet and up to our knees in slushy snow," Krenkel concedes, "we stood cursing the impenetrable, dripping murk. Letters, newspapers, spirit–all flew overhead and onwards to America."[31]

Its solidity is a deceit; in the play of brute forces, sea ice is an ephemeral raft. And beneath it lies a lethal bath. Only the instruments knew that imperceptibly the bivouac was being shifted. Erratically at first and slowly, the floe loitered; not until October–five months–did the camp drop south to 85°. Thereafter, its drift settled down, conveying the station yet farther from the motherland, ever farther beyond the pole. While the overall fact of this shift was known to the party, no special conclusions—for the time being, at least—had been drawn.

In the meantime, information was gathered hungrily. As conditions allowed, Fedorov "shot the sun" with a sextant to update their coordinates. His studies of atmospheric electricity and magnetic variations were granted sustained attention.[32] Shirshov, for his part, was engrossed in deep-sea studies using his water meters and bottles. Through an ice hole, this work was accomplished via line and winch, with a heavy weight lowered on a steel hawser. To divine the water column, Shirshov collected samples from various levels after each sounding—the number dictated by depth (in deep water, as many as twenty-five were taken). He then analyzed his specimens chemically. Thirty-three soundings were recorded by drift's end. One momentous finding was a thermal structure within the water column. Beneath cold arctic water, at a depth ranging from 248 to 745 meters (820 to 2460 feet), floats a layer of relatively high salinity and temperature. Detected at each lowering fully to Greenland, its presence demonstrated that Atlantic inflow spreads throughout the eastern Arctic Ocean and is not confined to coastal Eurasia—the then-accepted view.[33]

Maintaining an astounding level of performance, like machines the Papaninites displayed an almost pathological appetite for work. Relentlessly repeated, theirs was a routine of notes and readings, observations, housework, duties. Soundings (a group chore) were taken every thirty-five or so miles of drift track. When the plummet weight touched, an automatic brake stopped the drum. Crudely mechanized, retrieving cable and instruments demanded muscle power—and at least three hours of energy-stealing labor. When drift rate increased, exertions multiplied—because the cable lagged. After each lowering and retrieval, bodies ached to be relieved. In the teeth of these strains, regular soundings persisted data pinpoints in a maritime terra incognita.[34] "We established beyond all doubt that there cannot be any land in the vicinity of the Pole," Papanin exulted.

Bad weather bought time to work over observations. And to revive. (To further augment these sporadic reprieves, a rest day had been established—the 21st of each month.) But such respites proved malicious, nurturing gloom and boredom. And blizzards made a mockery of housekeeping outside of their sanctuary, all but obliterating camp and paraphernalia. An unrecognizable snowscape of drifts, rippled waves, and billows had to be cleared away from tents, huts, equipment, belongings, supply caches.

The rhythm of work ran down the adrenaline; yet somehow, the overworked four replenished their energy stores. Focus was total. "One dream possesses us all," Papanin said, "and we talk of it incessantly: that we may continue with our work to the end! No effort is too great for us; we *must* camp here till spring."

Every muscle was strained, then strained again, to gain the maximum of information and yet meet the requirements of their own immediate survival. A safe return lay quite beyond guarantee. Concerned for their priceless data payload, it was decided to transmit short, regular reports from the floe as they moved. Should misfortune strike, "the fruit of our labor would not persist unused, but would survive for science."[35]

To maintain energy, the classic pemmican and chocolate had been dropped. For their survival foods, the men dined instead on special concentrates cached in welded, forty-four kilogram cans—each enough for from four to ten days. The (frightfully named) Institute of Food Engineers had produced a nutritive and varied if disguised diet of staples.[36]

The pack is no place for the clean and tidy, for comfort and decorum. Domestic habits were necessarily casual. This included physical hygiene: the washing of face and hands, say, was largely ignored. Instead, the four simply sat down. "Only on big holidays and the twenty-first of every month, our jubilee day, do we get fixed up so we don't recognize one another. Then we shave, wash up, and even brush our teeth. Usually on days like these, we poke fun at one another: 'I bet you'll be quite a hit with the ladies!'" Though equipped with abundant fur clothing, they changed socks and linen twice a month. One vital commodity was fuel. "We liked to have our food hot with a minimum expenditure of our precious fuel," Papanin adds. "If successful in this, the cook on duty received good marks." Generating freshwater for anything other than drinking and cooking was a luxury. So crockery went unwashed.[37] As for baths, cleaning the body as well as toilet duties were rather more complex than when ashore. A real bath awaited retrieval.[38]

In this cold-blasted seascape, mid-July is the height of boreal summer. Under ceaseless solar attack, the floe was liquefying. During the period of maximum melt, ponds and wet snow overlay much of the ice—an unrelieved, impossible wilderness of light and liquid water. In all, about ten inches of underpinning would vanish. An engulfing dampness pervaded everything—cold, miserable, depressing. (In wet weather, everyone was more irritable.) The snow cover melted down, exposing food and gasoline stocks. Water dripped and impounded—pans of freshwater atop the ice that interfered with work and menaced belongings. Walking rounds were a soggy, dreary business. And the difficulties of keeping house were exacerbated. For one thing, the snow insulation encasing the living tent began to leak through. And each day, the shelters had to be refastened. Fogs draped the sea, as did wet snow, rain, drizzle. The laboratories had constantly to be shifted, the various depots becoming reachable only by rubber boat. Getting lost would be all too easy. Even had they had wished it, no aircraft could evacuate them now.

A new tyranny had intruded: wet clothes. Drying out was hopeless in rain or wet snow, so as much time as possible was spent under (cramped, wretched) shelter. To be shuttered indoors was not possible, however; so when soaking wet, the men had no choice but to change. The Primus stove was brought inside and lit and their clothing hung about.

Failing to plan in advance is planning to fail when adversity happens. As the occupation lengthened, the party had a constant companion: the threat of breakup. The seasonal cycle of freezing and melting of seawater produces a net excess of sea ice; eventually, this surfeit migrates out between Iceland and the Greenland massif and thence through the Denmark Strait into the far North Atlantic. The main "rivers" of ice-motion are defined by the Beaufort Gyre, a clockwise circulation centered off the Canadian Archipelago, in the western Arctic, and the transpolar drift in the eastern Arctic Ocean. The latter arcs from the East Siberian and Laptev Seas fully across the interior basin, into the Greenland Sea. This is the river the four were riding. And so, as their field shifted deeper into the American sector, the danger mounted. At first, wind seemed the sole engine of the ponderous drift. As observations piled up, the actual course (whose speed and direction the drifters knew) could not be explained by the deflecting effect of Earth's rotation alone. Calculations showed that, in addition, the ice was being driven by a constant 1.5-mile-a-day southeast current. Shirshov had found the Transpolar Drift Stream—the great midway debouching into the high North Atlantic. The Arctic Ocean, in other words, is dynamically coupled to the Atlantic.

Ice conditions had so far held relatively stable. No more. The frequency of floe interactions (collisions or sudden failures) intensified as the local ice field rafted south, loosening and diverging. In addition, the wind helped to bully, stir, and jostle the pack, raising up blocks of tortured ice. Ridges, fissures, and finely crushed white would attest to this insistent, incessant pressure. The nineteenth of July brought a warning, a precursor of the restlessness to come: with the camp's coordinates still hugging the district of the pole, a wall of blocks pushed upward along a pressure ridge.

21 July—two months' isolation. The pack lay covered in water, the interconnected lakes forming a large, improbable sea. On the 31st, in rising wind, under pressure, the floe edge could be seen breaking.

Polar warmth is evanescent. Within weeks, the quartet would be facing the raw, months-long moods of winter. "We know from experience that the long polar night is anything but pleasant. And to live through the frigid polar winter on an ice floe is exceptionally trying. But we are firmly resolved to hold out whatever happens and to take as many scientific observations as possible, and even more than possible!" The constant daylight and miserable warm soon would cease. And the nights grew long. "We began to prepare for the polar night," Papanin continues. "Though it is still a month off, we must put our household in order now. I collected lamp chimneys for our oil lamps, unpacked the cases containing our supplies, took out the lamps, and lighted them to estimate how much kerosene they consume. . . . All this testing and preparation gave occasion for much reminiscing about the polar night. Each of us had lived in the Arctic before, during the period of polar darkness; but we had been at large well-equipped Arctic stations, living in thick-walled wooden houses built on the land, on solid earth. Now we faced the prospect of spending the long polar night on a drifting ice floe."[39]

Extensive melting had ceased. And new snow lingered.

The Papaninites tired of powders and concentrates, delicious or dreadful, which satiated rather than satisfied. Interred in an ice pit, their perishable stock included a cache of veal. Early August found it edible, sort of. To enhance visual appeal and taste, a cleaning was conducted. "Of course, should such veal be put on sale in a meat shop, any sanitary inspector would certainly fine the store manager and prohibit him from selling it. But here on the ice floe we don't draw the line so nicely, on questions of this kind. Besides, we have our own doctor—he will agree, though our stomachs turn over on occasion from bad fare, there's nothing to be done about it. Fresh meat is essential for us here, especially for Krenkel. If the meat has a foul smell then we spice it."[40]

The party almost acquired a supply of fresh meat as August's first hours brought animal life. At 3 a.m., Krenkel crawled out of the tent, for a round. His alarm roused sleeping companions, "Hurry, get up! Three bears have come!" As the others dressed, a shot was fired. Thoroughly alarmed, the bears lumbered off, the men scrambling after. The quartet returned overheated—and disappointed. The flavor of fresh steaks would have roused their jaded palates. However, the visit was proof that Nansen had been in error when he claimed that life here was "nonexistent." A number of birds had been seen. Now the she-bear and her cubs. The great creatures were carnivores—living on sea animals, especially seals. So such had to be near. "The existence of life in the very center of the polar basin," Papanin records, "is now an established fact." As if to confirm the conclusion, a seal was spied two days later while canoeing in a lead.

The absence of professional medical attention further increased the party's malaise, as by drift's end each man would have suffered various maladies: headaches, vomiting, the insidious effects of nervous strain. With no doctor to attend sickness or accident, Shirshov's nursing—and his collection of knives, catgut, needles, lancets, and other surgical instruments—stood between the men and death. His skills held no allure: anyone under the weather manfully bore his affliction; operations would await Moscow. "Maybe Pyotr Petrovich [Shirshov] mastered the art of surgery in preparation for becoming our doctor," the leader offers, "but here on the ice I somehow do not feel like letting him practice his skill on me."[41]

The need to break routine, to reward oneself is altogether human. Camp life was made bearable by such luxurious touches as books, newspapers, a chess set, and other distractions. Offering relief in the form of domestic pleasures, the gramophone (a gift from the pilots) held special status; many off-hours were logged listening to records, especially jazz. Candies, tobacco and cigars were at hand, along with a modest cache of cognac. Following dinner, the men enjoyed settling in. Accessorized with fur rugs, the living tent was comparatively warm, cozy, communal—as pleasant a place as they could make, to smoke, to relax. Here, tuned to a concert or opera or news, the radio worked its miracles. Radiograms and messages streamed in—letters, greetings, congratulations. All were savored.

When light (and time) permitted, motion and still images were recorded. Meanwhile, newspapers and TASS, the official Soviet news agency, had special correspondents to appease. Cast as journalists, the four struggled to answer the demands of reportage, intercutting work with often vivid dispatches to fawning editors—a secret annoyance.[42] Much of this copy in turn was reprinted in the West—censored, of course. These reports from the ice, as Russians say, did not "carry the rubbish out of the hut"—air unflattering Soviet laundry. Unfavorable references had been deleted. Nor were specific data released.

Alone amid ice and sea (and, soon, the dark), their sole link to humankind was radio. Upon its set the ice camp relied absolutely. "The wireless means life itself to us, and we value it especially highly, tending the instruments constantly and looking after the wind motor which faithfully charges our batteries," Papanin recounts. "Our trusty receiver was an inexhaustible source of pleasure and encouragement," Krenkel affirms.[43] There was good reason for this. At electronic speed, their latest position was reported to shore—an incalculable comfort should emergency arise. In turn, mainland news (including messages from family) darted directly to their tent. If we can take Papanin at his word, a feeling of well-being was the rule. Despite the rough-and-tumble, "In general," he writes, "We live under very satisfactory conditions. The only thing we lack is time." Toiling sixteen hours daily, often much more, the quartet was shouldering a workload which, at a normal polar station, might engage a staff of ten. Sleep was scarce, hence prized.

August's third week—three months on ice. To celebrate, a special-occasion dessert and a wine glass of brandy were shared at dinner.

At mid-month, Levanevsky, pilot of the third polar transect, disappeared with his crew of five. Despite frequent fogs, a rescue expedition (Krenkel was told) might base at "North Pole." Abruptly, rescue-related work took over, and all but the most necessary observations were suspended. On instructions from Shmidt, a ski tour of the floe was made to find a site suitable for airplanes. Boundaries were staked out and work on an "airfield" closest to camp began in earnest. (The old strips were in a bad state, with holes and mounds.) Ice blocks were crushed and cleared away, mounds and ridges attacked with crowbars and spades. Within days, a place had been rendered smooth and even. The coming frost would leave it a solid sheet—a natural runway.

September announces winter. Surface air temperatures descend—the onset of fall freeze-up in the eastern Arctic. Ice ablation ceases as summer expires, defeated by the descent of boreal chill. In the average annual cycle, temperatures begin to drop rapidly early in the month. Meltwater as well as seawater begins to freeze, and the ice surface cools. A two-month transition to darkness commences, during which the light of a waning sun fails to a long, dark evening of hibernation.

The cycle commenced. As the icescape changed from puddled to frozen and snow-covered, the camp-floe of Papanin was again assaulted by pressure. Cracks, ridging and finely crushed ice attested to a mighty jamming in their sector. Dragged into the crush girdling northeastern Greenland, the raft now met severe ice deformations—the prospect of chaos and annihilation. The watch was redoubled. If necessary, the four could shift to a larger, safer pan. As latitude unspooled, however, the enclosing field would become more scattered, its component floes fewer. Meantime, surveys were conducted onto the surrounding pans.

In slanting light, the chill deepened. Inside their shelter, the below-zero air was scarce improvement on the cold without; the instant clothes were doffed, the men were obliged to enter their bags; otherwise, sleep would be snatched away as the men tossed about shivering. On 9 September, the kerosene lamp was lit for the first time. It would burn until February.

To mark the four-month anniversary of their isolation, Shirshov sacrificed five ounces of cognac.[44] "During our holiday meal, we drank a toast that our drift may continue to a successful finish. We are all in high spirits," Papanin records, "feeling refreshed, and glad that we all had shaved, washed, and changed our clothes. After dinner we relaxed and rested for an hour."[45]

Lowerings demonstrated that Atlantic water beneath had become decidedly warmer. The Greenland Sea—downstream—was nearing.

The waning month had its waning daystar: the barely lit sun. On the horizon blinked the first of the star—astronomical guideposts to help plot their course through blackness. In camp, a whole "street" of igloos materialized as, gradually, the observatory, stores, and other facilities were transferred from tattered tents. The disc of the sun now crawled their horizon. By early October the sun peeped from the rim of earth—its final performance for 1937. Frosty dark draped the ice world.[46]

Early in this drift, the floe had ambled sluggishly in the general direction of "south." October brought change. The movement (coordinates now confirmed) had grown increasingly stable. And rapid. Despite the frost, large areas of open water appeared.

Before the month was out, the bivouac-of-four stood closeted in snow-driven night. Darkness voided visibility, of course, but not the tether to home. One day, Krenkel woke the chief to report, excitedly, "Well, I certainly had a bustling night. I talked with ten American wireless amateurs. They passed me on from hand to hand, from one to the other. They know all about our expedition." Not calling any station in particular, the camp operator sometimes transmitted to anyone who might hear him. Krenkel would log contact with radio enthusiasts worldwide.[47]

Back home, a newspaper piece, "The Secret of Our Success," was published late in October. Thanks to their "most progressive equipment" and to Soviet science, Papanin informed his countrymen, the plan was proving successful. Earlier, tragic expeditions, he observed, had not enjoyed the embrace of the new Soviet system, and its party and leader.

Next day, powerful shocks–unlike anything the expedition had felt before–heaped ice. These heralded the floe's approach to Greenland–a zone of quickening ice convergence. The sector is permanently covered except (usually) during August and September, when the pack begins to melt and disintegrate. An aural riot commenced. Churning, shifting, buckling and regrouping, the ice fields had come alive, inveighing against the nocturnal map of stars (which simplified Fedorov's observations). "The message conveyed by this hellish din," Krenkel observes, "was clear: from now on we should have to be constantly on the alert." "We live as if on a powder keg," Papanin writes, "for at any moment an ice jam may occur, the ice floe crack apart and capsize, and draw us down with it to the depths."[48]

In the pole district, position plots had defined a looping, zigzag track. Now, as drift quickened, the track straightened–astonishingly so. In the central basin, the ice field had shifted southeast; losing latitude, by September it had met the Greenwich meridian–the line from which east and west longitude is read. Throughout the next two months, the drift path deviated but slightly east or west of 0°. And after early November, the pack pushed south-southwest–straight for the Northeast Foreland of the Greenland massif. The speed of its advance raised the likelihood of the camp getting caught (during boreal dark) in the Greenland Sea, a zone of breakup. Its floe was dashing for the exit (Fram Strait), the camp of Soviets bounding toward destruction.

On 8 November, alarmed functionaries debated having the expedition evacuate. Accordingly the chief of the Arctic and Antarctic Administration (then on Rudolf Island) advised that he would solicit–from Moscow–a December abandonment. Papanin could not agree: the party, he declared, was in no immediate danger. With everything yet "fine," the leader hoped for (and planned on) a spring retrieval. Any earlier, he added, would mean a withdrawal in dark. Agreeing (grudgingly) not to press his plan, Chief Shevelev emphasized he would keep the aircraft in constant readiness.

That November, the *New York Times* printed a portrait of life on ice. And before the month was out, articles described its half-year jubilee (the 21st), then Thanksgiving at sea.

On the 20th, Fedorov's observations had placed them at latitude 83°26′ N. Six months was celebrated next day. Closeted together, the quartet tuned in a jazz concert organized for them, radioed greetings, and sipped a cognac brew. "The real feature of the celebration," the *Times* reported, "was their exchanging messages with their families, with whom they talked almost as if they were face to face." Six days later, Fedorov's results announced that the floe had skittered 48 kilometers (30 miles) in 24 hours—a record. A month later (26 December), camp coordinates had shifted to 80°32′. Four days more had it at 79°54′. "Farewell 80th parallel!" Papanin jotted. "The drift of *Fram* came to mind, and we remembered that Nansen had been nervous and dissatisfied because the drift was too slow. But we, on the contrary, are alarmed because the ice floe drifts too fast."[49] An unseen river, the cold East Greenland Current, was bearing their pan to the high North Atlantic.

The expedition was drawing to a climax. Shoreside, there was an undercurrent of uneasiness. No explorer is ever guaranteed a return ticket. In boreal fall, the ocean freezes over and ice activity increases: the thick ice of winter develops huge stresses under the influence of wind and current. (When the canopy begins to thaw, activity diminishes correspondingly.) Off Greenland, the rumbling of hummocks forming was "our constant companion." The ice field was breaking up, and the resulting fragments were warring for space. In variegated chorus, as its floes pressed one upon another, they screeched, ground, and jammed in protest as they brawled for space. Fedorov discontinued his gravity readings: the "incessant shocks," he complained, shook the instruments.

A heightened, tense vigilance replaced routine. Countermeasures were installed and every effort made to prepare for emergency. Equipment-laden sledges, for instance, stood positioned at the tent. A night watchman now stood sentry, responsible for monitoring the state of their raft plus changes in weather. (Storms are particularly destructive to the pack.) And a quick-exit hatch had been cut into the top of the living tent.

As land drew closer, the condition of the floe—already precarious—deteriorated. (December saw the field shift 314 kilometers [195 miles]—3° of latitude.) Their field might slam into Greenland or jam and splinter. Nonetheless, near Christmas Shmidt announced that, barring any emergency, the four would float until spring. On 27 December, seeing Shirshov returning from his work tent, Papanin inquired about the depth to bottom. Rather than nearly 4000 meters (13,000 feet), the bottom had shoaled to 215 meters (705 feet)! Next day, the leader reported by wireless that his team "won't be ashamed to return to the mainland, since we have collected much valuable data." And to his private notes he confided, "Our safety, our further progress—everything depends on the radio station." In the two hundred days since alighting, the campers had floated 1200 kilometers (745 miles)—to sea, off northeast Greenland.

The last day of 1937. The four inspected camp before retreating to their "palace," to mark the moment. Each man shaved and washed his head. Celebratory caviar, salami, bacon, cheese, nuts, chocolate and candies were broken out. To assist the general merriment, they tapped into pulses from beyond the horizon: Moscow, with its Red Square traffic, the "International," the chimes of the Kremlin tower announcing 1938. Best wishes and kisses were passed round. Fedorov went out to make his observations, after which Krenkel transmitted a report. With a toast to Stalin, the four set to their meal—two hours. As they ate, Shirshov voiced unease: it was not the darkness he feared but the bears, he said, which furtively prowl the polar dark. The concern was justified; a carnivorous visitor might well appear. All the rifles were cleaned.

New Year's Day delivered a startling range of temperatures. The thermometer climbed to almost 43°—warmer than in many days. But by evening, a strong wind had squeezed the mercury downward, to minus 31°F. Cold, their worst so far, would torment. (On the 18th, a record low was recorded: $-52.6°$.) The dawn of the long daybreak appeared—a thin glowing streak along the horizon.

No rescue had been expected before spring. Amazed at the rate of drift (which exceeded all expectations), Shmidt announced that two ships carrying airplanes would steam to the embattled camp. The man's organizing skills were again asserted. Forces mustered. Radiograms to the ice told now of rescue preparations. (As late as 18 January, perhaps masking concern, an all but unflappable Papanin was advising a March removal.) On the 10th, the hydrographic ship *Murmanets* stood out from Murmansk, about 1610 kilometers (1000 miles) to steam. Alarmed accounts reached the West. "Soviet Now Rushes Aid to North Pole Party as Ice Flow Starts Drifting South Swiftly," the *Times* blared on the 11th.

Beset by ferocious (yet compelling) violence, the Papaninites watched the riot with a sense of marvel. "This morning [the 12th] I stood near the tent a long time, admiring the glow of the dawn," Papanin recorded amid the rumbling and erupting. "On a moonlight night our camp is like a painted fairyland. A mist of varicolored light lies on the ice floes all around us. Ridges of heaped-up ice blocks glint and sparkle in the soft moonbeams."[50]

The twenty-first of January. Their self-declared holiday had again come. The team shaved, washed, savored cognac and a special meal. "But today we are in no mood for celebrations or gala dinners. The jamming all around keeps reminding us that we are living on an ice floe and that at any moment it may crack asunder, leaving us at the mercy of the deep." Ten days later, the concern stood unchanged. "It feels as though we are living in some kind of huge sack, in the grip of some mighty, invisible hand which willfully gives us a sound shaking from time to time."[51]

An auxiliary motor-ship, *Murmanets* had hove anchor for the Greenland Sea. Its orders: to note ice conditions and to maintain radio communication with the imperiled research station. Krenkel established contact as she patrolled the metastasizing ice field, steaming its edge. That same day (the 26th), crimson clouds appeared to the south—a sign of sun. Instead, a hissing blizzard came then stormed on through a six-day night without stars. Hurricane-force winds buried the shelter entrance, noticeably tossed the floe and (save for brief excursions) confined the four to the thick, smoky atmosphere of the tent. Disrupted communications stole "our single pleasure"—listening in to Moscow broadcasts. With the sledges and canoe and depots snowed under, should jamming and the need to flee occur, death would be sure.

Retrieval, Shmidt had announced, would wait until March. But as January closed, constant, violent agitation persisted: heaving fearsome shapes, the plain of white was sweeping in against Greenland, jamming and twisting and piling up. The mitosis would undo the camp. A fissure appeared. Water welled up, forming a lead. Shocks were so jarring that snow slid off the living tent. The next day, along with spreading tremors, a new breach appeared. Then another wiggling black snake pressed toward and then *beneath* the tent. A hasty shift to provisional quarters proved timely; alive and moving, the crack yawned wide. On 2 February, the wireless tent was threatened. More shifting around on their flimsy, shrinking pan. "Soviet Polar Floe Splits; Rescue Efforts Hastened," the *New York Times* proclaimed. Amid gathering impatience, the government had ordered a rush to the rescue. On the ice, depots at the tiny camp were cut off: another blizzard. Disintegration well advanced, little remained of the original ice field. A reconnoiter disclosed only fine broken pack. The wind had proved stronger than the pack and had delivered a real jolt, though not yet a fatal one. Its assets and resources imperiled, the fate of the expedition stood very finely balanced.

As for drift, the speed was startling. On 1 February, after taking bearings, Shirshov had announced a latitude of 74°16′ N, longitude 16°24′ W. In six days, their raft had sprinted 193 kilometers (more than 120 miles) to the south-southwest. "Twenty miles a day!" Papanin penned. "Now that's speed for you!" From a nearby ridge, the ice that eight months before had withstood the shock and weight of four planes was but a mass of fragments rolling in the swells. Even a light-plane landing was impracticable: no adjoining floe floated large enough. And the property was near uninhabitable—a mere sliver detached from its neighbors. To these tightly defended square yards the four clung with a near-manic persistence. Pushed to the brink, the Papaninites rebounded; gathering itself together, the bivouac endured.

The rescue organized. Norway and Denmark offered aid. *Yermak*–fresh from duty on the Sea Route–was pressed into urgent service, assisted by the transport *Taimyr*. The *Murman* (specially outfitted) joined company. This was genuine theater. As preparations ended and the expedition sortied in great alarm, Western newspapers tendered almost daily reports concerning the rescuers.[52]

The operation called for brilliance–and plain good luck. Fog is common in that sector at that season. And the waters are crowded with drift ice; the current "delivering" the four conveys great fields of it. Fracturing, buckling and overriding, the white-ocean skirt off East Greenland is ill-behaved and, perhaps, the most heterogeneous anywhere. On shore, the ice sheet is rimmed almost completely by coastal mountains through which ice is discharged by outlet glaciers and ice streams. To seaward float countless bergs, winter sea ice from uncounted fiords, and thick multiyear floes exported from the Arctic Ocean.

Off ice-heavy Greenland, the sun peeped forth. Courtesy of amateur wireless enthusiasts, the camp heard alarming accounts of its peril. The camp raft was disintegrating, foreign readers were told, the ice around them shifting and breaking and piling up. "Soviet Scientists in Extreme Peril," headlined New York's own *Times* (on February 7). Tension heightened. Six days thereafter, another report:

> The Russians have shown that they have a flair for the dramatic in polar exploration as well as courage to attempt the unusual. As a result there has seldom been a more interesting finish to an adventure in the frozen north than that which is now approaching off the east coast of Greenland, where four Russians and a dog are drifting at the mercy of the grinding pack ice.

Taimyr, first to reach the target sector, scouted possible through-routes then nosed into heavy pack. Grinding a wake, the icebreaker artfully maneuvered the clearest course to shorten the distance, her conning officers exploiting leads and weak ice. Ship's master Ulyanov (according to *Pravda*) shouldered a particular anxiety: Had travail, stress, and seclusion, he worried, changed the four who, now, were his objective? On the 12th, Krenkel roused his comrades: a light dotted the horizon–the beams of searchlights. Lying to, with nearly impassable ice interposing, the would-be rescuers ordered aircraft aloft to reconnoiter, to help plot courses to the ice camp. "Everything," Vodopyanov writes, "depended on air reconnaissance."

The anxiety of the rescuers turned by degrees into elation. On the 16th, Pilot Gennadi Vlassov (from *Taimyr*) was first to succeed. Alighting upon Papanin's extemporized "airfield," he met the redoubtable *polyarnik*–the first new face since the pole. Vlassov then took off to continue his search, this time for comrade-airman Cherevichny, from *Murman*, whose seaplane had been forced down (safely). Meanwhile, a gift was savored in camp: a crate of beer and mandarins.

Guided by aircraft, *Murman* (in company with *Taimyr*) exploited the winding rivers of open water, negotiating the shifting, jostling floes. (*Yermak* had yet to gain the ice field.) Within forty-eight hours, *Murman* had bulled to within five miles, the red flag surmounting camp visible now through field-glasses. "We can see you quite well," the Papaninites radioed, "and welcome you with all our hearts. Best wishes." The night of 18–19 February was their 274th on the ice. From two miles, Vodopyanov recounts, "a light was seen to twinkle in front. The long-awaited moment had come. Ships' crews gathered on deck or climbed the shrouds. And there in the distance, on a tall hillock, stood four members of the drifting station waving torchlights."[53]

Ships' whistles sliced heavily—proudly—through the black chill.

The living tent was dug out, disassembled. "I walked all around the ice floe," Papanin records, concluding his documentary, "saying farewell to it." The leave-taking doubtless held an exquisite sense of relief: their awful bondage was done. A feral-looking group, the ice men were pallid, haggard, and filthy—unshaved in weeks, unshowered in months. For the final time, Krenkel left his snow-house station, having just broadcast "to all, all, all" that the work of the station had ended. Its snow walls then were knocked down and a sledge pulled away . . . to the bright and bustle of two ships. The date: 20 February 1938.

The "North Pole" station had been relieved near Scoresby Sound, off east Greenland. The expedition coordinates upon evacuation were latitude 70°03′ N, longitude 20°00′ W. During its nine-month passage, the rafted base had logged more than 2400 kilometers (1500 miles). Safely shipboard, the four luxuriated in an ecstasy of relief and extravagance. Krenkel, for one, dispatched exuberant word home. "Having a wonderful time," he telegraphed from *Murman*, savoring a newfound ease. "Have taken my first bath. Am eating oranges and smoking cigarettes. Life is marvelous."[54]

The party stood dockside on 15 March. All of Russia went wild; triumphalist rhetoric overflowed. The *Red Newspaper*, a Leningrad daily, reflected a genuine excitement, giving over an entire issue to the homecoming and festivities. Yet the Soviet Union was not a happy place. Its back page spat forth the names of the latest "enemies of the Soviet people" along with the obligatory charges. (Among the unfortunates were Leon Trotsky and Nikolai Bukharin, Lenin's "favorite son" and one of the original members of the Communist Party Politburo.) Stalin was consolidating his power: the arrests, denunciations, and purges, punctuated by paranoiac show trials, were rising to a grim crescendo. Where the czars had condemned the people in the name of the czars, the communists condemned the people in the name of the people—on a calamitous scale. No one was immune. In

October 1937, designer Tupolev, creator of the very machines to have gained the North Pole, was arrested.[55]

Justly celebrated, lions of the moment, the four–bedecked with flowers–were chauffeured down Kirov Street to the Kremlin. Along the route civic order was tested as delirious crowds set upon them. The men were decorated by M. I. Kalinin, chairman of the Central Executive Committee, who spoke of an air service to America within five years. (In the United States, the transpolar flights of the prior summer were cited as the year's most outstanding air exploit.) Fulsome headlines trumpeted a "Bolshevik victory," lauding the four as "conquerers" of the North Pole.[56] The making of heroes out of explorers and aviators, the glorification of technology and hyperpatriotic themes characterized Stalin's socialist society.

> The superiority of their deeds, the technology which they utilized and the conquering of the forces of nature, as Stalin seemingly guided them from victory to victory, all combined to convey images of order, progress and achievement, unity, and happiness. In all the media–press, graphics, stage, radio and film–the exploits of the polar explorers were used to create and present this image of Soviet society to its people. The image, in turn, fulfilled its function–it bolstered the legitimacy of Stalin's regime, helped him to implement his policies and enabled him to consolidate his power.[57]

Papanin ranks in the Soviet pantheon. That March, the *polyarnik* saw promotion–to head of the North Sea Route Administration, a position he retained throughout the war.[58] As authorized agent of the State Defense Committee in Murmansk, the world's most northerly port, the former *polyarnik* would supervise delivery and acceptance of a flood-tide of Lend-Lease supplies (1941–45) from the Western Allies, to help sustain the USSR against Nazi Germany.[59] Well before that crisis, word of a stamp engraved to honor the expedition was released. And more drifts, the party announced, were planned. Privilege and esteem showered down–every possible reward and honor. This had an element of unreality. "People whom I had seen only on portraits before," Fedorov marveled, "were seated at our table and asked us, still stunned, about our work, life on the ice floe, and our families in a friendly manner."[60] Adulation aside (an acquaintance confides), in the next chapter of his life Papanin "could not be in earnest about it, because he realized that he was far from being an outstanding scientist. During his life-time he was as popular in this country as Richard Byrd was in your country."[61]

Taking risks and making choices, the four drifters had survived. It may be difficult to truly appreciate how dire their situation was, in part because the Soviets were so successful in dealing with it. Ivan Papanin's (and Shmidt's) superb leadership, sound planning, grit, skill, forbearance, luck—each had contributed. Exploiting clumsy technology in service to the state, this one venture had verified a new pilgrim trail. Many would follow. The experience, Shmidt remarked, had demonstrated the great possibilities of using the airplane as a research tool.

The North had exacted its price, certainly. Through brutal months amid Zen-like deprivation, each man had been tested. Often, the party had known danger. Yet the four had absorbed all, conducting themselves with uncommon self-discipline, tenacity and confidence. And the joy of discovery had been theirs. Disappointments had come, of course. The rations and their gear had sustained the Papinites well enough. Yet the fare had been largely unappetizing, sickening the party more than once. As for equipment, the silk tents had proved less than adequate, as had the design of the hurricane lamps. One inexplicable omission: the failure to make provision for a radio beacon to assist any hunting aircraft.

Success had been nurtured ashore by means of superb planning, organization and logistics. At a time when Alaskan waters were unplumbed, the Soviets had thrown an expedition across fully half the central Arctic Ocean. The north, some claimed, had been "conquered"—an exaggeration. Still, a beachhead of hard-won data *had* been set for all that was to come—teams of scientists from different disciplines and different agencies, sharing resources and a common purpose. P. A. Gordienko is a legend. "The oceanographic, geophysical and meteorological work of [the Papanin camp and *Sedov*] into the heart of the Arctic Ocean," he writes, "combined with the data issuing from the network of fixed stations ashore, began to yield an entirely new understanding of weathermaking processes of the Arctic—and of the earth as a whole—even before the outbreak of World War II."[62] Soviet achievement in the north casts a very long shadow.

Results were canon-smashing. The drift had yielded the beginnings of real scientific understanding of the maritime polar depths. Radical discoveries stood logged, tidy theories had fallen. Compared to temperate-sea broth, for example, the central Arctic had been dismissed hitherto as a nutritional desert. (Nansen had failed to observe much animal life.) Its extreme seasonal cycle notwithstanding, the interior ocean ecosystem is robust—the Papaninites had seen gulls, seals, bears. The sea itself had proved alive with zooplankton—microscopic plant life thrives in the upper layers throughout a limited, intense growth period. "Evidently sufficient sunlight penetrates the polar ice in the summer time," Papanin writes. "And if vegetable plankton is present, animal plankton, which is the ultimate source of sustenance for all sea creatures, can also develop in fair quantities"—an analysis that proved correct.[63]

As revelations were teased from the data, further wisdoms fell. The Northern Ocean fills a deep crustal basin, as Nansen had shown. A regular, predictable system of surface currents and circulation seemed verified, a fact hinted at by the remnants of *Jeanette* and *Fram*'s own drift. Meteorologically, the notion of a constant region of high pressure in the central Arctic had been replaced by evidence that lows born in the Atlantic frequently overrun the polar area—a cardinal discovery.[64] And the camp's readings made it possible to better correlate near-pole conditions with those prevailing at lower latitudes, particularly along the Sea Route.

In sum, the pathbreaker camp of Shmidt and Papanin is a defining exercise, an enormous act of preliminary analysis, a bold stroke whose importance is great but not obvious. Knowledge builds upon itself. The future would see more systematic coverage with a long-term series of "North Pole" stations installed to deepen this reconnaissance-glimpse, multiplying and corroborating its revelations by painting in the finer-grained detail.

Months before hostilities, another Soviet initiative pressed to seaward. Jumping off from Wrangel Island, off the northeast Siberian coast, pilot Cherevichny and his aircrew flew three long-range probes into the central Arctic. It was March 1939. The aircraft was a USSR N-169, reengineered as a "flying laboratory."

Airlifting mobile stations onto the pack for spot observations at selected positions, Cherevichny verified the value of brief tarries at predesignated coordinates. Compared to setting up semipermanent ice-borne stations, this technique was active, simple and inexpensive yet productive.

Reconfigured for essential comfort (even including a kitchen), the flying "laboratory" was all but independent of mainland support. (Exceptions were fuel and engine oil.) With a four-man flight crew and three researchers—hydrologist, geophysicist, meteorologist—three sites were occupied in the area of 80° N, 175° E. The first, begun on 3 April, had N-169 on the pack for nearly five days recording weather and ice conditions, soundings, currents—the "lab" itself a base of operations. The pilots again set down at 78°31' N. Following a third "jump," the expedition retired, having logged 25,760 kilometers (more than 16,000 miles). Meteorological observations had been taken, currents defined, ice conditions probed, soundings recorded in sectors little known and difficult to access.[65] One revelation was that in-sweeping Atlantic waters circulated throughout the basin. And depths were much shallower than those recorded by Wilkins.

War flared, sweeping the map of Europe. Further expeditions were suspended. The circumpolar north—a distant, novel realm for battle—was to be assailed in force. For the Soviet Arctic, flying would spotlight the Sea Route, weather reconnaissance—and the dispersal of Lend-Lease aircraft inserted via air-ferry steps to Alaska into Siberia (thence the eastern front). Leningrad would revive offshore work only in 1948, three years following defeat of the Axis.

Aeronautics had assumed an overarching priority through the requirements of war and survival.

CHAPTER THREE
FLETCHER'S ICE ISLAND, T-3

> You can observe a lot just by watching.
> –Yogi Berra

FAR-NORTHERN SEAS AND AIRSPACE PROVED vital during World War II.[1] "Cold" war would accelerate operations and adjunct military research on, over, and under the Arctic Ocean–the center of the circumpolar north.

Russia had gained first-rank status; by 1945, Soviet power stood immense. But the Soviet-Western alliance against fascism had sown bitter seeds, and Moscow turned from the West. A continued coalition with the United States and Britain would interest Stalin only if it could be maintained while he spread Soviet influence through collaboration and aggrandizement.[2] Each side now indulged its worst suspicions about the other. In Soviet domestic propaganda, a powerful sense of "capitalist encirclement" obtained. The U.S. War Department had offensive plans. And Washington's polar strategy–in Alaska, Greenland, Iceland, and Canada–was manifestly militant.

> No, it is not serving defensive purposes! The talk about the "defence of the Western Hemisphere," about the danger alleged to be threatening it "from the North," i.e. from the Soviet Union, serves simply to fool the naïve and mask the real intentions of the American imperialists. . . . The measures taken by the USA in the Arctic are aimed entirely at inflaming a war psychosis, at war propaganda.[3]

To counter the "expansionist plans" of the West, Moscow would remain on guard–xenophobic, proud, obsessive of attack.

Washington nursed anxieties of its own. The Russian bear was on the march, "alarming the Western states into believing the worst about Soviet intentions."[4] The problem of Stalinist power and the thwarting of Soviet ambition shifted to deliberate national policy. Communism, in short, had moved center stage. In the logic of the time, any encroachment of Soviet power, any thrust of Moscow's influence—diplomatic, economic, political, ideological—was endangering. Strength plus forceful action alone could check the Kremlin. This logic was to imbue—and debase—conduct both at home and abroad.[5] Resistance to Soviet imperial power (the "implacable challenge" of diplomat-historian George F. Kennan) would shape Anglo-Soviet relations and the contours of the postwar geopolitical world for two generations.[6]

From 1945 to 1990, the history of the world was that of two competing visions of social equality and economic justice.

The 1950s were rich with geostrategic confrontation; in northern high latitudes, an escalating militarization was manifest. "The 'eventual enemy,'" a Russian geophysicist remarked, recalling the stakes, "was on the other side of the pole and the Arctic Ocean could have been the center of a great war; thus the leaders of the country did not grudge the expenses [of programs and operations]."[7] Offshore, Soviet initiatives resumed. (Ice reconnaissance, weather forecasting, and general support along the Sea Route had in fact never ceased.) Sorties into the central Arctic basin would revive as well, though not immediately.

Remote-environment research is not practicable without backing from big institutions or government. U.S. political, military, and intelligence planners would have to work very hard and very fast to match Soviet polar initiatives. Beneath the pure science lay strategic policy. The containment doctrine—and the sheer momentum of events—impelled an aggressive militarization of the Northern Ocean. As an instrument of policy, then, Washington's initiatives were less a strategic vision than a response.

What had happened? The combatants had dispatched countless personnel—along with material support—into the Arctic and sub-Arctic. Far-flung infrastructure was installed, enlarged, improved upon. Logistical obstacles had been assailed, all with urgent timetables. Knowledge of polar conditions exploded. As airborne surveys had multiplied (for example), so too had operational experience. Cold-climate construction, communications, flying, navigation, forecasting, search-and-rescue, and survival had sustained near-revolutions. Flood-tides of know-how, data, and observation advanced a host of sciences and disciplines—each having immediate application for the postwar international order.

Still, of themselves, these tremendous events explain little. The new international order was key. In the span of five years, the north had become strategic space. A peripheral realm hitherto on the outskirts of military notice, the Arctic held immense geostrategic value after 1945. Eurasia and North America front directly on the Arctic Ocean—the only waters shared by Russia and the United States. There the lines of latitude gather. Accordingly, the shortest arc between the newly born superpowers transited maritime boreal wastes.

> During the middle of the 1940's it became evident that the course of any future war would be further northward. Implications of arctic geography on world strategy and national security are evident from an examination of a polar orthographic projection made of the Northern Hemisphere. Our approaches must be guarded in the Aleutians, in Alaska, and in the complex of islands of the Greenland Sea—Iceland, Greenland and Spitsbergen.[8]

Against this backdrop, "north" to policymakers now embodied an exposed flank, a theater of competing interests as well as a *direction*. Thus defined, the high latitudes have retained an irreducible geopolitical significance.

ICE ISLANDS

August 1946. Ladd Air Force Base, near Fairbanks, in east-central Alaska, sits near the arc of 65°–100 miles (161 kilometers) shy of the circle. Here, the rigors of cold-weather flying are a U.S. Air Force (USAF) specialty.[9] The 46th Reconnaissance Squadron (Very Long Range [VLR]) Weather, Alaskan Air Command (AAC), has an established and exacting routine. The men rise early, don flight gear, and mess. A briefing follows. On the tarmac, specialists attend the aircraft. The WB-50 is the reconnaissance version of the Boeing B-29, the four-engine giant designed to bomb the Japanese home islands. Stripped of armament and turrets, this Superfortress is fitted with additional fuel tanks. Its mission—to collect weather data—will be a tiring twelve- to sixteen-hour probe into nameless nothingness.

The forecast holds no particular threat, only a hovering gray overcast. At these latitudes, dreadful summer visibility is normal—of scant importance for this aircrew. The north polar sea is a vacant waste devoid of landmark. Lodged in perennial ice pack, the North Pole holds little mystery, certainly no romance. It is but a turn-point on the charts, the place at which the pilots' return leg thankfully commences. In terms of scenery, the geographic pole is unremarkable.

Ranging easily, the WB-50 had inaugurated long-range penetrations into the central Arctic, opening its white reaches to airman, soldier and researcher. Hardly routine, these "Ptarmigan" missions—preassigned flight lines through the void—were nonetheless unexceptional. This particular sortie, however, would shift the course of science in the maritime north. The bird lifted away, assumed a northerly heading, ascended to 10,000 feet. It was 1730 miles (2785 kilometers) to the geographic pole. The Yukon basin, the Brooks Range rose up (unseen) then fell astern. The North Slope and the coastline were crossed.

About three hundred miles out, the navigator saw it on his radar. (No visual sighting was possible.) An immense mass was filling his screen, an apparently dark object against the brilliant returns of the ice cover—a signal characteristic for low, ice-bound terrain. Startled, the airman checked his equipment. The returns, he concluded, were genuine; this was no electronic gremlin, no deceiving "angel." The signal was unmistakable—land, where none was known to exist. It seemed to be a large island, roughly fifteen by eighteen miles.[10]

Its crew alive with conjecture, the plane returned. Word flashed to Washington did not fail to impress: a fury of excitement ensued. Never before had such an enormous ice feature been reported in that sea. The object—whatever it was—was designated Target X (later T-1) and its existence classified "Secret."[11] Subsequent sorties would realize further sightings—and, with each plot, a different position. At first these were thought to be errors—understandably given the fiendish demands attendant to northern air navigation. But a specific rate of movement was finally noticed. The air force, apparently, was tracking an exceptionally large island of ice. Speculation as to its possible origin and probable future soared.

The mass was finally seen, visible from seventy-five miles: a low-floating tabular iceberg. Rather than the chaos of the pack with its tortured ridges and zigzag of leads, this mass showed a ribbed texture of parallel ridges and troughs running from edge to edge. And its pale blue contrasted with the enfolding pack. Unlike bergs but like floes, ice islands are more or less horizontal. Over T-1, low-level runs confirmed an amazingly smooth tableland.

Could T-1 be exploited as a vehicle—for a weather station, airfield, a geophysical observatory? These and related questions garnered intense scrutiny.[12] Why ponder such nearly unreachable objects? Maxwell E. Britton, Ph.D., is a former director of the Arctic Research Program, Office of Naval Research. "The air force was interested in landing places," he points out. "Landing is important. Lots of ocean up there in the Arctic getting from point A to point B."[13]

In 1947, regular WB-50 overflights had commenced with the mission of general reconnaissance and the collection of weather data. Code-named Ptarmigan, these sorties kept track of T-1 and, to pass the time, hunted other islands and undiscovered land. (By 1951, daily missions were being made to the geographic pole.) Often, T-1 escaped detection.[14] When it *was* reported, however, the mass had shifted slightly east. Its shape, though, was unchanged—implying great thickness and strength.

An aircraft cleared an Alberta runway that April. Its mission was a test of low-frequency Loran as an aid to navigation in the north. As well, its crew would photograph ice conditions and land features. Off the archipelago, the flight log records an unusual sighting: "Observed a great slab of 'shelf ice' lying most conspicuously among the pack ice. Its size was about 4 by 6 miles, perhaps larger...." One of the airmen was Wing Cdr. Keith Greenaway, Royal Canadian Air Force. This ice, he muses, was different—yet somehow familiar. "We'd seen the [Ellesmere] ice shelf before," he recalls. The berg's rolling surface "was immediately associated by us with that of the ice shelf along the north coast of Ellesmere Island." Great sheet-aprons of ice gripped that far shore, quite like those in Antarctica. But how had they—and the T-1 fragment—formed? "We had no idea of the mechanics of the formation of the ice shelf, or undulations," Greenaway adds. "Our guess was that the ice was a remnant of the last ice age."[15]

The island slab was photographed. Published in 1948, this is the second known image of ice island T-3.[16]

Muscling the pack, T-1 was following what appeared to be an erratic, clockwise drift track. By late 1949, the target lay beyond routine flight tracks; "lost" in the back ways north of Greenland, it had shifted some 1500 miles (2415 kilometers). Average rate of drift since discovery: approximately 1.2 miles per day.

SEVER AND SPS

The Soviets—in secret—were resuscitating offshore campaigns. Reviving Cherevichny's "jumping" technique, the Arctic Institute fielded its first "high-latitude air expedition." Leningrad would all but run a shuttle over the Arctic Ocean. Known as *Sever* ("north"), and deploying mobile research groups, these hit-and-go airlift operations acquired as much data as was practicable. Every spring (sometimes fall), sometimes basing at ice camps, these aerial fleets would sortie–detachments taking spot observations at stations over large sectors. (As well, *Sever* would relieve and resupply the semipermanent *Severnyy Polyus* [SP] stations soon to come *and* replace camps closing-out.) *Sever* airmen were the best. No mere ferriers of men and cargo, many were veterans of polar aviation, having (for example) ferried U.S. warplanes from Nome to Siberia. Dr. Zalman Gudkovich, a hydrologist, accompanied the pioneer *Sever*. He recalls the pilots as singularly able and confident, gods of the air on whom all depended. "The most important thing," he said of that expedition, "they had good experience; they were well-trained, experienced and broadminded."[17]

Six or so weeks in 1948 saw men air-deployed at interim stations at eight positions. On drift ice, the technique is simple but not easy: at predesignated coordinates the pack is assessed, a suitable floe chosen, a touchdown made. Always hazardous, ice landings demand a great deal of skill. Deeply experienced, the Soviets hold a near absolute command of the procedure. In all, eight semipermanent stations would see deployment before the decade was out. Still, the institute had chosen to emphasize its mobile air detachments: by onset of the International Geophysical Year (IGY), Leningrad would log more than 550 of these "jumping" oceanographic stations, taking thousands of readings.

The deepwater geometry and circulation were not yet known.[18] The inaugural *Sever* had recorded a depth of 1290 meters (4231 feet) 200 miles off the geographic pole—impossibly shallow. Supplementary measurements from *Sever-2* (in 1949) confirmed the sounding. The Lomonosov Ridge had been detected. Named for the philosopher, poet, and scientist, this transarctic highland bisects the Arctic Ocean into the Eurasia and (smaller) Amerasian basins, crossing near the pole. Knowledge of deep-ocean geology is based largely on geophysical studies, with some direct information (e.g., cores). In the inevitable way of science, seismic and echo soundings, gravity, magnetic, seismic refraction plus ocean drilling were to flesh out the crustal structure and bathymetric map of the abyssal ocean with its rifts and ridges, seamounts, plateaus, and fringing continental shelves.[19] For the Northern Ocean, measurements were to reveal the influence of subsea topography in determining the major components of water-mass motion and mixing. For reasons known only to itself, no Kremlin announcement of discovery was made.[20]

Thus rewarded, more expansive *Sever* air expeditions sortied, to plumb the most intriguing and understudied sectors. "These 'laboratories,'" Gudkovich remarks, "allow us to cover simultaneously vast areas of the basin." On station one to three days, "jumping" detachments took soundings; measured temperature, ice thickness and snow cover; sampled the bottom; and recorded magnetic and astronomical as well as meteorological observations.[21] Further, a long-term camp would deploy. Although versatile, "jumping" campaigns were restricted to spring-season ice: *Sever* are darting probes, not sustained appraisals. As a scientific tool, then, it had become "absolutely clear" (Gudkovich) that *Sever* did not—indeed, could not—offer a *continuous series* of observations. The harvest from overflights and brief on-ice tarries, useful to be sure, were not synoptic. The observations lacked continuity. "And therefore, it was necessary to organize long-term, so-to-say, drifting stations in the Arctic"—that is to say, a Papanin-type drifting station.[22]

A decade had passed since the audacious "North Pole" expedition. Why not revisit the ice, reinvent the Shmidt/Papanin technique of drift? A protracted loiter would lengthen the period of measurement, granting regular observations throughout the boreal year. As well, such data would supplement air-expedition results. Shifting with the pack, moreover, a semipermanent station could traverse a large (if limited and unpredictable) area. Drift track, moreover, would decode near-surface circulation. "No other people," George Kennan writes, "has such a thirst for knowledge, such a zest for intellectual and artistic experience." With apparent ease and aplomb, America's number one adversary would set the terms of postwar scientific enterprise above, beneath, and atop the boreal canopy. The technology may not have been revolutionary. The concept was. The *Sever* and SP programs represent the (complementary) twin pillars of Soviet ice-based investigation in the postwar period.

Something has to drive such pushes. The objective in this case was improved forecasting (ice, weather) for the Sea Route, thereby bettering conditions for shipping. "All stages of Arctic research during the years of Soviet rule have been linked with the problem of charting and mastering the Northern Sea Route. A network of polar stations guaranteeing exhaustive observation throughout the year of basic natural processes taking place in the Arctic expanses" was, and would remain, the Soviet Union's basic policy.[23]

For the 1950 field season, equipment was made ready, supplies stockpiled, *polyarniks* selected: glaciologists, oceanographers, aerologists, geophysicists—sixteen men in all. The diaries of Papanin and his drift comrades were studied. Polar work is a vocation rather than simply a job; these men do and dare for the sheer adventure of discovery—and for Russia. Granted the sunlight and relative warmth of spring, *two* field parties were airlifted from Leningrad and its celebrated Arctic Institute—the trailhead of Soviet polar science.

Not since Ivan Papanin had men dared to land—and to linger—on the restless canopy of white.

Leaping time zones, the expedition was conveyed to the Russian Far East. Northeast of Ostrov Vrangelya (Wrangel Island), an advance team was installed (from a coastal airfield) onto a frozen-over polynya. Atop multiyear ice about half a mile from the point of touchdown, a foundation for "Eastern Drifting Ice Station" (1950–51) was chosen. Its position on 2 April was 76°03′ N, 166°36′ W—a point about three hundred miles northeast of Wrangel. At ten feet, the floe seemed amply thick. A pioneer camp thus installed, tons of equipment plus a full station staff were shunted to seaward. Led by oceanographer and sea-ice expert M. M. Somov, the party manhandled everything over a ridge of hummocks to the site, over trails worked into snow. (In a few days, dogs were airlifted out—to help reduce the stubborn pile at the airstrip and to warn of polar bears.) The work was brutal. "How strong my reluctance to crawl out of the warm bag!" G. N. Yakovlev, a glaciologist, writes of his first awakening. "My hands and feet ached from working like a horse the day before. Today, we had to do more of the same. Shivering with cold, I pulled on my warm, down-filled pants, and left the tent. It was –22 degrees (C). A blizzard was blowing."[24]

On ice, work is as much physical as scientific. One day is peaceful; the next, all is pressure, motion, deformation. Almost immediately, this group found itself beset when a crack materialized, interposing between the camp and the airstrip. So the nascent station was reestablished on a less risky floe. The shift put gear out of action; as well, decamping devoured the clock: sleeping and work tents had to be repitched, instruments relocated and remounted, through-shafts prepared (using ice spikes and explosives), stakes set out, benchmarks frozen. Luckily, the double-layer KAPSH-1 and KAPSH-2 frame tents were readily shifted fully assembled.

Order restored, defenses in order, team Somov prepared for observations.

At *Severnyy Polyus-2* ("North Pole-2")–as the station later was designated–the process of inquiry commenced. Work was to persevere for thirteen months. For its part, the other team did not encamp: their floe cracked and broken, A. F. Treshnikov and his party had had to abort. Returned westward, the would-be colonists resumed "on-shore duty." Reticence is a mania, very much Russian. The fact of SP-2 would not be disclosed until 1954, its data withheld longer.[25] With Soviet research policy a cipher, the West was left to speculate. "Believe me," Joe Fletcher frets, "had we been able to show our authorities the information [regarding Soviet initiatives], it would not have been so difficult to get authorization and support for Arctic Basin investigation."[26]

Hidden well away, SP-2 probed. With both time and manpower limited, the regimen was unrelenting, the grunt work endless. Readings of depth (wire soundings), of water temperature and salinity, of current velocity at various levels, and

UNKNOWN TO THE WEST, THE SOVIET UNION REVIVED THE SHMIDT-PAPANIN TECHNIQUE OF DRIFT. IN 1950, *SEVERNYY POLYUS-2* ("NORTH POLE-2") WAS DEPLOYED OFF EASTERN SIBERIA BY A *SEVER* "HIGH LATITUDE AIR EXPEDITION." SP-2 WAS THE ARCTIC INSTITUTE'S SECOND SEMIPERMANENT STATION. ITS SCIENTIFIC MISSION: TO IMPROVE ICE AND WEATHER FORECASTS ALONG THE NORTHERN SEA ROUTE, BY TAKING SYSTEMATIC MEASUREMENTS OVER A FULL ANNUAL CYCLE. (*ARKTIKA MUSEE*)

of bottom samples comprised the hydrological research. Especially labor-intensive, this work demanded all off-duty comrades.[27] With time, the complex relief of the abyssal ocean floor (reaffirmed by "jumping" soundings) would yield. Not far off the pole, a great underwater rise was discovered: the Lomonosov Ridge. And the Chukchi cap was detected—a flat-topped prominence rising from abyssal depths to less than one thousand feet of the surface. These were major payoffs. As before, surface and airborne atmospheric readings (the latter via radiosonde balloon) were compiled then radioed south. Scientifically, the new camp was traversing a void; these data were its most practical service, helping to heal a blank on Soviet synoptic charts. Wandering its trackless, tortuous course, the scientists studied gravity and geomagnetic measurements, ice studies, and various components of the heat budget.[28]

Position is crucial to datasets—and to rescue. As the line of drift unscrolled, camp coordinates (latitude and longitude) were calculated, new data-points plotted, the results radioed from deep isolation to the sweet south.

A six-month stay had been planned, the station provisioned accordingly. "But after this [initial] period," Gudkovich recounts, "it became clear that it would be very good if this station could continue its work. And this station continued . . . , although the program was a little reduced."[29] By close of business, indeed, its work would fill four volumes. Three would be allowed to leave the Soviet Union. The most sensitive results—gravity measurements—were consigned to use by Russian researchers and withheld absolutely.[30]

The north is an *air* theater in great part, dependent on aircraft for logistical support and transportation.

> Since the Second World War, changes to the map of Canada have been made from air photography. Geologists, archaeologists, and other scientists have reached their destinations by airplane and helicopter in a few hours, rather than by traditional means in months. They have been able to bring much of the south with them, replacing the satisfactions the north provides. They have lost more than they gained, and I do not envy them.[31]

High-latitude research is hostage to the available logistics. As the West's doctrine of containment took hold, experience favored the Soviets. The Arctic-focused scientific and survey programs of the USSR outmatched all others for one reason: the northern air and sea-transport service conferred unrivaled know-how on those who conducted it; playing to its history and strengths, quietly and without fanfare, the Soviet Union was advancing the use of aircraft (and ice-capable ships) for polar missions, accustoming men and gear to the operational demands of an extreme environment.

The value of aviation in this vast area is much greater than the number of its aircraft or the scale of its operations would suggest. Though it carries only a very small percentage of all freight moved in the North it is of vital importance in the movement of passengers and high priority freight and for much of the year it performs a unique function, for its operations are carried out on a year-round basis—the only interruptions apart from weather occur at certain airfields during the breakup period. It is also of great importance to the administration of the North, providing rapid all year transport within the region, to Moscow and to other areas of the USSR.[32]

High-northern logistics were fast maturing.[33] And Washington could hardly ignore the challenge. In the heady days just after the war, federal patronage of science continued to soar. Created by Congress in 1946 to help stimulate peacetime science, the Office of Naval Research (ONR) was the most significant organization for research in the federal establishment immediately after World War II.[34]

ONR was charged to plan, originate, coordinate and support basic and applied research in a joint effort with civilian institutions and scientists by funding them through contracts and grants. The ideas would be those of the scientists, and their research would usually be open, unfettered by security restrictions, and freely published in the world's journals of choice. Not until four years later did the National Science Foundation [NSF] come into being with much the same philosophical base as ONR, and with a much broader mandate to serve the public scientific interests of the United States.[35]

TO SEAWARD OFF ALASKA, SKI-JUMP

The Arctic Research Laboratory (ARL) was established in 1947 at Barrow—Alaska's northernmost point (75° N). ARL provided a support facility for contract-research. Operated for ONR under contract, the lab was to become a major field logistic support activity.[36] Little more than a Quonset hut initially, assets for the transport of personnel and gear—landside, aloft, and to seaward—were wanting. The navy is a seagoing outfit; during the summers of 1950–51, indeed, icebreakers were operating in the eastern and western Arctic. How best to get government, industrial and university researchers *onto* the pack?[37] "Most of the work was, in fact, carried out by land-based scientists," Dr. Britton records, "for, until 1959, ONR/ARL capability for supporting research far from shore was confined to occasional opportunities for penetration of the pack ice by Navy or Coast Guard icebreakers during the summer season. It was not until it became possible to operate on the drifting ice itself that work at sea could really progress."[38]

The U.S. Air Force was similarly occupied. At Eielson Field, outside Fairbanks, Lt. Col. Joseph O. Fletcher, USAF, commanding officer (CO), 375th Reconnaissance Squadron (VLR), organized a special radar and visual search along the Ptarmigan route for drifting ice formations similar to T-1, now known as "ice islands." In 1950, Fletcher was an opinioned, dedicated and enthusiastic officer capable of total immersion in a project. His advocacy and zeal would realize the first drifting research station for the United States. Meantime, John F. Holmes, a civilian oceanographer at the Woods Hole Oceanographic Institute, was flown to Barter Island, off North Alaska. Its auxiliary airfield lay 350 or so miles east of Point Barrow. Here the Tenth Air Search and Rescue Squadron had a detachment. Joining Dr. Albert P. Crary, a geophysicist from the Cambridge Research Center (U.S. Air Force lab), the pair was deployed 135 miles to seaward of the Alaskan shore. The mission was to investigate the possible use of aircraft for landing oceanographic parties onto pack ice. "On the first landing [approx. 71° N, 145° W]," Holmes reports, "the landing gear was bent, which made it necessary to lighten the plane [C-47] by leaving all of the scientific equipment behind us. This equipment remained out on the ice for more than two weeks before the weather was clear enough to rescue it."[39]

No further landings were logged in 1950.

Casualty aside, observations were in the logbook—the first oceanographic data for the United States taken from pack ice. In Washington, Holmes and colleague L. V. Worthington proposed obtaining oceanographic data using a *navy* airplane as the working vessel. With the concept approved, the Bureau of Aeronautics was directed to provide a ski-equipped R4D. That December, BUNO 12417 was delivered to Quonset Point for reconfiguration, to support an ONR-sponsored arctic oceanographic mission code-named SKI-JUMP I. In January, Lt. Cdr. Edward M. Ward, U.S. Navy, received orders to report to the naval air station as officer-in-charge.

In their joint report of the 1951 project, Holmes's own introduction outlines the objectives of SKI-JUMP that long ago springtime:

> The purpose of Project Skijump was twofold. The primary object was the procurement of oceanographic data about the Arctic Ocean. The secondary purpose was the operational experience of a flight crew on the pack ice.[40]

Ship and crew reached Barrow on 17 March, whereupon Ward and copilot Lt. E. C. Woodward, USN, flew long-range reconnaissance. The navigator was Lieut. D. L. Moorhead, USAF. The pilots' ability to distinguish ice types from the air was basic to success. Unlike new and first-year ice, multiyear ice has undergone deformation and cracking, hence is of variable thickness and presents a comparatively hummocky surface. "Practically all the [Beaufort Sea] ice was old hummocky ice laced with pressure ridges," Ward recalls. "There were however a few freshly frozen-over leads. I decided the old ice was tantamount to suicide and we'd try our luck on the frozen over leads." With Holmes and Worthington rounding out an eight-man crew, twelve landings (eleven oceanographic stations) were logged that March–April off North Alaska. These were the first successful deliberate landings by a navy plane on uncharted ice. Farthest-north station: 76°30′. Though ski-equipped, the pilots elected to wheels-down all but one set-down. "My greatest concern," the former pilot admits, "was landing on unprepared or un-surveyed ice floes. The question uppermost in my mind was how to determine ice thickness?" Snow cover offers clues, as does ice color and texture. Ward soon developed "an eerie ability" to estimate thickness from the air. At ARL, skepticism vanished. "Now we were besieged with requests for a days' outing on the ice."

Independent of SKI-JUMP, a party consisting of Bert Crary, Robert D. Cotell, and Jack Oliver was shunted into that general sector. Its pilot was J. R. Whitmer. That April, between latitudes 73° and 76° N and longitudes 138° and 151° W, a half-dozen airlifted stations were made. Loiter time: four to six hours per site. Though rivals, the two operations shared identical objectives. The services were going to school, testing the feasibility of aircraft on sea ice. Although equivalent to the Soviet's "jumping" surveys, there were no borrowings for SKI-JUMP. The United States knew nothing of the Soviet technique.[41]

On 24 April, the navy secured for the year. "The weather [Holmes held in his report] was too risky, and the ice too soft, to venture out again." Basic hydrographic data had been won: seismic soundings (in which an energy source imparts acoustic waves into the subsurface, then measures travel times for analysis); magnetic declination; ice-drift and surface-wind data; gravity readings. Pack movement, researchers learned, correlated with near-surface winds—the apparent engine of pack-ice drift.[42]

SKI-JUMP II comprised a three-plane armada: the specially equipped R4D plus escorting P2V Lockheed Neptunes, to act as tankers. O-in-C of the 1952 operation: Cdr. Vernon "Jack" Coley, USN, with Ward again commanding the R4D. At Ladd Air Force Base, where it arranged for hangar space, spare parts and sundry matters, the flying sailors crossed trails with Joe Fletcher, now O-in-C of the Ice Island T-3 expedition. An "O" club meeting was held with both SKI-JUMP crews and Fletcher's entourage. "We discussed the respective missions and the possibility of mutual aid if necessary," Ward says. "It was generally conceded that once the Air Force was established on T-3 they would lose little time jumping off for the pole. That however was the prize we coveted."[43]

The R4D reached Barrow on 14 February. Following tests, local ice landings and at-sea reconnaissance, Ward rendezvoused with one of the escorts on 11 March. A search for suitable ice culminated in a touch-and-go north of 76°, followed by touchdown. The cargo door was flung open and the oceanographers jumped out, to hole the ice. Using a five-horsepower gasoline-driven post-hole digger, a five-foot floe was bored in about ten minutes. (The chainsaw and chisel method used in 1951 had consumed as much as two and one-half hours for similar ice.) To make each hydrographic station, Nansen bottles were cable-lowered via winch (mounted forward in the R4D), to record water temperature and salinity up to 3200 meters (10,528 feet) in 1951, to 2950 meters in 1952. On the first station of SKI-JUMP II, a radioed "All clear and plenty thick" brought the thirty-five-ton escorting bomber alongside, the first heavy warplane to attempt a sea-ice landing. "The refueling operation went smoothly and we took on enough fuel for two more working days on the ice." Altogether, seven oceanographic stations were logged. A final unfortunate takeoff, on 27 March, made explicit the risk attending ice-as-runway. Throttles to the firewall, the ship started to lift away from rough ice. Suddenly a sharp joint was felt, and the copilot was amazed to see the port propeller "walking" across the ice ahead of the plane. At the same instant the left wing dropped and the ship slithered sideways to a stop.

Fearing fire, Ward sang out, "Get everybody out!" All hands made for the hatches and doors. The R4D had hit an uplifted block hidden by drifted snow; this had collapsed the port landing gear, allowing the prop to make contact, shearing it off.

> Had a fire followed the crash the situation would have been grim. As it was we were fairly well off. The hull was intact albeit a list to port. Survival rations were plentiful. And as the heaters fed off the main tanks, we had 700 gallons of fuel to see us through. But the first thing to be done was get a "May Day" off to Pt. Barrow, inform them of the damage and stress the fact that there had been no injuries.[44]

Airlifting men and gear to effect repairs exceeded its value, so the aircraft was abandoned for salvage. Rescue, though, called for a much-improved runway. "It must be remembered we were 810 statute miles (1304 kilometers) north of [the logistics base at] Barrow," Ward reminds. "The P2V would require fuel aboard for the 1620-mile round trip plus a reserve. So crucial was the situation that every foot of useable runway would have to count."[45] Three days of attack using shovels and ice axes produced a strip marked with red flags and hydraulic fluid poured at its approach end.

Retrieval by Colby–all throttles forward–was effected on 31 March. The

A NAVY PARTY AUGERS A "HYDROHOLE" DURING SKI-JUMP II, MARCH 1952. ONCE THE DESIRED OCEANOGRAPHIC READINGS ARE RECORDED, THE R4D WILL JUMP TO THE NEXT TOUCHDOWN COORDINATES TO SET UP ANOTHER SPOT STATION. NO CAMPS WERE SET. (NATIONAL ARCHIVES AND RECORDS ADMINISTRATION)

oceanographic phase of SKI-JUMP concluded, its team having eighteen floe-ice oceanographic stations. Its results suggested that surface circulation "is more complicated than the cyclonic system suggested by Nansen. The chief feature [Worthington wrote] is a large anticyclonic eddy north of Alaska." Further, "some evidence" of a submarine ridge crossing the basin's center had been recorded.[46] Fieldwork logged, ice reconnaissance and a landing at ice station T-3 remained. Ordered to Antarctica, Ward became the first CO of VX-6, the U.S. Navy's support squadron on the Southern Continent.

Sea ice offers a natural yet dangerous runway. Landings demand well-schooled judgment and the ability to "read" the ice surface. Unlike the turbine (e.g., HC-130), piston engines take a beating in the cold. This image was taken March 1952, during Project Ski-Jump II. (National Archives and Records Administration)

—•—•—•—•—•—

The first orders to establish a prefab ice station ("experimental base") north of Barter Island had been issued in mid-1950, by the AAC. (By late 1951, the project became known as Project Icicle.) When deployed, the mission would evaluate weather and communications problems and, not least, the feasibility of airlift support. That July, the search for "floating islands" stood duly rewarded: a "target" 200 miles (322 kilometers) from the geographic pole. T-2 was huge: a three-hundred-square-mile tableland. Nine days later, T-3 was sighted by radar three hundred miles from Wrangel Island, drifting poleward.

Three such "targets" were known. Even so, weather and other factors caused delays. Meantime, the substance of the air force's results went public. In December 1950, Joe Fletcher and Maj. Lawrence S. Koenig, his ice project officer, presented a paper at the First Alaskan Scientific Conference. The desirability of "semi-permanent geophysical observing stations on secure positions in the Arctic Ocean" was stressed.[47] In contrast to the pack, which is thin and everywhere subject to buckling and cracking, ice islands (the pair argued) offered relative invulnerability. The paper was well received. "Afterwards," one attendee remembers, "several of us sat up late into the night conjecturing about the origin, composition, and probable future of these paradoxical floating 'lands'. We were especially curious as to where they had come from."[48]

Headquarters Air Force approved continued flights by the 375th. The siting of any floating camp required good visibility, so March and April were chosen–that slice of the boreal year granting both daylight and minimum cloud. But no new masses were found. Meanwhile, special "Ptarmigan" missions investigated Ellesmere; photo reports had suggested that ice islands might originate from the shore-fast shelves gripping that underexplained coast–an analogue to the floating shelves fringing Antarctica.

Braving severe icing as well as adverse winds, Ptarmigan Special Roger reached Nansen Sound, the seaward edge of north Ellesmere. Fully to Ward Hunt Island, its shore-fast ice had a distinctive ribbed texture; the resemblance to the ridge-and-trough relief of T-3 was striking. On the return leg, a small mass was sighted off the Queen Elizabeth group, provoking another probe. That mission would rediscover T-1.

—•—•—•—•—•—

In the Soviet sector, summer's first "day" at SP-2 was marked by snow, a sullen rain, fog and drizzle, and a temperature about freezing. Speeded by the fog and rain, ice loss proved intense. Instruments requiring precise installation began to "swim." Mast guys loosened. Puddles, ponds, and streams overspread a dank, darkening white. (The puddles which melted through were especially dangerous.) High rubber boots became uniform; wet footware and clothing festooned the sleeping tents. Now airdrops were the sole resupply option.

October to April is boreal winter. And autumn is crucial; in addition to essential research, staff must prepare for the polar night. Intense students of the North, Russians have the gift of making themselves at home anywhere they put down. This household–a team of sixteen–had adjusted to life on ice. In mid-August 1950, the Main Administration to the Northern Sea Route radioed: Would the personnel agree to extend the drift? And so, in the weeks before dark, additional supplies were shunted out, selected men retrieved. A ten-man contingent was left to continue the (winter) work of the station. Racing the twilight, a new campsite was chosen. Snow is a superb insulator. Relocated, the tents (which dissipate heat) were winter-proofed against wind and heat loss using "brick" cut from compressed snow.

October first marked the half-year jubilee–no mean prize. On 4 November, the last plane screamed away. The pilots' ability to fly brilliant no-light landings

The Ward Hunt Ice Shelf, Canadian High Arctic. Note the pattern of troughs and ridges. The dark streaks are melt lakes. These floating shelves—nonsaline ice of freshwater origin—are nurseries for the most massive ice features in the Arctic Ocean: ice islands. The Soviets have deployed five drifting research stations onto ice islands, the Americans two, Canada one. (Royal Canadian Air Force, courtesy George D. Hobson)

amazed Yakovlev. "What other pilots," he asked, "would risk landing on pack ice during the polar night?"[49]

Dawn's glowing strip would not cut the drape of night again till January, the first peek of the sun until late February. In star-pocked dark, the station inscribed its meandering night-passage. Beyond its circle of lights, young ice in compression sang protest.

Through storm and dark, investigations—including signal work on the heat budget (prize data)—persisted.[50] The sheer physicality of intense cold assails the human body, transforming the simplest chore. At SP-2, observations at distant work areas proved particularly draining. The winter of 1950–51 delivered the daily realities of life on ice: a Stygian dark, hummocking, blizzards, and an unsettling diminution of their floe-foundation, the edges of which sliced closer. The hazard is plain: that the tortured plain of white will crack, taking the camp or shortening its airstrip. "Drifting ice remains drifting ice," Yakovlev notes, "and the winterers do not for a minute forget that they are not standing on solid ground. They must always be prepared to have their work and lives disrupted."

Ice camps endure at nature's indulgence. Early February 1951–ten months at sea. Yakovlev felt a jolt accompanied by a shot-like report. For some reason, "my feet started to drift apart. I looked down and was struck dumb: directly below me there gaped a black, progressively widening crack." Grabbing on, he pulled himself to one side of the tent. Instinctively, he shifted the galvanometer to a safe corner. "I took a knife from the top of my boot," the Russian continued, describing the blow to the camp's foundation, "intending to cut through the tent, when suddenly I heard someone's scream and the sound of tearing canvas. The crack slowly widened pulling the tent in various directions, until the canvas—stretched like a bowstring—tore. . . . I . . . saw Kolya Milyayev . . . on the other side of the crack, jumping about in the snow in his underwear, barefoot and hugging a theodolite and tripod. Frightened, half-dressed men were running out of all the tents."[51] An ensemble of welded-together fragments, their floe had fractured traumatically.

—·—·—·—·—·—

In Alaska, plans continued for an experimental *pack-ice* camp. As *Severnyy Polyus-2* struggled to revive and restore itself, AAC reconnaissance located what appeared to be a compatible floe about 115 miles out from Barter Island. On 20 February, test landings by the Tenth Air Search and Rescue Squadron found conditions to be "perfect." For three days, the ski-equipped version of the C-47 (Ski-47) ferried gear and personnel (eight men) to establish a station—the first U.S. experimental camp on boreal ice.

Sound preparation does not ensure excellent results. Wind brings the canopy alive. This debut continued without incident for two weeks, until high winds began shoving the pack. The ice began to fissure, opening a lead on the runway's far side. No emergency was declared because, unless the strip was severed or the camp itself riven, cracks held no immediate threat. Three weeks on, however, a breach opened close aboard then closed under immense compression, quaking the sea-ice cover. The camp-floe splintered and, with it, the bivouac. The men—brave but no fools—radioed for aid then fled to safer ice, to await a risky emergency retrieval. Camp was obliterated as its staff watched. Evacuated on 10 March, it had narrowly survived. Three days later, a return sortie found . . . nothing. Ground to oblivion, even the leavings of habitation had vanished.

Had it been worth the carnage? The air force's notion of a floating, semipermanent ice base clung stubbornly. Still, six years would pass before it chanced another pack-ice station.

In the Eurasian sector, SP-2 was approaching the center of the basin–the only functioning thing amid emptiness. Resupply so far out would be daunting and costly, so closure was ordered. On 11 April, near 82° N, 164° W, journey done, station staff decamped by airlift. The drift time was 374 days. The distance traversed was 1612 miles (2595 kilometers).

Three spring seasons would lapse before another SP station would deploy.

FLETCHER'S ICE ISLAND, T-3

The Air Weather Service, for its part, hungry for observations, was driving the matter of a U.S. ice-island station (Project Icicle) to decision.[52] The Alaskan Air Command would establish (the April 1951 Operation Order reads) "an experimental camp on a floating ice island" north of Arctic Alaska. Its purpose: to maintain a manned weather station; collect information relating to arctic communications; assess the potential for rescue, and operate a base for "the support of scientific investigations." In the eastern Arctic, the air force initiated construction of a major air base at Thule, in northwestern Greenland. USS *Atka* (AGB-3) led a convoy of 120 ships to anchorage off the site. Massive surface and air support of the construction would persist for two years.

Preparations under way, aerial reconnaissance again probed the ice shelves fringing Ellesmere. "Suppressed excitement," a passenger recalls, infused the preflight brief. On 1 August, a B-29 roared off in glorious weather. Moored to and shelved from the coast, an ice-apron extended coastwise fifty miles or more, its freeboard looming perhaps twenty feet above the pack. Ice shelves (and the bergs they spawn) are common in Antarctica. But no true shelf had been reported near this pole. "There were distinct furrows on it," Maynard Miller of the American Geographical Society reported, "making it look like a great plowed field, with clean snow on the ridges and blue ice and meltwater showing as dark lines in the depressions."[53]

Was this the "target" source? A tabular mass was spotted. Having all the characteristics of the land-fast shelf, it likely was a piece that had calved–"the very evidence we needed." About fifty miles north of Disraeli Bay, the plane chanced upon a very large fragment. Shuffling photographs, Major Koenig (lending his seasoned eyes to this mission) confirmed the find: "That's it! It's T-1, all right. What luck!"–the first sighting of T-1 in twenty-two months.

Fifty feet off, the aircrew skimmed a monochromatic tableland–a remarkable unrippled plain much flatter than its known companions. "A smooth runway of any length could have been selected in almost any direction," Miller recalls. Estimating its freeboard, a thickness of two hundred feet was calculated. The overflight of six and a half minutes had covered twenty-one miles. "Target Two" was the largest such feature ever detected in that sea.

On 3 January 1952, the director for research and development, USAF, chaired a Pentagon conference. Fletcher, project officer for ICICLE, represented the AAC. Among the agencies attending were the Air Weather Service, the Air Rescue Service and the Cambridge Laboratory. The centerpiece of discussion: Should SKI-JUMP, a navy operation, be combined with the air force project? Fletcher, teeth in the project, urged independence; in his view, both services would benefit if each did not rely on the other. (The navy proved equally opposed to a joint plan.) Nonetheless, the AAC *was* encouraged to support the navy with supplies, "so long as the Air Force project first received all that it needed."[54]

Colonel Fletcher returned to Alaska. Having won scattered, cautious support for his notion, Fletcher concluded months of "wheedling and nagging" (*Time*'s characterization). His project for a military ice base was under way.

A trial camp was erected (to test equipment), supplies assembled, personnel assigned. That February–March, ICICLE was shunted to the new air base at Thule. Because the target lay closer to the eastern Arctic, Thule would serve as staging base for the C-47s and C-54s slated to support the project. Authority was solicited from the Royal Canadian Air Force to use Alert, on Ellesmere Island, for emergency and staging operations in connection with a "special arctic project." The C-47s had limited range—and Alert station sat halfway to target ice. (A similar request was received from the navy for SKI-JUMP II.) Ottawa's response was lukewarm, due in part to the potential impact on its northern facilities and, as well, to possible U.S. interference with spring resupply operations. In a terse reply, the U.S. Air Force reiterated that its needs were emergency only, adding, "Project Icicle will proceed as planned with staging at Thule, Greenland."[55]

Denied Canadian runways, the air force positioned fuel drums. T-1–the planned depot—could not be found because of poor weather, so a depot was established on sea ice in Robson Channel.

A final reconnaissance located the errant slab. (The sun had reappeared for the first time in five months, on 13 February.) A low-level inspection disclosed a fairly even surface blanketed, apparently, by a hard crust of snow. Though a site for the larger planes was wanting, a ski-landing by C-47 seemed feasible.

T-3 CAMP

The expedition lifted away in the early hours of 19 March 1952. At the controls of the 10th Rescue Squadron C-47 was Capt. Lewis Erhart, USAF. To his right was

Gen. William D. Old, Commanding Officer, Alaskan Air Command. Fletcher was navigator. Aboard as chief scientist was Dr. Kaare Rodahl, a Norwegian expert on arctic medicine and nutrition. Enthusiastic and social, Rodahl was known to revert to a Norwegian phrase for effect–usually for comic relief. Also along were Capt. Michael Brinegan, USAF, and George Silk, on assignment for *Life*. More than 1000 pounds (454 kilograms) of gear were aboard; the escorting C-54s lofted additional supplies. Following a refueling at the cache, Erhart steered for the T-3's last known position.

The target proved elusive. As the hours added, tension rose. "I opened the door to the navigator's compartment," Rodahl records, "to find out what was going on. I could see that Curley and Mike were frantically checking their calculations; Joe was sitting in the corner with his earphones, apparently trying to get in touch with the leading C-54. The pilots up forward were eagerly peering down at the ice for anything that might resemble the target." When the engineer was questioned, Rodahl was told, "We're lost!"[56] Thanks to ground haze, white against white, there was no horizon–opaque flight conditions. "We were flying through a kind of milky twilight," Silk writes. At last, the leading C-54 radioed: it had the target. The bird circled, to home the trailing planes in.[57] On flight decks, headings were adjusted. Erhart gained the place ten hours out of Thule. In a race entirely unofficial and unannounced, the air force had beaten the archrival navy. At separate altitudes, the C-54s began orbiting.

On board the C-47, gear was reorganized, snowshoes pulled out, cameras checked, outerwear donned over flight gear. Below, conditions seemed short of ideal–the surface appeared rougher than hoped and a strong wind was whipping snow. In a blinding "nightmare" of white haze, snow, and glare, a landing area was chosen. "We are going in. Hold on to your teeth!" Erhart shouted.

The intention had been to land *along* one of the troughs. Instead, the pilots elected to land *across* the rolling relief (horizontal change in elevation) of the island. A ski-drag verified deep snow. Seeking thinner cover, another site was probed, with similar outcome. After nearly an hour and several tries, an area close to an edge was chosen. Twice Erhart touched down, dragging his skis. On the third pass, he landed–then pulled up again after a long run, the skis slapping the surface "and the aircraft skipping from crest to crest," Rodahl writes. "On the fourth attempt, a complete landing was made, and the aircraft was quickly slowed down in the deep snow." Under full throttle, the C-47 trundled up the nearest rise. Misled by the roaring engines, Fletcher, aft, growled, "If they're going to take off I'm going to jump out." Encased in billowing snow, the ship taxied back and forth. Finally, a site chosen, the engines were cut. Instantly, the skis froze to the surface. Silence. It was 1:00 p.m. local time–eleven hours since Thule.

No one rushed the hatch. Silk (first off) plopped into knee-deep snow to take photographs. General Old "took one unbelieving look around and trudged off down the landing [ski] track. He didn't like what he saw. He came back and told Fletcher 'I don't see how any man can live on this thing. We must take right off if we can.' Fletcher had expected this and had his arguments ready. He and the general went off by themselves to talk."[58]

Mocking a brilliant sun, the chill (they estimated) stood at about –60°F. Scudding along at ten to thirty miles per hour, the wind swept the island undulations, lifting the granular snow, reminding one of the party of a "restless ocean." Overhead, the C-54s orbited.

Eight men tramped to a ridge to orient themselves and look around. Meanwhile, the camp-to-be was debated out of earshot. Walking surveys and photographs were taken, the aircraft dug out. (Snow cover averaged about three feet.) The slab itself appeared solid. The temperature was appalling. "Never before on any of my arctic journeys," Rodahl writes, "have I experienced such cold." At length, Old walked off–to plant an air force flag. Fletcher trudged back. Managing a slight, frozen smile, "I've gotten permission to stay," he announced. "Brinegar and Rodahl and I. The rest go back." With snow too deep for wheeled landings, the general ordered the orbiting pilots not to try.[59]

Twelve hundred pounds of equipment were manhandled out. Another 3000 pounds (1362 kilograms)–tents, tarpaulins, rations, clothing, stoves, oil, gasoline–were air-dropped off the C-54s. Fuel and equipment for at least forty days were collected at the campsite (Old assisting)–installed on one of the ridges half a mile or so from an edge. (The mission was to assess the possibility of establishing a permanent base. Because the bivouac was necessarily experimental, no gear was along for scientific work.) A camp emerged from dishevelment. While the debut shelter–a four-to-six man double-walled mountain tent banked with snow–was erected, the C-47 was refueled from drums and JATO (jet-assisted takeoff) bottles attached.

After nearly four and a half hours, tension resurfaced: the experiment would degrade to failure if the Ski-47 could not escape. "Before I got in the plane," Silk writes, "I brought out a bottle of Scotch from the bottom of my bag. I gave it to Fletcher and climbed aboard." The colonists were bid farewell. As Silk leaned out for a final shot, the C-47 wobbled off. When the JATO was cut in it surged ahead, broke ground, zoomed steeply up. A final dive, and the fleet faded south (refueling at Alert). Singularly alone, a mere splash of humankind, the colonists sat about one hundred miles shy of the pole. "Now there is no turning back," Rodahl mused. "We were probably as isolated as anyone can be in this world."

The trio organized. A large tarpaulin was spread on canvas, cold-stiffened mattresses inflated, sleeping bags pulled inside. Each man had two: one for normal usage plus one beneath, as added insulation along with blankets. After the men wiggled into two-piece "mummy bags," a pair of kerosene heaters and the primus stoves were coaxed to reluctant life, a case of field rations opened, a meal begun. "Sitting on ration cases in a circle around the tent pole, warming our hands over the blazing stoves," Rodahl recalls, "we chewed some frozen sausages, biscuits and fruit, and drank hot chocolate which we had prepared."[60] Stoves extinguished, each man–half-dressed and weary–addressed himself to his cocoon. (The three had not slept for a full day.) To check the fiendish chill, arctic helmets would remain in place. "Already in bed," Fletcher writes, "we drifted after dinner into sleep interrupted by sharp sounds, like the popping of pistol fire, caused by the cracking of ice in the extreme cold."[61]

Grit was required to vacate sleeping bags: rime frost and icicles framed the openings. And partly frozen zippers proved irksome. "We lit the primus stoves before we got out of the bedding in order to raise the temperature as much as possible before dressing."[62] Still, the special, all-down bags had ushered in a "night" of restorative sleep. And so, amid silence and emptiness, a snowshoe reconnaissance was made. "It was a wild panorama–grotesque white ridges as far as the eye could see," Fletcher recounts, "one large lead of open water in the distance, immense blocks of ice lying in 30-foot-high masses"–products of ice-to-ice collisions.[63]

These explorations tendered reassurance; the island appeared solid–and seemingly unassailable from pack-ice pressure. Further, a veneer of good snow (a wonderfully adaptive material) lay to hand for building shelters.

His party radioed: no retrieval desired.

Entirely suitable as a raft, T-3 offered longevity *and* a margin of safety not to be had on the thickest drift ice. Unmolested by ice cracking and ridging, virtually a national laboratory, the base here begun (enlarged and improved upon) would serve more than twenty years–a triumph of persistence. The environment was punishing, to be sure, yet (exercising care) nonlethal. Just now, moreover, the berg floated intermittently accessible: athwart Ptarmigan tracks to the pole, thus ensuring reliable contact and resupply.

26 March–one week on. "By now," Dr. Rodahl continues, "we were well organized in our camp, and life began to be comfortable and easy, and the work a matter of routine. We slept well in our cozy snow cave, our appetites were excellent, and no one had any physical complaints, save backaches caused by lifting and pulling the heavy equipment in the deep snow."[64] Fletcher–practical, resourceful, companionable–was similarly at ease. "Rank disbelief," he remarks, would often greet his assertion that the men were comfortable. "Our supplies–clothing, equipment, food–were the best. Our permanent station had everything we needed."[65]

Defenses tightening, the bivouac promptly boasted two tarpaulin-roofed snow-block huts plus snow caves. Connected by an unroofed passage, one served as

Tents and Jamesway hut on T-3, a large ice island first detected in 1946 by airborne radar and occupied in 1952. Its status exploratory, the first camp was small. The decision to remain marked the beginning of systematic at-sea research—geological, geophysical, meteorological, oceanographic, biological. Note the transit—for sun shots of exact position in summer, star shots during polar dark. Its life-support system: airlift resupply. (Hazard C. Benedict)

kitchen complete with oil-burning stove–their central heating plant–the other tent as storeroom. A hut slated for emergency shelter was occupied "bag and baggage the day after the first real storm"–a choking blizzard. On the 29th, instead of vapor trails and soundless sky, engine roar announced a Superfortress. It circled, during which a radio conference proceeded. Two planes, Fletcher was advised, would reach him next day, weather permitting. He was to stand by, to home-in the visitors should it be necessary. The project, in other words, was continuing. "This was probably our happiest moment during the entire expedition," Rodahl exults. "The uncertainty and doubt that had prevailed the past few days gave way to a spirit of overwhelming optimism." The strivings to come here, at this singular place, might unlock mysteries of lasting scientific worth.

Early on the 31st (continuous daylight had returned eight days before), a C-54 reached T-3's then-position. Within thirty minutes more, a C-47 also was circling. A paradrop from the larger bird blossomed forth: fuel drums, plus containers filled with equipment and supplies. Despite a stiff wind, all arrived undamaged. As one ship skimmed over, ration cases were tossed out. The C-47 pilot flew a low pass then skied in. Thanks to whiteout conditions, he missed the runway (which was marked) and landed roughly, amid whirling white.

A scientific party had arrived. It was led by Bert Crary, a seasoned geophysicist—one who applies physics to the study of geology.[66] Robert D. Cotell, an electronic specialist, would assist. A superb researcher, Crary was soft-spoken, experienced, and an indefatigable fieldworker—a one-man expedition, a colleague said. "He was a very modest person," ONR's Max Britton remembers. "But a very steady, very hard-working, very reliable, a very *comforting* sort of person. A good man to have around in a pinch, you know, 'cause you could depend on him."[67] Norman Goldstein, a geophysicist as well, was a (later) Crary associate aboard T-3. "Absorbed in his work," Goldstein remembers, "you could see that he was keen on doing everything he could to advance his research."[68] Crary's curriculum vitae of contribution would result in his being loaned to the National Science Foundation and, in 1957–58, to appointment as deputy chief scientist of U.S. IGY Antarctic Programs.

Also disembarking was communications officer Capt. Paul L. Green, USAF— and Tundra, Fletcher's dog. Newly arrived as well: radio gear, a small meteorological station and geophysical equipment, replenishment supplies, a power plant. C-47 pilot Erhart produced a pole surmounted by a large sign. White letters against a blue background announced: "Fletcher Island." (As unloading proceeded, the plane's engines were started at short intervals.) The C-54 dropped two drums of gasoline, both of which were sled-hauled to the idling aircraft. At about 1:00 p.m., Erhart, throttles forward, pulled away in a cloud of white. Rodahl was returning to Thule, to assist in the organization of supplies slated for T-3. The venture was written in the snow. "We banked," Rodahl records (pleased), "and as we passed the camp on our way home I saw four men and a dog standing by the camp site surrounded by the scattered equipment. When we had landed thirteen days earlier the snow was unbroken, but now there were tracks and footprints in all directions."[69]

Two days later, another bird skied in. The island camp was fifteen days young. George Silk found "a smoothly functioning outpost. . . . The expedition, which had stayed to struggle against great odds, was snugly going about housekeeping." Fletcher—loyally supported by acolytes—had harnessed his air force career to the notion of a floating Arctic research station. And won.

> The first occupation [Cotell would recall] employed tents as living quarters. USAF sleeping bags were quite adequate and during the day the usual complement of parkas, etc., coupled with work made the environment adequate. Later, two Jamesway huts were erected. One was the kitchen, the other was

sleeping quarters. The oil-fired range in the kitchen made a comfortable place to work over data and to dine.[70]

Though facilities were spare, the most basic function—shelter—had been addressed. The United States held an incalculable northern asset: a long-haul base from which to conduct systematic field investigation. Here science was to persevere to 1974; indeed, a slab remnant endured to the 1980s. "It was a wonderful program," Max Britton summarizes. "Joe was really the patron saint in the air force for that activity."[71]

Researchers view the world through a lens of interlocking disciplines. Eager to exploit the berg, the talented and the trained would come to T-3 to encamp and to query—biologists, geophysicists, geologists, meteorologists, oceanographers, more. On the ice, knowledge was happening: assumptions made, hypotheses posed, the air-ice-ocean system probed by East and West. Each observation gave contribution. "Every time you did a sounding," a geophysicist told the author, "it was like a new discovery."[72] Late in 1952, a technical paper using results derived at T-3 saw publication, the first of a library of papers and reports.

Meteorologists, oceanographers, biologists, and geophysicists were the earliest to exploit T-3. Broad in scope, military in rigor, their ice-rafted investigations yielded basic, foundational data and observation. As well, the structure and past (origin) of T-3 was examined. To help correlate natural ice-drift patterns with atmospheric and ocean dynamics, the berg's wanderings were plotted. So wind and current measurements commenced—among the earliest of U.S. data sets concerning sea-ice motions, physical properties, and general ice-related problems. Readings of the gravity and geomagnetic fields (the first of thousands) were recorded, to help quantify the regional (but changeable) gravitational pull of the sub-bottom bedrock and its magnetic strength.[73] Not least, the challenges inherent to under-ice navigation, detection, and tracking of submarines brought classified acoustics research onto the berg.

Meteorology was vital. On 1 April, the inaugural readings were recorded. Next day, the following report addressed to the CO, U.S. Air Weather Service, was transmitted out: "Temperature 17 degrees below zero, wind calm, and visibility 20 miles." Surface-wind velocities were taken at six-hour intervals. In addition to regular surface observations, upper-air readings of pressure, temperature, humidity, and winds were initiated (in June), using radiosondes lofted by balloons. This was rich information; altogether, these data-from-nowhere helped to improve forecasts for the whole of the hemisphere. Soon, upper-air readings were to help revise notions of polar circulation and temperature.

Before satellites, to study the ocean meant going to sea. Riding the canopy, oceanography was built-in: the nature of the water column along with soundings, marine life, the geography and geology of the seafloor—the latter derived from seismic experiments.[74] Solar radiation studies began. Crary, with much to do, had settled in—until October. He would log a prolific stay. In ceaseless daylight, the colonists pursued a "kind of timeless existence." The clock was irrelevant; the men worked until exhausted, then slept. When hungry, they ate. That April, *Life* summarized the situation: "The gamble for T-3 was paying off. The Air Force was prepared to stay through the summer—and perhaps permanently."

The original camp was exploratory—necessarily small. But with the decision to continue, a second plane delivered additional gear (on 30 March) plus two men. "Facilities were increased and the scientific program enlarged, as it became clear that it was feasible to maintain an operation on an arctic ice island." A small wall tent served as kitchen, a hexagonal tent consigned to supplies. A two-section snow-block house was erected, to serve as radio room and sleeping quarters. That April, three "Jamesway" shelters were dispatched. Functional and economical, two of the Quonset-like structures were set end-to-end. Airlifted out in sections, one served as a combined kitchen and community center. It boasted a modern oil range, washing machine with drier, florescent lights, electric mixer, phonograph—and "comfortable furniture." The second sheltered quarters for sleeping, the third a laboratory. Here sat the weather station, geophysical equipment, radio gear. At one end were arranged messing facilities, with scientific gear and communications at the other. A third hut—sleeping quarters—was erected a short distance off. Cobbled together, the island camp had begun to resemble a true research installation.

An unforeseen problem now started to get bigger: the local population. Requests to visit T-3 had piled in. The commander of the 7th Air Weather Group, frustrated, was heard to remark, "Individuals here in Alaska, with no valid reason for going to the island, will go to no end to get to the place."[75] The first guests arrived—unbidden, unexpected (on 5 April) when two P2Vs from SKI-JUMP II arrived, to compare notes. But an engine failure brought an emergency return-landing for one P2V. The offending engine was a complete loss, the ship itself undamaged. Abruptly, thirteen names had joined the ice-island roster.[76]

The Commander Alaskan Sea Frontier arranged an engine change—cargo deliverable only by C-54. Eleven days later, two aircraft materialized, one of which flew a slow inspection-pass. An M-129 Weasel (a tracked vehicle indispensable to operations) had prepared a packed surface; hard snow, everyone had reasoned, could support a wheeled landing. A C-54 swept in; its load of nearly seven tons of treasure included a replacement engine, an "A" frame along with a prefabricated working shelter and aircraft heaters.

A vital design feature for landing on snow and ice is strong landing gear. A conventional tail-wheel configuration can absorb great abuse from soft snow and drifts that blanket runways. A tricycle landing gear, in contrast, has a nosewheel–a weak point for snow-landings. This was the first big plane to chance an ice island. Wheel met snow. "Once the C-54 touched the crusted snow," 1st Lt. Robert D. Hale, navigator, commented, "I knew we would never take off again without building a runway." The nosewheel had broken through. "The plane buried itself in the snow like an ostrich on a windy day," skidding to a stop barely three hundred feet following touchdown.[77] Having intended to remain only two or three hours, ship and crew were stranded, swelling the habitation to thirty-three souls—with accommodations to hand for six. (A companion C-54 air-dropped its load then sped home.) The bird was shoveled out. A takeoff attempt failed. Only a runway scrapped down to the ice could accept C-54s. Fletcher radioed for a D-2 tractor.

A hardened strip required four days to clear; on the 24th, the C-54 lifted away. Engines exchanged, the P2V escaped as well. Visitors gone, population slashed, island routines settled back to work.

> Fletcher and Green [Cotell continues] worked on the household chores to make the place livable. Crary and I helped, obviously, but tried to maintain the scientific program that Crary had planned. We would work all day, skipping lunch, and return at night for a social meal. Crary would then work on the day's recordings until midnight and after. He was up again at 4:30 or 5 AM to complete the record analysis. Crary really only required four hours sleep at night.[78]

The encampment (quite habitable now) offered a forward logistics base. Early that May, a jump-off was made for the pole, a mere hour's flying away. The navy had been publicizing its intention to deposit the first U.S. airmen at earth's northern axis. The air force was no less eager for the honor.

TO THE POLE

The third of May saw an off-island sortie. The pilot was thirty-four-year-old Lt. Col. William "Wild Bill" Benedict, USAF, a superb airman. Copilot Fletcher was in overall command of the nine-man expedition. (Two public-information types accompanied Crary, Cotell, and the aircrew.) By virtue of this mission and his later Antarctic work, Bert Crary would be the first man to stand at *both* poles. Benedict enjoyed himself, showing "what could be done with the Ski-47." The sortie was allegedly for science. But esprit held equal consideration. "A fun thing to do," Cotell concedes, "there was not much scientific justification."

The geographic pole is a mathematical abstraction amid limitless white. On the arrival of the airplane, an old, thick floe was found roofing the place—sheer dumb luck. Below, a half mile or so area free of pressure-ice floated. A low pass checked for ridged, broken ice and tested for thickness. The procedure for ski-landing a C-47 was simple if startling: footprint pressure. The plane approached a flat site, the gear was dropped and, with an observer in the bubble aft, the ship plopped down hard and bounced, pulling away at full power; the observer then checked for cracks. "Barring their appearance," Cotell explains, "the pilot was expected to circle around and land in the same spot." On this approach, the copilot-observer (Fletcher) was aft: he would drop a smoke flare to reveal wind direction. Benedict, though, dispensed with a ski drag.

> Obviously Benedict was determined to be the pilot who landed at the Pole. Therefore, on the initial pass instead of bouncing and taking off, he landed. Joe F. who was back beside the door was disgusted and threw out the flare anyway. As the plane trundled to a stop it filled with the yellow smoke from the flare with Joe grumbling about Benedict's ploy.[79]

"Shooting" the sun, 1st Lt. Herbert Thompson, USAF, verified their position. "We took position readings [Cotell continues], gravity, measured ice thickness, water depth, and took ice samples." Of themselves, these data were not especially significant, save for the novelty of location. The party lingered for three hours ten minutes in minus-three-degree air, constant communication being maintained throughout.[80] (Every fifteen minutes, the flight engineer restarted the engines.) After a gravity reading, a ten-foot hole was drilled and TNT exploded to measure depth. Seismic echos put the bottom at 14,150 feet–4316 meters of water column. This result was comparable to that at T-3, floating 135 miles off–and agreed nicely with that obtained from "North Pole-1." To help assess currents, six sample bottles were left behind.[81]

Returning, the party rested briefly then pressed east, in search of T-1. About forty miles off Ellesmere, the C-47 alighted aboard Target One. Crary and Cotell ran a survey line to ascertain surface relief and cored two holes. The ice surface and dirt layering proved similar to that of T-3. Departure now was taken for Ward Hunt Island and its ice shelf, to compare it with the then-known island bits. Rodahl was convinced. "There can no longer be any doubt," he concluded, "that this is where [the ice islands] originate." During 1953–54, complementary investigations by small parties under the leadership of Geoffrey Hattersley-Smith, Ph.D., a Canadian glaciologist, added much to the knowledge of Ellesmere's shelf-ice fringes and the floating islands they spawn.[82]

—·—·—·—·—·—

Maritime research–Yankee style–was flourishing. Strain gauges had been set into T-3 and at the island's interface with the pack. Anemometers were installed five, ten, and fifteen feet above its ice, to probe wind gradients. Off-island, regular current-meter measurements exploited a chiseled through-hole. Soon, twice-a-week soundings were being taken and the water column was being plumbed via a temperature-recording bathythermograph. Samples of surface dirt and boulders, flora and fauna were collected. As for surface biology, the only recorded wildlife was eight birds. (Tracks of bear and arctic fox were noted, however.) Sea life seemed to consist of shrimp and jellyfish. "The oceanographic data, bathymetric data, and the drift positions versus winds were of very great value," Cotell affirms. "We occupied stations that had never been manned before."

Rotation of field personnel is a seasonal rite. On T-3, racing the thaw, spring brought airlifts trucking all possible freight before the runway softened. Decamping in June, Fletcher returned to Washington. The exchange brought Lt. R. Derrickson, USAF, aboard–the new CO. On the 20th, geophysicist R. F. Sexton joined the staff. Cotell rotated out. "Our technology has grown so that we can go and come in the Arctic almost at will," he told reporters of his ride. Asked if he had had any adventures, Fletcher hesitated. "I guess not," the officer replied. "We were comfortable; it was always interesting; and all we had to do was to keep busy. That was easy; there was much to learn and to record. But it was all relatively easy, the quietest Arctic adventure on record, I guess. It was very pleasant. I'd be happy to go back."[83]

Snowmelt began on the ridges, collecting as dark patches and puddles. (Air temperature hovered just above freezing; during the first two years, the warmest reading recorded on T-3 was 36°.) Water began to impound, the impoundments to interconnect. Erected on one of the ridges, the shelters stood safe from the pools, ponds, and channels overspreading the island. By mid-August, it was water-saturated and gray in color, with chains of slender lakes daubing the decay in blue. (Eventually, these drained.) Off-island, open water cut sea-ice concentration (fractional cover).[84] T-3 itself lost one to two feet of upper ice that season–to odd effect. As unprotected ice ablated, camp buildings, drums, and even misplaced tools pedestaled. For staff, landmark headings–"north" or "south"–became unreliable, since their foundation now rotated relative to the disintegrating pack.

The sun settling hard, the late-summer decay ceased. Open water began to refreeze. The canopy metamorphosed–from scattered, puddled, and broken floes to a snow-covered plain, which reattached itself to T-3. Concentration approached ten-tenths. (In the interior pack, typically, coverage exceeds 97 percent.) In dying light, final supplies and equipment were airlifted out and stockpiled. That October, Maj. H. G. Dorsey Jr., USAF, relieved Derrickson as officer-in-charge.[85] As well, Lt. R. R. Shorey, USAF (a geologist), relieved Crary and Sexton, both of whom packed out via the same plane that had shuttled in their replacements.

On 9 October, during one of the last resupply missions, C-47 tail number

315665–the first U.S. aircraft to touch the north geographic pole–crashed on take-off. Damaged beyond repair, the airframe was stripped of useful components then abandoned. The carcass became a fixture of station life. "It was a tourist spot and a walk destination for the natives," Gerry Cabaniss, Ph.D. recalls, "as well as an azimuth reference for measuring island rotation–until [in July 1962] it slid off its ice hummock."[86]

On 1 November, two more joined the winter crew: L. V. Worthington, oceanographer-veteran of SKI-JUMP, and Charles Horvath, a marine biologist and graduate student. "[Horvath] went to T-3," according to his faculty sponsor, "with little time for planning, with equipment gathered hastily from several laboratories, and into winter darkness."[87]

The sun disc faded. The final resupply sortie–a paradrop–was logged on 11 November. The long northern twilight had faded, leaving its flight crew to locate the drop-place with some difficulty.

For the wintering-over, adrift were a station commander, one cook, radiomen and meteorologists, two geophysicists, a marine biologist, and an oceanographer. Utterly isolated, the T-3 base–a candle in boreal dark–shifted through a numbed void. Harangued by storms and the deepest chill, its rounds of observations, routines, and chores persisted, the work of its self-immured inhabitants eased by a few rude diversions.

Polar night is a cold slap; bereft of sun, the air itself seems weighed down. Wind chill speeds cooling; in strong winds, work and exposure is minimized. Even Nansen was not immune to the discomfort of sustained cold and lightless, isolating winter. His heirs in exploration are no less tested. "Eternal night, crowded houses, snow, work that goes on hour after hour and day after day. It is quite monotonous, even oppressive," *polyarniks* concede. And the quality of comradeship may ebb. "The monotonous polar landscape, the unchanging order of the day, the same faces. If you have an argument with somebody there's no place to go to get away from him, and tomorrow something will bring you face to face with him again. It is not surprising that guests who spend several days with us board the airplane with no particular regret."[88] An entire race has sustained life in this environment with no modern-day equipment whatever. A bit of equanimity can sustain a sense of adventure.

> An initial point to appreciate is that severe cold is not as bad as many people believe. In many respects, the psychological effects of low temperatures are as important as physical ones. Sustained, extreme cold promotes depression in some people and encourages others to give up. Consequently, it is important to have a positive attitude toward dealing with the cold.[89]

Regular communication—the thread that binds field teams to the land—is critical. On ice, one concern is safety—are the men all right? Another is morale. At any outpost, the hunger to exchange information is extreme. Radio nourishes the spirit—a form of therapy. (At SP-2, the operators' tent—"our holy of holies"—burned to the ice, silencing the station for anxious days.) Regular traffic for American drifters would be sweetened by contacts with radio "hams" worldwide. And when atmospherics obliged, amateurs helped to patch through telephone calls to families.

When support of the T-3 camp was taken over by the Northeast Air Command (spring 1953), two "Mark-3" huts supplanted the Jameswa120s—a big improvement. One became sleeping quarters, the second for messing, communications, and weather-station equipment—thus separating working from sleeping spaces. Fire spreads swiftly in cold: *two* units enhanced safety. The *heated* spaces were welcomed—the lowest temperature of 1952–55 was recorded that February: –60°F. The layout of the main camp (plus shelters off-island, on attached pack ice) would alter little until 1957, when BRAVO camp was airlifted out to accommodate the International Geophysical Year.[90]

Arctic-related news for this period recounts the Canadian-U.S. expeditions hunting the source of "giant ice islands"; the air base nearing completion at Thule; the T-3 occupation; the West's expanding network of airfields. Fletcher (the *New York Times* reported) would urge the use of T-3 as a forward support base for fighters in the event of an attack upon the United States. Concerning Soviet initiatives, the Western press offered little. Inciting no alarm, drawing scant notice, *Severnyy Polyus-2* passed in secret.[91]

In May 1954, programs at T-3 were suspended: the peripatetic berg had straggled from 88° N, 156° W (March 1952) to 84°40′ N, 81° W—a position off the Canadian Archipelago in the neighborhood of the weather (and listening) base at Alert. Given the rigors attending airlift resupply, proximate stations are a costly redundancy. So the air force quit the outpost. A workable camp was left in place for reoccupation and reuse.

SEVER, SP-3, SP-5

As the Americans all but retired, the Soviets reengaged—with a special audacity. The USSR was reinvigorating its commitment to ice-borne science, and on a scale undreamed of by Western investigators. Experience breeds confidence. Leningrad was about to inaugurate an imaginative investment—a *continuous* campaign. In what was more of a *process* than an event, Soviet *polyarniks* were to dauntlessly inhabit the polar sea. The mechanism would be *two* semipermanent research stations, operating concurrently. If its ice outlasted its shift, outgoing personnel were relieved via *Sever*—the springtime air expedition. Also, wintering staffs were enlarged to between twenty and thirty men. Floe ice is expendable, longevity the anomaly. Should active ice threaten, leading to abandonment, a replacement station would follow.

The surge opened in April 1954, as planes deployed to forward bases along coastal Russia: Dickson in the west, Tiksi off the Laptev Sea, Mys Shmidt in far eastern Russia.[92] The spring offensive, termed *Sever-7*, included a PE-8 transport, the IL-12 and LI-2 (the Soviet-built DC-3 with ski undercarriage), and the AN-2– the latter a workhorse biplane. In overall command: I. P. Mazuruk. "The Arctic had never before seen as much air traffic as it did that spring," a triumphant Yakovlev records. Seeking compatible multiyear ice, parties reconnoitered well to seaward in the hunt for targets. Young floes are unsuitable for long-haul stations; young ice is much preferred for *landings*, however.

> Now [1954] pilots of polar aviation have learned very well to choose a landing place from the air. They can land the LI-2 on skies almost anywhere. But they don't choose old, thick ice field, as Vodopyanov, Molokov and others had done. Now they look for iceleads recently frozen over. Such ice is bound to be even. The whole thing is to judge the thickness of the ice correctly. . . . Now our pilots not only criss-crossed the ocean, but also landed where and when it was necessary. The planes were not little U-2s or aging R-5s, but the big LI-2s into which we could load everything we needed. Common sense and the group commander's experience were the only factors restricting the flights.[93]

In the face of fog and low ceiling, which cursed all flying (the season was exceptionally warm), two stations were installed as, concurrently, mobile research groups did hit-and-go, gathering spot data. One group, jointly led by Cherevichny and a magnetic expert, logged a number of landings in the vicinity of the pole; there, soundings were taken. Another group deployed to study the continental slope in the northern Chukchi Sea. On 9 April, Hero of the Soviet Union I. S. Kotov circled for a long time at 86° before setting down, so as to assess the floes upon which to deploy SP-3. Oceanographer A. F. Treshnikov tells us

> As leader of this station I was aboard the aircraft and had to inspect the region of the landing and select a floe on which to set up the camp of our party. The floe we selected stood out from the surrounding ice by reason of its size and thickness. It was surrounded by high ridges. Its thickness was about 3 m., and its horizontal dimensions 2 x 3 km.[94]

A constant shuttle from shore delivered instruments, equipment, stores. "There was soon a little village of tents and instrument huts on the floe. Radio masts were erected." In mid-May, *Severnyy Polyus-3* (1954–55) was declared operational. The foundation was an old floe 250 miles off the top of the world, at coordinates 86° N, longitude 175°45′ W. (The nearest flat expanse capable of receiving aircraft floated eight kilometers away, however.) North of Wrangel Island and nearer the mainland, a mobile research group unloaded onto multiyear ice. Supplies were shunted, living quarters erected: SP-4 opened for business under the meteorologist Ye. I. Tolstikov.

Though the standard dome tents remained, collapsible wood huts were replacing canvas. Manufactured in sections with plastic-foam insulated walls and floors, the huts (or cabins) were air-transportable and quickly assembled. Heated by a two-element gas stove, each held four bunks. SP-3 saw their first use for drift operations. (That October, five more were installed, after which "all personnel moved into them from the tents.") For logistic support, its aircraft, Leningrad had decided, would deploy more regularly and more frequently, conveying fuel, fresh food, equipment, and mail to the ice. Comforts were realizing a quantum jump. *Severnyy Polyus-3* even boasted a piano! Reportedly, the instrument was a tremendous success, with V. G. Volovich, camp physician, the life and soul of many a party.[95]

These long-haul camps operated in concert with the annual spring resupply and its mobile research groups (*Sever*). When deployed onto a suitable floe at preselected coordinates, the mechanic and hydrologist (with auger) deplaned. Engines were kept idling, ready to taxi off should the ice prove unsound. A hole was drilled; the raft declared sound, measurements commenced. The "jumping" party loitered for from hours up to a day or longer, after which it was conveyed to the next oceanographic station.

The entire *Sever* force retreated, having spent about a month in the field. The Soviet Union would log about 750,000 miles (1,207,500 kilometers) that research year. "It was a pleasure to work in the Central Arctic again," Fedorov remarks of his tour, "and to see with my own eyes how the scope and rates of the Arctic Ocean's study and development had exceeded our most daring dreams during arguments in our only black tent lighted up by a smoking kerosene lamp on the North Pole 1 drifting station." Soviet operations, Fletcher observed, were not only on a larger scale but planned with "meticulous care and superbly equipped for scientific work." Leningrad would have agreed. "Never before," Vodopyanov explains, "had such large-scale exploration been undertaken in the Arctic, and with such first-class equipment, as in recent years. The country supplies Soviet explorers of the high latitudes with all they need."[96] The flying (and the science), the airman could have added, was being carried out by the Soviets' most seasoned aircrews and experts in weather, ice, and other fields with years logged at polar stations and in various expeditions.

Meteorology, oceanography, geomagnetism, and studies of the ice itself were the institute's standard on-ice repertory. Starting with SP-3, drift teams were joined briefly by specialists shunted out on the regular flights. An operational flair was added as well: helicopters. SP-4 (1954–57) was first to boast rotary-wing support its full service life. Clattering off, helos made it possible to press well away from the main encampment. (Heretofore, *polyarniks* had confined themselves to floes that could be reached on foot or with skis.) Checks on conditions as well as scientific forays were thus made practicable within an ice radius of about eighty miles.

The violence of the pack is a major stressor. One's ice raft can mutate—rapidly, unpredictably, lethally. SP-5 (1955–56) would fracture 111 times. Over one twelve-month span, SP-8 (1959–62) was to report fractures on twenty-two occasions, obliging several relocations of both camp and runway. Hypervigilance obtains. Assessments of the ice along with accounts of work performed and other matters of consequence were radioed out. Adrift, state-of-emergency instructions and regulations presided throughout each twelve-month "shift," round-the-clock. Underlying these rules was a single anxiety: should disintegration begin, each man must know precisely what to do.

Before earth-orbiting satellites, major marine events passed largely unobserved; being in the right (or wrong) place to record phenomena was a matter of happenstance. Ice stresses are considerable during winter. Late in November, Treshnikov's command tasted near-catastrophe. No quivers had warned of approaching pressure. When ice breaks in the dark, the awakening is abrupt, the threat compounded, the reaction primal.

> The end of November brought many alarms [Treshnikov writes]. Deep depressions passed over the station causing strong changeable winds. Movement and hummocking of the ice started near the camp; shocks were frequently felt and the noise of hummocking heard. On 24 November, a crack in the ice passed through the camp. Most of the men were asleep and only the man on watch heard the noise. Suddenly a blow was felt and the floe shuddered. Everyone woke up quickly and ran out of the huts. All went to pre-arranged places for "ice alarm." The crack passed between the huts of the meteorologists and started visibly opening. It passed beneath the tent housing the magnetic instruments. The edge of the tent hung over the water . . .[97]

Within a quarter-hour, the break had opened to 50 meters (164 feet) of open water. In darkness and 40°C of frost, the huts and tents were hauled onto a neighboring floe. SP-3 had passed over a volcanic eruption, near the Lomonosov Ridge. As the floe approached, it had taken a series of strong shocks then fractured as sulphuric gases took the life of one man. "One soon learns to live with the threat;

the unpleasant expectancy subsides, although not the vigilance," station leader P. A. Gordienko [SP-4] records. "I learned," he continues, "that the ice field will break apart on sunny days as well as during storms. Fortunately the crack does not always go through the camp!"[98] As well, SP-6 would be jarred by sulphurous fumes from volcanic activity. And in August 1960, SP-8 reported another active submarine volcano in the Chukchi-Beaufort Sea area.[99]

As wind and current shoved, squeezed, and twisted the indolent pack, *polyarniks* shifted from danger, maintained their assigned rounds of investigation. The cycle of observations persisted. In the warm of 1954 Leningrad established a first: a runway prepared *during* the thaw. At SP-4 that July, aircraft delivered fresh vegetables, fruit, and letters—plus the specialists. And in August, a group of doctors deplaned.[100] They found the strip on relatively thin but even ice, the bivouac sited on a thick, hummocky field.

> The station consisted of about 15 wooden assembled houses, some to live in, others holding laboratories. There were many black expedition tents to either contain reserve equipment or intended for temporary inhabitants. . . . I

SPRING REPLACING WINTER AT SP-2 AND SP-3, 1954. WITH BREAKUP, THE ICE TRANSFORMS INTO A BROKEN, SODDEN COVER. HERE (SP-3) MELTWATER—A PRODUCT OF TWENTY-FOUR-HOUR INSOLATION—THREATENS TENTS AND EQUIPMENT. DRYING OUT IS A GENUINE VEXATION: WHEN THE FLOE CONVEYING PAPANIN AND HIS TEAM FLOODED, SUPPLY DEPOTS WERE REACHABLE ONLY VIA RUBBER BOAT. (*ARKTIKA MUSEE*)

IN SPRING 1954, THE SOVIETS DEPLOYED TWO DRIFTING RESEARCH STATIONS—COMMENCING A CONTINUOUS OFFSHORE PRESENCE UNTIL 1991. THIS IS *SEVERNYY POLYUS-4*, THE FIRST IN CONTINUOUS OPERATION FOR THREE YEARS (THREE SHIFTS OF PERSONNEL). NOTE THE HELICOPTER, FOR WORKING THE SURROUNDING ICE— AND FOR SAFETY. HERE AN OCEANOGRAPHER INSTALLS HIS ALL-IMPORTANT WINCH. OUTMODED INSTRUMENTATION AND EQUIPMENT WOULD PLAGUE THE SOVIET SCIENTIFIC COMMUNITY. (*ARKTIKA MUSEE*)

saw the spirals of the radio-sounding radar antennae and a tractor towing one of the houses, which were all on runners so they could be transferred to another site in case of danger. . . . The electric power plant had three diesel engines, one of which was always running to supply the settlement with power. Telephone wires were strung up. The number of personnel at the station was about 30 people, but there were also several dozen more expedition workers, pilots and newsmen. The station had a helicopter and a small AN-2 plane to work the environs.[101]

In the Eurasian Basin, SP-3 nudged to within thirty-one kilometers (nineteen miles) of the pole. Its floe field then shed latitude; conveyed on the Transpolar Drift, within eight months a peripatetic camp was fast approaching Greenland. In light of its inevitable egress (plus a winter's roughing up), evacuation was ordered. On 20 April 1955, the last man was airlifted to the nearest ice able to receive transport aircraft (forty kilometers off). Drift days: 376 total. A week later, SP-4′ was relieved by the next deployment cycle, brought by *Sever-8*. Concurrent with the closing of SP-3 and the relief of SP-4, a new floe raft was chosen at 82°04′ N, 157° E–a point more than twenty meridians west of where SP-4 floated. *Severnyy Polyus-5* was declared operational on 21 April. The station head was N. A. Volkov.[102]

SP-5 would approximate the track of *Sedov*, yielding useful, comparative observations as it moved. Like its predecessors, SP-5 would be supplied and, ultimately, evacuated by air. Pursuing good geographic coverage, however, Leningrad continued to emphasize air-mobile detachments or "jumping" stations.

Unlike the short-lived SP-3, SP-4 survived three annual cycles. During 5–17 April 1955, the first "shift" welcomed its at-sea relief. For the first time on this ocean, a science camp was advancing its programs into a second year. That springtime, cargo-laden aircraft shuttled out almost daily, hauling provisions, equipment, fuel. Station staff–ten to thirty-five men–were accommodated in tents and collapsible cabins. Altogether, its floe-ice foundation rafted a village of thirty structures.

Building on existing knowledge is the engine of science. Wandering the meridians, working to common cause, each expedition pushed forward–one piece in an ensemble–building upon the preceding body of work. (In these years, for example, assumptions based on data from SP-2 were vigorously attacked.) Sustained Soviet performance on and over pack ice was no aberration, success no lucky stroke. "The lesson should be plain," Fletcher would caution. "For such specialized operations there is no substitute for continuity of effort and for experience."

Putative enemies are not in the habit of endorsing one another. How did the Americans know of Leningrad's investigations?

> In sharp contrast to the secrecy surrounding Soviet activities during the last decade, a large amount of information has been released about the Soviet High Latitude Arctic Expedition to the central polar basin during the summer of 1954 and about the two drifting stations established by this expedition. Those taking part were continuing a tradition of previous Soviet studies in this area. I.D. Papanin's North Polar Drift of 1937–38 inaugurated this tradition. . . . The two main stated objects of the expedition in 1954 were to increase knowledge of the central polar basin, with special emphasis on the study of ice movements, and to obtain information about the weather of this area.[103]

T-3 CONTINUES

In the non-Eurasian sector, hostage to the gyre, T-3 wandered its meandering way. By 1955, it floated off Ellesmere, at the edge of the continental shelf and edging westward–another useful shift. Under air force sponsorship, a five-man reoccupation was airlifted out, arriving on 25 April. The old camp stood encased in snow. This field season, Bert Crary again was directing the program, accompanied by Goldstein. S.Sgt. Hazard C. Benedict, USAF, a technician from the Geophysics Research Directorate, was the sole military representative.[104] The marine biologist was a returnee: Charles Horvath, now under contract to the Arctic Aeromedical Laboratory. An observer from the U.S. Weather Bureau, Clifford E. Goodall, completed the ice-island contingent.

The men settled in; routines and rituals started afresh. This field season, the work would range from meteorology, snow depth, and lake-formation studies to solar radiation and magnetic studies, along with oceanographic measurements. The latter included below-ice ambient noise. The berg itself was probed with thermocouples, cores, and stakes; the latter for ablation measurements. Test plots were treated with powders (to study differential melting) and surface dirt collected for carbon-14 dating. Off-island, a shelter was erected for biological investigations. (As a precaution, it held a life raft plus emergency rations.) "We were busy," Norm Goldstein summarizes. "We had science to do and we were busy doing it."[105] Cloud deck or fog obliging (often they did not), vertical elevation angles to the sun were measured with an engineering transit, for computation of geographic position–after which the coordinates were radioed out. (A few stars appeared in September, allowing stellar shots.) Goodall, as weather observer, was fully occupied.

> The surface weather observations continued to be taken every six hours including local noon and midnight until 1 September. From 1 September to 24 September surface weather observations were taken every three hours and transmitted to Thule Air base via Alert Weather Station. Included were wet and dry bulb temperatures, surface pressures, wind speed and direction, sky conditions and visibility.[106]

A seismic shot was used to calculate ocean depth and dip of the bottom horizon(s). For this, a small chemical explosive was detonated. The energy returns from the ice, water, and bottom were precisely detected and recorded. From these, properties of the sea ice, water column, and seafloor (and sub-bottom) were ascertained.[107] Additional soundings were taken while collecting cores and dredging bottom samples. A transit magnetometer–obtained through the U.S. Coast and Geodetic Survey–was used to measure the horizontal strength of the magnetic field and declination. Over three months, forty-eight observations for horizontal intensity were recorded. In that span, T-3 shed less than two degrees of latitude as it shifted from near 88° to 98° west longitude. In Washington, Irene Browne, an air force colleague, reported on Bert Crary's results to the American Geophysical Union meeting, at the National Academy of Sciences. Soviet teams had found the Lomonosov Ridge. Now, based on echo soundings, another large chain had been detected south of the Soviet discovery. There bedrock rises to within 5000 feet (1525 meters) of the sea surface.[108]

If hardly idyllic, life on ice was agreeable enough. More or less self-sufficient (the last plane escaped on 21 June), domestic duties such as cooking, kitchen patrol, and sawing blocks (for drinking water) were shared among the party, as at Soviet stations. Goldstein enjoyed (and still enjoys) cooking, so he had assumed much of that chore. Camp fare included steaks, ham, and chicken; from a Dutch oven, pies, cakes, and doughnuts were conjured. In the main quarters, heat came courtesy of a fuel-oil stove—tended constantly. The men elected to sleep in unheated spaces, however, thereby eliminating a second fire watch.

Diversions were few. Still, every expedition logged its own moments. And satisfactions. Delights this drift included Goldstein's menu, Horvath's enthusiasm for his specimens—and a Soviet plane that roared over, its crew waving, then dropping chocolate. The reoccupation, Goldstein observes, "went very smoothly. I think we all enjoyed it."[109] For U.S. polar research, the year was a comparatively quiet interlude. Ashore, meanwhile, a sled-mounted party investigated the ice shelves fringing Ellesmere Island, the nursery for T-3 and sister icebergs.[110] Helping to celebrate the Fourth of July, Crary detonated a ten-pound charge of TNT. Concluding the celebration, gasoline on piled trash produced a bonfire. Crary, as it happened, would be recalled to head the air force's contribution to the IGY in Antarctica. In 1957, the geophysicist led the wintering party of scientists and a naval supporting party at "Little America V."

The worth of observational data is dependent in part on *where* it is taken. Accordingly, navigation accuracy—true and correct position—was cherished. ("Navigation" implies control. Attempting to influence drift is absurd.) Coordinates were obtained by "shooting" stars or the sun. Three stars can be shot in minutes, with excellent result. Thanks to overcast or fog, sun-shot calculations often suffered. Given its low elevation, moreover, an uncertain correction was required for atmospheric distortion and for ice drift *between* shots. Storms proved vexing: T-3 might drift seven miles a day in open water in strong winds—at a time when solar or stellar shots were unattainable. And as the pack loosened, the slab itself rotated; during these weeks, location data were generally poor.[111]

Ashore, a complete record of position and overall conditions was maintained at each nerve center. At sea, accuracy increased the likelihood of resupply as well as rescue. To assist hunting aircraft, homing beacons were installed. But they were weak—and useless if the power failed. "So everyone, from pilots to camp commanders to replacements to 'natives,' felt much more comfortable when we knew where we were."[112]

The season was dying; dark would soon again drape this paralyzed sea. Wanting the T-3 party off as quickly as possible, the air force insisted on evacuation.

(Ample food lay stored for wintering; "I wasn't the least bit concerned," Goldstein told me.) On 1 September, the runway was declared usable. Aircraft from Thule sortied for the errant camp. But high winds and blowing snow foreclosed a rendezvous. And continuous bad weather meant no recheck on camp coordinates. Horvath saw retrieval on the 16th. Eight days later, a navy radar aircraft "rediscovered" the roving island. A pair of C-47s were directed in. "Swooping down out of the snow storm" in one-mile visibility (the *Air Force Times* insists), the aircraft spent more than an hour collecting the three scientists and more than 3500 pounds (1589 kilograms) of instruments and samples. Engines had to be started and warmed each thirty minutes, to prevent them becoming "cold soaked." In near-zero air and a twenty-knot blow, cargo and fuel were man-hauled to the waiting aircraft via sled and caterpillar-tracked weasel. Finally, igniting JATO bottles, so as to boost off, the two-plane squadron climbed hard away. One of the left seats held Maj. C. Bedrick, who, three months before, had piloted the last takeoff from a softening T-3 runway. Inside of three hours, the research team for 1955 was restored safely to Thule Air Base.

What was it like living and working on ice, journalists inquired?

"Well," Mr. Crary said, "it was not bad, not so bad. It didn't hit more than 10 below all summer." However, he conceded with a smile that in winter you had to slow down a bit when the temperature reached minus 60 degrees or so.

Soviet natural scientists, also, are interested in the Arctic region. Mr. Crary credited them with a greater knowledge of the area than anyone else possesses.

They have a greater interest in it because of their vast Arctic territories, he explained. He implied that this is probably the main reason for their work, rather than primarily military considerations, although these cannot be completely discounted.[113]

Investigations tailed off. Having disengaged, an extended hiatus ensued: the Americans would not return to the ice until the surge attending the IGY with its logistic support of remote science stations.[114]

IGY PRELIMINARIES

When 1956 opened, a pair of Soviet camps were operational: SP-4 and SP-5. The springtime expedition—*Sever-9*—comprised thirty-five aircraft under the scientific control of M. Ye. Ostrekin. A member of Cherevichny's 1941 party, he now was deputy director of the Arctic Institute. Altogether, about 150 oceanographic stations were occupied. That April, *Severnyy Polyus-6* (1956–59) was installed north of Wrangel Island, near 74° N, 177° W—coordinates near that of SP-2 when closed. This ice raft was an ice island—a first for Leningrad.

Ice is deformable. Camps are lucky to log uninterrupted months before pressure erupts, requiring a shift or full abandonment. Immeasurably more useful than floes, ice islands are unassailable in this respect. This sixth camp was to survive longer (1243 days) and migrate farther (8634 kilometers/5363 miles) than any SP camp, prior to SP-16 (1500 days operational). In all, Leningrad would exploit these rare rogues five times. Nikolai Vinogradov was a station leader at SP-22 and at SP-31. What is the platform of choice, the author asked. "An ice island is preferable, of course," he replied, smiling. "Certainly, the work is much easier . . . than on a thin ice floe."[115]

Measurements derived from *Severnyy Polyus-6* and its sister, SP-7 (1957–59) would constitute Moscow's promised contributions to the IGY program for the Northern Ocean.

Sustained by airlift, *three* Soviet stations rode the ice in April–October 1956. SP-4 and SP-5 were each relieved and resupplied during the latter half of April— each for the *second* time—by the *Sever-9* expedition. Relief staffs encamped, station business held fundamentally unchanged: oceanography, meteorology, glaciology, geomagnetism. Starting the year moving but slowly, SP-4 gathered speed as it approached the pole. On 5 July, it passed within thirteen kilometers (eight miles). Now shifting southward toward Greenland, "4" would be evacuated by *Sever-10* within about three hundred kilometers of that island-continent. Less happily, "5" was nearing the Siberian shelf and, giving way to the next in line, abandoned its site by airlift on 11 October 1956, having logged 536 days of useful operation.

> High latitude air expeditions are becoming an important annual event in the Soviet Arctic [*Polar Record* reported]. Groups of scientists are transported by air to a number of points on the ice of the Arctic Ocean, and there carry out observations. The basic principle is that knowledge of the area is so slight that even single observations of, for instance, ocean depth, water temperature and salinity, geomagnetic elements, and ice thickness are of value. At the same time the opportunity is taken to establish or relieve the drifting stations. The operation is mounted each spring and normally lasts about six weeks.[116]

Synergetic yet rival plans for the IGY (July 1957 through 31 December 1958) entered their final phases. For the Arctic Ocean, two ice-based Soviet IGY camps were in prospect. Moreover, the Soviet Union planned to designate thirty-four of its *shoreside* stations for the IGY (operating polar stations, mostly). In all, Leningrad would field twenty-two expeditions during 1957. The high-latitude air expedition, *Sever-10*, led by M. M. Nikitin, mobilized 270 men and 17 aircraft to close out "4," relieve SP-6, and deploy *Severnyy Polyus-7*. In addition to oceanographic work, mobile groups deployed fourteen automatic weather stations plus eleven radio beacons–the latter to track ice movement. Further, an expedition based aboard *Lenin* would investigate water exchange between the Polar and northern North Atlantic basins within the Greenland Sea.

As IGY neared, then,

> Soviet investigators had carried out a very ambitious program. . . . The basis of their success was continuity of effort, careful planning, and the accumulation of experienced people both for scientific work and for air operations. This continuity of effort was provided by the Arctic Research Institute on the scientific side and by the Arctic Air Service on the air-operational side.[117]

To counter these initiatives, no U.S. ice-based stations were operational. The match-up lay tipped against the Stars and Stripes. Was the Soviet Union's stakeout a spur? "Well, I can assure you it was," Dr. Max Britton states. Washington felt the prod. "You talk about things that will stimulate interest in your program. And certainly, if other people are doing it and you can't, then something needs to be done. Sure. You work that to the hilt. Everything that I could learn about the Soviet Union–what they were doing–got put to work right away, by pointing out how backward we are, in not accomplishing what we could do."[118] Recruited from the Navy Department's Office of Polar Programs, Cdr. Ronald McGregor would succeed Britton at ONR. The polar regions, McGregor told me, were works-in-progress. "Finding the pole–the day of the explorer–was done. Now it was the scientist who was an explorer."[119]

Chapter Four

INTERNATIONAL GEOPHYSICAL YEAR

> I see [the IGY] as the most significant scientific effort in Arctic history.
> –Dr. Kenneth L. Hunkins

THE INTERNATIONAL GEOPHYSICAL YEAR (IGY) of 1957–58 was a direct descendant of earlier collaborations. When the U.S. Congress authorized President Rutherford B. Hayes to establish a temporary Arctic outpost for science, two stations were secured to help support the International Polar Year (IPY) of 1882–83. An unprecedented consortium of nation-participants, the IPY was the first coordinated international cooperation in polar science. In 1932–33, the jubilee of that inaugural "year" was celebrated with a repeat–and an extension–of the first. The Second IPY suffered an anxious gestation–nearly succumbing to international financial chaos. "The acute economic crisis that had struck the capitalist world," radioman Ernst Krenkel remembers, "hindered scientists from different countries in their desire to work together as a single united family."[1] The Depression put an end to a number of projects, including that of Nansen for the establishment of a drifting station near the pole.

The seeds for a third polar year had been planted in 1950, during an informal gathering at the Maryland home of Dr. James K. Van Allen, a leading researcher in upper atmospheric phenomena and space science.[2] As a physics student, he had assisted the Second IPY. "My participation was minuscule," he later wrote, "but it made a deep and durable impression on me."[3] Rapid advances in scientific technique, the instigators reasoned, allowed the interval between polar years to be halved. The twenty-fifth anniversary of that IPY, moreover, would coincide nicely with an anticipated maximum of solar activity. The notion was brought to the International Council of Scientific Unions. As officialdom considered, the view developed that the enterprise should deal as much with the tropical and southern nonpolar regions as with circumpolar extremes. The venue was therefore enlarged to the whole of the planet. Late in 1952, the name of the enterprise was changed to International Geophysical Year.

The general objectives of the IGY were broad, but the major focus lay on securing new and better observations of poorly understood natural phenomena. Resources were directed especially to high-altitude rocket observations and artificial-satellite monitoring of Earth from space, Antarctic and Arctic observations (including ice drilling in Greenland), echo soundings of the abyssal ocean basins, and studies of long-period waves in the solid earth.[4] Among the scientific specialties, geophysics and oceanography in particular would gain in stature.

Superpower rivalry and nationalism, certainly, were ingredients of this collective push for world data:

> The IGY with its unprecedented funding was energized by a mixture of altruistic hopes and hard practical goals. . . . The government officials who supplied the money, while not indifferent to pure scientific discovery, expected the new knowledge would have civilian and military applications. The American and Soviet governments further hoped to win practical advantages in their Cold War competition. Under the banner of the IGY they could collect global geophysical data of potential military value, along the way gathering intelligence about their opponents, and meanwhile enhance their nation's prestige. Some saw the Cold War the other way around, hoping that the IGY would set a pattern of cooperation between the rival powers—as indeed it would.[5]

The "year" held eighteen months (1 July 1957 to 31 December 1958), with programs persisting two years more. A coordinated global cooperation to understand planet Earth, the IGY stands as the most ambitious and comprehensive scientific assault on the polar regions ever undertaken. In Antarctica, a dozen countries established sixty research stations. In the opposite hemisphere, the United States set up or modified existing facilities at seventy-six locations for continuous observations, including thirty-eight stations manned in cooperation with Canada, Sweden, and Denmark.

The IGY embraced all the major nations. Russia had assisted both IPYs; formally invited to again cooperate, the Soviet Union agreed. As endorsements multiplied, IGY planning engaged government agencies worldwide. "World data centers" were established along with a global communications network to coordinate scientific observations and to ensure that the hard-won information would reach them.

Planning and direction of the U.S. IGY program was placed under the National Academy of Sciences. Inside the USSR, a coordinating committee had been formed (in 1955) under the auspices of the Soviet Academy of Sciences. The nation's primary sponsor of basic research, it served to channel government funds to numerous institutes. IGY projects received highest priority: scientists and technicians, facilities, equipment. By early 1957, twenty-eight institutes (Leningrad's Arctic Research Institute among them), observatories, committees, and societies were hurrying preparations for the extravaganza.[6] Emerging from its self-imposed isolation following Stalin's death, Soviet science, so to speak, was going public.

The IGY was the largest coordinated international-research effort yet undertaken. Sixty-six countries would participate, including all in the Arctic region. For its part, the U.S. National Committee for the Geophysical Year formulated an Arctic Ocean Study Program. The Department of Defense, meantime, had been directed to provide the logistical support necessary to deploy scientific teams, many of which would be working in remote, comparatively inaccessible locations. Early in 1956, the planning committee requested logistical support for establishing and sustaining a *pair* of drifting stations. The U.S. Air Force agreed to put IGY research into the basin by supporting "an extracurricular project" within the basin. In October 1956, Headquarters, USAF, instructed the Alaskan Air Command "to establish a scientific observation station on the drifting pack ice of the Arctic Ocean prior to 1 July 1957" and to "logistically support this station until after 31 December 1958." Executing similar orders, the Northeast Air Command began to land men and equipment for the refurbishment of T-3—that roving fragment, which still was opening sea sectors to the study of geology, oceanography, climatology. The staging base was Thule Air Force Base (AFB), Greenland.

The Arctic Ocean was no navy playground. Diesel boats had logged a handful of experimental cruises. Now, though, the exploratory phase of the transarctic submarine was about to open, courtesy of nuclear propulsion—embodied in *Nautilus* (SSN 571) and *Skate* (SSN 578)—progenitors of a fleet to come. Waldo K. Lyon, Ph.D., had joined the U.S. Navy Electronics Laboratory years before the advent of nuclear propulsion. *Nautilus* changed everything. A pioneering naval physicist, tireless in his advocacy of an arctic submarine *program*, Lyon "was shoving and pushing and yelling and what not, to do something in the Arctic." The fight was uphill. Robert E. Francois was active for decades in navy-sponsored programs for underwater acoustic tracking and sonar research and development. "The navy was very slow," he concedes, "to recognize the importance of the Arctic to the defense considerations of this country."[7]

Under-ice operations are strictly the province of submarines. And nuclear boats are ideal for extended survey work—and for Arctic war-fighting. For Dr. Lyon, then, Arctic-capable boats were the *primary* platform for development work, ice-based stations the secondary element. The physicist's notes for a 1956 IGY Oceanographic Panel meeting are emblematic:

> ... we made no commitment to directly take part in any observations on any ice floe stations in arctic basin. Air Force Cambridge indicated they were going to make seismic bathymetric, ice physics measurements at both T-3 ice station and ice floe station; however, there appeared considerable confusion about where personnel were to be obtained. They are officially committed to conduct the work. Our position stands that we would only consider brief stays on ice floe by our people, say 2 months at most for heat budget thru ice work, otherwise cost too great in time and personnel.[8]

Offshore, meanwhile, Project ICE SKATE reestablished T-3 as a forward support base. (It was acceptably positioned, hence useful again.) Late in the winter of 1957, ski-equipped C-47s and a C-54 field were deployed to the air base at Thule. At Eureka, on Ellesmere Island, fuel was cached for the smaller planes. The "big push" was flown on 7 March, in twilight. With the C-54 (which landed first and set out flares) as escort, five C-47s skied in on Eureka's snow-covered strip. "We then proceeded to refuel the C-47s from the cached fuel drums," Ron Denk remembers, "first with the aid of a small engine-driven fuel pump which failed after ten minutes. We then resorted to a hand-pump operation which was really slow. We kept the C-54 engines running the entire time since we had more than enough fuel on board. . . . We then took off and found T-3 with no difficulty."[9] As the C-54 orbited, a survey party of six specialists deplaned and unloaded. The camp commander was Maj. Willie Knutsen, USAF. The mission: to survey a base for IGY work. The armada was escorted back to the base nonstop with just enough fuel. An accompanying newsman, deposited at T-3, fought down a sense of abrupt isolation on the leave-taking.

> We watched as the propellers of the five brightly painted C-47s churned up miniature blizzards; as the planes jerked to life, their ski landing gear scudded over the rough snow. Then a roar told us that the first plane had ignited its JATO, boosting it into a steep climb above the whirling clouds of snow. Another roar, then another and another, until all five Gooney birds were airborne, winging south across the polar pack ice to their base. . . . Now we were alone. . . .[10]

On a routine two-week assignment, heavy weather stranded the newsman seven days.

Norm Goldstein was there. "The old camp," he says, "would certainly not have been suitable for the IGY." Blighted and derelict, buried to its roof in drifts, the outpost had survived six field seasons of wear and tear. Untrodden for eighteen months, the hut floors lay sheathed in ice. Further, the site was ringed by the detritus of hard occupation. Anyway, the old camp would not suffice, as it was undersized for IGY-related programs. Airlifts brought supplies and mail and, in April, (from 18th Air Force *Globemasters*) paradropped D-4 tractors. Despite breakdowns (which required drops of replacement parts), the D-4s plus exertion resulted in a two hundred- by five-thousand-foot airstrip. Nine hundred tons of cargo followed, along with a thirteen-ton grader. Commercial house trailers comprised the essential base—now redesignated IGY Arctic Ocean Station "B" or BRAVO (T-3). Forty-two and a half feet long, about ten feet wide and nine feet in height, the shelters weighed in at 21,000 pounds (9,534 kilograms). Each held a private stateroom (for the camp commander), a staff lounge, kitchen, and lavatories—unheard-of amenities.[11]

———·—·—·—·—·—

In its sector, the ice rafting SP-6 had drifted tortuously its first year. On 20

POLAR ZONES ARE EXTREME ENVIRONMENTS—LETHAL, OPPRESSIVE COLD; SLICING WIND; MONTHS OF LIGHTLESS WINTER. DARK-SEASON STORMS ARE PARTICULARLY VEXING. HERE A *POLYARNIK* BRAVES THE DARK SEASON AT SP-6, THE FIRST SOVIET STATION DEPLOYED ONTO AN ICE ISLAND (1956–59). WORK AND EXPOSURE ARE MINIMIZED IN STRONG WIND. (*ARKTIKA MUSEE*)

April 1957, the high latitude air expedition *Sever-10* arrived, a relief crew accompanying. The island-raft continued its slow shift to the northwest. A week following that relief operation, SP-7 was air-deployed, with V. A. Vedernikov as station leader. The floe position was 82°04′ N, 164°50′ W—somewhat to the north of where *Sever-2* had ended its run. The winter crew at each station: about fifteen men. Both crews would rotate out in April 1958—the standard twelve-month shift. Until then, the pair was to function as part of the Soviet contribution to the IGY.[12]

—•—•—•—•—•—

What of that other IGY drift camp promised by Washington? In accordance with instructions to establish "a scientific observation station" on pack ice, the Alaska Air Command assembled men and equipment at Point Barrow, ready to sortie when a suitable floe had been chosen.[13]

An air search commenced in March 1957. Operating from Ladd and Eielson runways, WB-50s sought a "possible"—a large multiyear floe, preferably located about midway between the Alaskan coast and the pole. Weeks of almost daily sorties proved fruitless, however. Candidate ice *was* spotted but proved elusive; it was lost to view and never seen again.

Summer softens the pack, precluding heavier planes. So the search was shifted northward. On 4 April, a C-47 skied onto prospective ice—a floe three miles in length, half a mile wide at 79°29′ N, 168° W. Drill cores confirmed eleven feet—a satisfactory foundation. A second Ski-47 set down. In addition to Maj. Richard E. Freeman, commander, ICE SKATE, the advance team consisted of Col. Joe Fletcher and Tech. Sgt. Pat Garrett, an instructor from the AAC's Arctic Survival School. Four civilians had joined them: Father Thomas Cunningham, an Air Force Reserve chaplain,[14] and three construction specialists. Their primary mission was to install a radio beacon for resupply aircraft and to begin construction of a runway, using heavy equipment paradropped in. "As the last two planes headed south, they dipped their wings as if to say good luck and farewell to the courageous men they were leaving behind."[15]

Six construction workers deplaned on the 15th. Additional military personnel were airlifted in—three communications men, two cooks, and a medic among them. Late that April, heavy equipment was airdropped, including (as at T-3) a D-4 bulldozer—the largest single piece of gear. "Within five minutes of its uneventful descent, which was slowed by four 100-foot parachutes, it was working on the ice floe." Livability installed, the permanent staff turned to other construction, the most important of which was leveling a runway.[16] When the scientists gained camp, on 21 May, the strip was largely complete though yet short of specifications. Most of ALPHA's Jamesway huts had been set, wired, and modified for purpose. "The [permanent] camp (a report summarizes) was equipped with a continuous electrical power supply, oil heating, weasels, caterpillars, and a grader for transport and runway maintenance, electric washer, dryer, and shower facilities, and an excellent radio station for both official and amateur frequencies."

The gross weight of this newest bivouac was five hundred tons; most of this consisted of fuel flown in primarily by C-124 *Globemasters* during May–June. Routine resupply would rely on C-54 and C-47 aircraft (See Table 2).[17] Inescapably, June realized channels of runoff and melt-throughs, leaving the runway unusable—closed until August. Strip construction and maintenance in this formative period represented *the* on-ice chore for the U.S. support teams. Such was emphatically the situation at Station "A."

> This was primarily due to the Air Force requirements for sea-ice runways to be used by the C-124 type aircraft, namely 5,000 x 200 ft. with minimum thickness of 54 in. of solid ice, also with 500-ft overruns and 1,000 ft from the point of touchdown to the nearest lead. These last requirements were demanding and ALPHA was, for the most part, serviced by C-124s under a special waiver.[18]

RUNWAY CONSTRUCTION, MAINTENANCE, AND REBUILDING (FALL SEASON) INTRODUCED MAJOR LOGISTICAL HEADACHES DURING THE 1950s AND 1960s, AN ERA WHEN CAMP SUPPORT SYSTEMS RELIED ON LARGE AIRCRAFT (HERE, A C-124), LONG AIRSTRIPS, AND HEAVY EQUIPMENT. AT T-3 AND THE ARLIS STATIONS, MANY DAYS WERE EXPENDED DRAGGING THE RUNWAY. FRACTURES CAN STILL TERMINATE OPERATIONS IF LENGTH DROPS BELOW SPECIFIED MINIMUMS. (DR. KENNETH L. HUNKINS)

When a plane was scheduled, almost the entire camp turned out if the runway had been drifted out of operation. And from June to August, virtually all supplies arrived via parachute, including heavy drops by C-124.

Following hard upon ALPHA, aircraft deployed *Severnyy Polyus*-7 into the central Arctic on 23 April. The coordinates were near 82° N, 164° W—somewhat north from where SP-2 had surrendered. Located on floe ice, "7" would be sustained by airlifts throughout a 718-day career. A permanent staff of about fifteen would log visits by Soviet, Czech, and Swedish journalists, and, later, a Moscow film crew. This same April, on the 19th, 322 kilometers (less than 200 miles) off north Greenland, planes carried out the final evacuation of SP-4. The first station in continuous operation for three years (1007 days), "4" had transected the basin east to west, meandering more than 7000 kilometers (4376 miles) under three station leaders—a bravura performance. In the meantime, SP-6 and "7" had been designated the USSR's Arctic IGY stations.[19]

—•—•—•—•—•—

The IGY realized a munificent harvest. Tens of thousands of researchers were deployed—literally from pole to pole. The work proved heady, fruitful, chaotic, career-making. Geophysical research was stimulated worldwide, especially in polar regions. The legacies affect the landscape of geoscience still.[20] At ice-borne ALPHA, a balloon launch commemorated the IGY kickoff—zero hours Universal (Greenwich) Time, 1 July 1957. The sphere was tracked to 80,000 feet (24,400 meters), the results transmitted to Point Barrow for relay to IGY headquarters in Washington for archiving. Mere dots upon boundless waste, four ice-borne pockets of humanity floated. Their mission: *science*—no guns, no bombs.

But the West had underestimated the Soviets. Dead on time, the *Sputnik* shock would annihilate its conceit—and realize a massive escalation of cold war. A shift in the balance of scientific power, the West feared, could realize a shift in military power.

Knowledge of currents had been derived largely from land-based expeditions, from Soviet drifts (including involuntary freeze-ins), and from the track of T-3. In broadest outline, ice movement *was* known. *Sedov* had confirmed a persistent pattern in the Siberian sector. Still, near-surface circulation is rather complex. Off North America, the pattern is clockwise—as T-1 and T-3 had shown. Near the pole, however, drift tracks diverge: ice fields may shift southward, to debouch (as Papanin's had) with the Greenland Current or recirculate for another orbit. The cycle duration is about ten years.

A major contribution to the Arctic Ocean Study Program of the IGY planning committee, ALPHA floated its useful, wandering way. The floe would receive (as did T-3) logistical support—camp construction, maintenance, supply—

courtesy of the U.S. Air Force, staging out of Ladd Field.[21] ALPHA would host world-class research. Heading the ALPHA team was Norbert Untersteiner, Ph.D. A specialist on thermal budget, Untersteiner would codevelop one of the earliest thermodynamic models of snow and sea-ice. His colleagues: Kenneth Hunkins, Ph.D., Director of Studies in Oceanography at Columbia University's Lamont Observatory; Arnold Hanson, Ph.D., a meteorologist from the University of Washington; and Lt. Thomas English, USAF. A marine biologist assigned to ICE SKATE, English was part of the Arctic Aero Medical Laboratory at Fairbanks–the sole military member of the science party.

Adopting sensible routines, research programs embraced gravity, seismology, aurora and airglow phenomena, geomagnetism, oceanography, ice drift, meteorology and microclimatology, sea-ice physics, biology, underwater sound, and marine geophysics. Beneath a nonsetting sun (which disturbed sleep and hindered a normal routine), the research programs began. The original layout for ALPHA camp would endure a full year before ice pressure intruded, disrupting day-to-day rhythms of domesticity and work.

On 16 June, the Kremlin disclosed that a pair of its aircraft had landed at the geographic pole on 16 May; this was the first such touchdown since Fletcher and company had jumped off from T-3. This mobile group (part of *Sever-10*) lingered six hours, setting up an automatic weather station.

The IGY underway, Soviet plans for a satellite became a regular boast. On 27 August, Moscow announced that it had a proven intercontinental ballistic missile; on 1 October, the frequency of the prospective orbiter was disclosed. Forty-eight hours later, *Sputnik I* burst upon the world–the first man-made object to orbit the planet. Surprise–indeed shock–was universal. Inside the Soviet Union, the most spectacular event of the IGY startled its news-starved peoples and elated its governments. Nikita Khrushchev hailed the launch. The USSR, he boasted, "has now entered the decisive phase of its economic race with capitalism." A shattering blow had been delivered to U.S. prestige. *Sputnik*'s data-chirp on the ionosphere and the orbiter's own temperature held a darker message: Soviet technology was not to be underrated. "We didn't suspect we were behind," an engineer with the *Redstone* (missile) group recalled. "My reaction was shock that they had that much launch capability."[22] Instantly, Washington was frantic to reclaim the lead in space–an alien realm abruptly in vogue.

Boris I. Silkin, Ph.D., was a junior researcher on the Soviet Committee of the IGY.

> In the West, much was written about the American program of forthcoming space researches [he recalled]. So much greater was the sensation on October

Geophysics at ALPHA. Study of under-ice phenomena included seismic methods and underwater acoustics, the latter conducted by the Navy's Underwater Sound Laboratory. USL work here consisted primarily of using explosives for measuring low-frequency transmission loss. (Dr. Kenneth L. Hunkins)

4, 1957! I shall never forget this day . . . in the official government report it was said: ". . . according to the program of IGY." In reality, everything was secret and we were not admitted to these secrets and so it was difficult for us to answer numerous questions of public and mass media. But we also could not admit the fact and so the natural pride in our science and technology was combined with embarrassment. And when I took the risk of telling one of the Soviet journalists the name of a person who could have answered all his questions, I had trouble.[23]

THE INTERNATIONAL GEOPHYSICAL YEAR OF 1957–58 HELD A STRONG POLAR RESEARCH COMPONENT. THE AIR FORCE AGREED TO ESTABLISH AND SUPPORT TWO FIELD CAMPS ON THE NORTHERN OCEAN. THE SOVIET UNION ALSO HAD TWO DEVOTED TO IGY GOALS, SP-6 AND SP-7. ALPHA WAS THE FIRST U.S. RESEARCH CAMP ESTABLISHED ON AN ARCTIC FLOE, ESTABLISHED IN APRIL 1957. HERE, BIOLOGISTS PREPARE TO EXPLORE AT ALPHA, JULY 1958, TO EXAMINE THE UNDER-ICE SURFACE AS WELL AS BIOTA ON AND NEAR THE BOTTOM OF THE CANOPY. (DR. KENNETH L. HUNKINS)

In November, *Sputnik II* gained orbit. (On T-3 that autumn, signals from the *Sputniks* were monitored.) "It was becoming all too apparent," *Life* lamented, "Russian scientists are as good as any in the world—or better." The Soviets (of all people!) had inaugurated the Space Age. The absolute belief in the supremacy of American science and technology lay in tatters. Awash in self-doubt and recrimination, the nation engaged in an urgent, emotional, bitter debate. "We reacted hysterically," one observer conceded.[24]

America had forfeited being first. With IGY plans well under way, an urgent interest had been expressed by the National Academy of Sciences. But the academy had had no money and turned to the National Science Foundation. Its resources proved unequal to the task, with Congress disinclined to appropriate more. In the spring of 1955, the classic option had been taken—a scientific advisory committee. Asked if a launch would someday be feasible, the panel responded with an equivocal "yes"—provided personnel, facilities, and funds were made available on a high-priority basis. But the committee was stymied. The president had directed the Department of Defense to develop the necessary vehicles and to provide appropriate logistic support for the conduct of the firings. Then rivalry and interservice parochialism bred delay. Inexplicably, the services viewed satellites as a nonmilitary item; none wanted to divert resources spaceward.[25] In sum, a U.S. satellite had never been granted the push crucial to orbiting first.

—·—·—·—·—·—

At ALPHA, navigational fixes revealed a generally northwesterly drift—a segment (if completed) of a crudely elliptical orbit.[26] By October 1957, the estimated position would plot the floe mere miles from the International Date Line and Soviet waters and a few hundred miles off the pole. (A southerly drift, though, soon commenced.) A busy lieutenant was seen to shiver in a thin coat. "Of course, biologists are always cold in the Arctic," Untersteiner said, "because they have to mess around in the water." English was perhaps the first to scuba beneath the pack, to study its inverted underside. Spring, he discovered, brought plankton blooms. (But the nutrient-poor sea cannot sustain them; this explosion of life quickly collapses.) Curious as to fish, English's nets yielded . . . nothing. Frustrated, he shot a seal, to examine its stomach. The officer was thus amazed when a by-catch floated up—a cod inebriated by the antifreeze keeping his work-site instruments operable.[27]

At any outpost, aircraft—a lifeline—rank supreme. Or rather, the supplies, mail,[28] gossip, and fresh faces that arrive in them do. Landings are an opportunity to learn firsthand what is happening in the sweet south, within the earth's green cover, a world away. "The entire station," a *polyarnik* records, "except for on-duty and watch personnel, awaits the arrival of the airplane at the strip and flares are sent aloft to meet the sound of the engines." Less fortunate than their American counterparts, Soviet drifters were provisioned from afar, and less frequently. "If you only knew," another observed, "how dear . . . is every word from the continent. Every day, while sitting at home, thousands of words pass by our ears. But when they are few and far between you suddenly begin to notice something in them which earlier you completely ignored. It is as though you have discovered a gem in the darkness."[29]

News commentator Lowell Thomas debarked at ALPHA on 7 September. Foraging for images, his party exploited photo opportunities that afternoon and evening. "The visitors left the following day and the camp returned to the normal routine."[30] Already, the temperature had commenced its seasonal decline, and the short summer calm was fraying. On 30 September stars were sighted; a week later, the sun began its tour below the horizon. Its glowing disc was not to offer full face again until 1958. "It was not long before the total darkness began to have an effect upon nerves," an air force history records.

Ice duty is no cakewalk. Still, privation and discomfort aside, amenities were to hand. Motion pictures for one. ALPHA's first had arrived in July 1957; before long, a screening was being run in the dining hall six nights a week. At BRAVO, films were a nightly ritual. Movie distractions proved equally helpful at Soviet stations. At *Severnyy Polyus-19* (1969–70), "sacred time" was set aside daily for films. "Only a very unusual event," an SP-19 *polyarnik* accedes, "could break that tradition."

In January 1958 *Explorer I* was heaved into orbit. (A descendent of the V-2, the army's *Redstone* rocket sustained these earliest probes and would loft America's first astronaut.) Its primary payload was a cosmic-ray investigation designed by Van Allen and his team. On 5 April, ALPHA observed the anniversary of the first aircraft landing. That spring, the floe hosted its third group of personnel, under Maj. Joseph P. Bilotta, USAF.

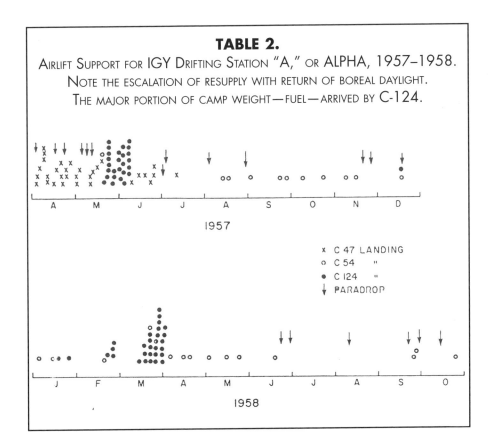

TABLE 2.
AIRLIFT SUPPORT FOR IGY DRIFTING STATION "A," OR ALPHA, 1957–1958.
NOTE THE ESCALATION OF RESUPPLY WITH RETURN OF BOREAL DAYLIGHT.
THE MAJOR PORTION OF CAMP WEIGHT—FUEL—ARRIVED BY C-124.

TABLE 3.
Air Temperature Data for the Anomalous June–September 1958 Ablation Season, IGY Drifting Station BRAVO (Ice Island T-3).

Period	Maximum	Minimum	Average
Weekly:			
8 to 14 June	43	22	32.5
15 to 21 June	39	29	34
22 to 28 June	39	28	33.5
29 June to 5 July	37	27	32
6 to 12 July	35	28	31.5
13 to 19 July	40	25	32.5
20 to 26 July	38	26	32
27 July to 2 August	36	31	33.5
3 to 9 August	38	26	32
10 to 16 August	40	25	32.5
17 to 23 August	36	26	31
24 to 30 August	35	24	29.5
31 August to 2 September	35	30	32.5
Monthly:			
8 to 30 June	43	22	33.2
July	40	25	32.4
August	40	24	32.1
1–2 September	35	30	32.5
Season:			
8 June to 2 September	43	22	32.5

1. Based on data from the U.S. Weather Bureau station installed at the IGY camp on T-3. Values are in degrees Fahrenheit. In comparison with records of preceding summers, the 1958 season on T-3, and apparently throughout much of the Polar Basin, was unusually warm and long. Effective ablation ended on 2 September.

Source: David D. Smith, Ph.D., "Sequential Development of Surface Morphology on Fletcher's Içe Island T-3," Geophysics Research Directorate, U.S. Air Force Cambridge Research Center, Scientific Report no. 4, 1960.

LIFE ON ICE

Isolation incubates psychopathology, and weather affects mood. One might assume that dark and chill are most vexing. Not so. Warm is more blight than blessing—a perverse irony. In what passes for summer, temperatures range from the low twenties to the high thirties. A monotonous overcast and calm prevail. By mid-July, snow has vanished and the ice itself is ablating. The voice of fluid water is again heard. Conditions for living and for working degrade (as Papanin had found): in ceaseless sun, everyone confronts a miserable, soggy seascape. "Summer means fog," a Soviet drifter noted, "a little sunshine and water, water everywhere." Air drops sustain riders on a maritime cover whose surface is hostile to aircraft.

The ablation season for 1958 was both long and intense (See Table 3). On 8 June, air temperatures rose above freezing: thawing began, snow density increased and compaction of the snowpack began. Temperatures held above freezing almost continuously thereafter. On T-3, the runway became inoperable on 10 June because of thawing. The melt period there ended in very early September—after roughly seventy-two days of thaw.[31] Water levels rose until channel or stream flow was attained. And an average of two and one-half to three feet of ice ablated.

T-3 CAMP TRAILERS AT SUMMER'S END, 1958. PARACHUTE CLOTH IS DRAPED OVER THE MELT-SEASON "PEDESTALS" IN AN ATTEMPT TO INHIBIT ABLATION. COMPARE THE IGY BASE WITH THE INITIAL (1952) CAMP ON P. 88. (DR. ARTHUR COLLIN)

At SP-7, the snow was damp and settled by the end of May, with the first puddles evident about ten days later. Often, it rained. The Soviets recorded the highest temperature in five years: a blistering 40°F. At ALPHA, perhaps a third of the floe lay under water. This was life amidst a floating lake. The rate of decay was quite remarkable. As August neared, almost a half inch of ice was vanishing daily. A low albedo augments melt. So intense was solar weathering that buildings at ALPHA had to be moved or jacked up and reset. Clothing and equipment were dank and heavy; electrical gear especially was hard hit. Long-lost articles resurfaced, as did accumulations—the debris and discards that ring all encampments. "Good hip boots (no leaks) were in high demand for most summers," Cabaniss says. (The wet eased, he adds, in 1959–60 because snowfall proved slight and drainage was much better organized, allowing runoff to gain the edges of the island rather than ponding.)

One unexpected headache was refrigeration. Cooks had assumed that meat could be safely stored outside, but found that it spoiled. Complications multiplied. Dark objects—trash, tools, dirt—left on snow or ice vanished in minutes, drilling themselves in. "A pebble no larger than a pea," the air force notes, "would eventually make a hole several feet in diameter, and a drop of oil from an engine, a hammer, or any tool left on the ice would make a similar crater."[32] Hydraulic fluid and fuel oils are fatal to ice surfaces. Runway discipline therefore called for drip pans under the engines of each arriving plane and the policing of each landing and takeoff path. "During the summer, when the only contact with others is via radio, morale reaches its lowest ebb," one researcher records.

> Feelings of isolation, helplessness, and hopelessness become predominant. The long hours of daylight make sleeping difficult, and time passes very slowly. Personnel problems build up and tempers become short. The arrival of the first plane in the fall becomes the most discussed item of daily business.

Soviet counterparts voice similar reactions. "He who has lived in the Arctic," a *polyarnik* writes, "and has waited . . . for infrequent flights from land understands how important human kindness is for the soul. That is why we are so fond of our aviator comrades."[33] Wives of SP personnel endured private tortures of their own:

> There is a problem of communication because letters can be delivered to the drifting stations only when planes fly there. It is possible only in spring and autumn. In winter and summer you can only send cables. It is possible to send only brief information in a cable and it is very hard both for men wintering at the station and their wives staying here because you will never write

all details about your life and in-time things in a cable although sometimes it might be necessary. . . . Our institute has never planned to make wintering period shorter due to difficulties with transportation, preparation of the runway, etc. Furthermore, you should understand that it is very costly to transport people from an ice floe. In this case nobody will think that there are psychological and moral problems for people to stay on ice for a year or more.[34]

Which season is cruelest? "Summer," Nikolai Vinogradov replied. Thaw, rain, meltwater, and humidity breed an inescapable wet. In a realm of lengthened dark, who does not embrace the salutary effects of sunlight? Or agreeable weather? *Polyarniks* (I was assured) do not suffer the cold: the huts are well equipped, the mens' clothing superb. And winter-dark landings had become more or less routine. "In summer, the work is much more difficult and the situation is much more unpleasant," Vinogradov continued. Water penetrates the huts, vandalizes equipment, destroys food. All air communications are severed; aircraft-provisioners cannot land. "Winter is the best time; no melting, night—you can sleep day and night," he smiled.[35]

Prior to the microchip, Arctic communications were beset by a vast geography and radio blackouts, the latter due to the aurora borealis.[36] Regular human-to-human contact confers an inexpressible boost, a kind of psychotherapy. Radio operators are therefore key—because they manage all communications, for example, exchanges of data, phone patches with families. (Today, handheld satellite-navigation devices and telephones provide a welcome sense of connectedness.) Disconnections spawned irritations "almost as variable as the number of individuals," Hanson notes. "The reaction runs from passive to violent—from 'that's life' to kicking a hole in the wall. A period of poor communication tends to depress morale, which largely manifests itself as exasperation and shorter tempers."[37]

In sum, field parties require a package of minimum comforts. "[*Polyarniks*] are not staying on the ice for a day or two, but for a year, and conditions must be such that trivial things will not interfere with life and work."[38] Dispensations once deployed? None, given the logistical and fiscal challenges. Quite naturally, a hankering after those persons and places and scenes left behind is keen. "We think about our families. We want to embrace our wives and children. We are lucky when sleep, the only real machine of time, takes us back through the weeks and months to our distant homes. But then the alarm clock rings. It is time to get up, open the door, face the snow and do the job for the hundredth or thousandth time."[39]

Exile from the familiar calls for adjustment. Cables from the mainland—the sole indirect communication with family—brought the briefest of details of home along with the ritual "good drift and firm ice." "For the first six months," one

Soviet writes, "you remain close to what you left behind you, and you live by its interests. But gradually the details fade and disappear, you remember only the important things, and a lack of information only promotes this process. You become involved in the routine of like and only events like a mail drop occasionally break through the shell you have built around your world."

In the rut of routine and staffs closeted together, a heroic attitude is difficult to maintain—unless getting through twelve unbroken months with one's self-possession intact is heroic. As tours edge toward end, the most welcome transmission is: "replacements on the way." The men are weary of sequestration, some bored, others—beyond anomie—depleted to a spiritual core. As the last calendar day draws near, wardrooms are awash in excited confabulations of home. "I have had enough, I'm not coming back," members announce. Knowing smiles attend these proclamations. Within this culture, this cult, each man knows the sweet pain of entering the white world, the seduction-affliction known as "Arctic sickness." Driven by an ineffable hunger, *polyarniks* will somehow succumb, returning yet again to the ice.[40]

Heads of camp are responsible for the physical, social, and psychological well-being of all hands. Hazard is companion to every drift; the potential for casualty haunts every manager. This is no place for fragile temperaments, the undisciplined or the unsteady. Inhabiting a remote working environment, researchers and support staff are unreachable during storms. Radio blackouts are hardly unusual. Bears, heavy equipment, firearms, and the handling of explosives amplify the probability of genuine emergency.[41]

Reactions to life on ice—deep isolation at close quarters—vary among individuals. Here, camp life is at once a privilege and a predicament. A tour can be an adventure—or a season in hell. "I can say that prior to the [SP-19] expedition to the drift ice I never imagined all the hardships of wintering above the Arctic Circle. . . . But the physical hardships, in spite of the fact that they were quite severe, were easier to overcome than morale problems. Isolation from the world, from our families and living in a small group for a long period of time are much more difficult to tolerate than the most difficult physical labor."[42]

Group expeditionary living is a bonding experience similar to combat experience. Unfortunately, an infallible method for gauging resistance to isolation has yet to be devised. At BRAVO, each airman-volunteer was selected for technical proficiency only after passing a rigid physical exam. "Anyone with a history of marital or disciplinary problems, with neurotic tendencies, or any kind of troubled past, was eliminated. Even the health of a man's immediate family was taken into account. Finally, each man had to have at least five years of creditable military service." Arctic experience? Hardly essential.[43] Most drifters measure up, adapting readily, responding to the demands of their craft, living brightly and altogether at ease for weeks, even months. A few, though, become depressed. Communal living under harsh, confined conditions can stoke resentments among the most indefatigably social. Leaders have to be alert to morale and behavior—the abrupt welling up of trouble—*before* a tour degenerates into an ordeal.

> With all personnel morale is a large problem. Some men can spend months on a station with no problem, whereas others become depressed in a matter of days. There seems to be no positive method of evaluating a man's ability to withstand periods of isolation. A man unfit for isolated camp living must be removed, as he can disrupt the whole camp and bring morale to a dangerously low ebb. At best life in isolation is unreal, and it takes the cooperation of all to keep morale high and life enjoyable."[44]

This creates a potential for self-destructive behavior, which is compounded by the presence of guns and the handling of explosives.

At *Severnyy Polyus-7*, physical and mental tasks were the best prescription for easing crankiness and depression. At ARLIS II (chapter six), game-playing—baseball, football, rugby, soccer, volleyball—was exploited, as were walks to leads—comparable to the ski trips prescribed by Dr. Korolov for his Soviet *polyarniks*.

Radio was particularly valued. At ALPHA camp, the surest antidote was radioing home. In addition to the official equipment, amateur stations were operated (call letters KL7FLA for ALPHA, KG1DT at T-3.) "Hams" in the States contributed hours of their time, connecting isolated team members with home. When conditions permitted, usually late at night, operators could link the local telephone directly to their transmitters or receivers, allowing the families to talk directly with loved ones.[45]

Thule Air Base sits 931 miles (1500 kilometers) from the geographic pole—America's most northerly garrison. Its principal mission was supporting the Strategic Air Command and providing air defense. Conceived for retaliatory bomber attacks on the Soviet Union, the introduction of the intercontinental ballistic missile shifted emphasis (in 1959) to an early warning role. In terms of supplementary missions, Thule was serving as the resupply and jump-off point for the U.S.-Canadian weather stations at Eureka and Alert—and, as long as it floated within range, for Fletcher's Ice Island.

In 1943, the Army Air Force had established a weather station at Thule. A gravel airstrip followed. By 1951, after a bilateral agreement for the defense of Greenland, the air force initiated construction of a major air base. A detachment arrived that summer, as well as a naval task force. A vast construction program was finished late in 1953. To support its aircraft, a 10,000-foot (3-kilometer) runway had been leveled and large hangars erected plus smaller ones for fighter aircraft on alert. Near the center of the base were 125 barracks and six mess halls. To help dispel the isolation, a gymnasium, service club, officers club, hobby shop, library, exchange, post office, theater, chapel, and hospital were to hand. There also were a television station and clubs and softball and volleyball tournaments—complete with trophies—and for readers, the *Thule Times*, the "Northernmost Newspaper in the World."[46]

Thule base is a sprawl. Until it was heated (insulated pipes were completed in 1960), heated tank trucks brought lake water and hauled sewerage away—"different [vehicle], one hoped." Heating and electrical requirements were met by ten diesel generating plants. The size of a normal late 1950s winter population was between five thousand and six thousand officers and men. In the summers of 1958 and 1959, with new construction under way, Thule's population probably peaked, with approximately ten thousand airmen and support personnel.[47] Installations are unprepossessing—rectangular, one-story insulated barrack-like structures set on piers. Windows were double-paned Plexiglas—all of which (the joke ran) were installed scratched and dirt-fogged. But inside, a transient marveled, "all was bright and warm . . . and my VIP room was surprisingly spacious and furnished in excellent taste." Also to hand in each building: emergency rations. (Windstorms off the ice cap—called "phases"—are notorious; at phase three, over seventy-five miles per hour, no one is allowed outside.) Access to the officer's mess and officer's club mandated a sport coat and tie, but "was inexpensive and the slot machines were generous."[48]

In Boston and Washington, the 1958 field-season party for T-3 had begun to assemble. Following a rigorous physical and psychological examination, each man was issued an ID card and a set of orders. C-124s belonging to the Military Air Transport Service then shunted these individuals northward—most from McGuire AFB in New Jersey, but a few from Dover AFB in Delaware. After refueling stops at Goose Bay, Labrador, the men were delivered to the sprawl that is Thule AFB, the largest military installation in the eastern Arctic—and, in 1958, *the* supply point for BRAVO camp.

Lingering, the group received safety and survival instruction, an issue of air force arctic gear was drawn,[49] further preparations concluded. Field equipment—seismographs and geophones, reels of seismic cable and electrical line, surveying equipment, a gravimeter, stakes, a power auger and corers, sample bottles, current meters, water-level recorders—had been assembled, checked, and packed for transport. When all was ready, a C-54 lifted off for T-3. The commanding officer of this season's support detachment was Lt. Col. James J. Giles, USAF.

In mid-1958, roughly 400 miles (644 kilometers) interposed between the two U.S. IGY encampments. However, their respective personnel jumped off from opposite sides of North America. (Soviet scientists and shift changes were deployed from intermediate coastal bases, jump-off points dependent on station coordinates.) ALPHA was reached via Ladd AFB by way of Point Barrow. Those T-3-bound gained camp via Thule.

The year's first party reached the BRAVO strip 1 May. "We landed [C-54] on the ice runway," geologist Charles "Carlos" Plummer recalls, "and my first impression and lasting memory is the smell of diesel smoke coming from the generator room. The station consisted of trailer-like buildings built like refrigerators arranged in a large rectangle with a large vehicle and maintenance building on one side. The recreation/movie building was in the center of the rectangle." The men met with Colonel Giles, a quiet but firm commander and camp manager. Of twelve to fifteen volunteers, Giles was the lone officer.[50] Arthur Collin, a Canadian scientist, was the sole non-American. His special domain: the "ice hole," an off-island site on drift ice welded to the slab. His workstation was little more than equipment for oceanographic observations and a frame tent over an access hole.[51]

Geophysicist Gerry Cabaniss describes a day in the life on ice. On T-3, the alarm (he recalls) rang at about seven a.m. In summer, the trailer was cool—not cold—as he started the furnace. Though daylight prevailed for the full cycle of twenty-four hours, typically, a deck of low cloud masked the dome of sky. And fog signaled nearby open water. Dressed in waffle-weave underwear, blue air force shirt and pants, and turned-down hip boots, he paused atop rickety wooden stairs to look for polar bears before clambering down onto dirty ice and the mess hall. "The trailers and supply building look ridiculous, sitting on the tops of four- to six-foot high rounded ice mounds, partially swathed in old parachutes to protect them from the sun."

> The smell in the mess trailer is wonderful—some of these cooks are really good, even with C-rations. Fortunately, we have plenty of food for the summer, even though fresh vegetables are long gone. After exchanging pleasantries with the Japanese scientists, the radio operator and the "cat" driver, I dig into a bowl of oatmeal, three eggs straight up, grits, and toast, washed down with adequate reconstituted orange juice and excellent coffee. The tables each have at least two bottles of Tabasco sauce, labels long gone. Some of the men put it on almost anything. Most of the enlisted men, came from the south it seemed—rather ironic to be at the Air Force's highest latitude station (until it drifts to a latitude south of Thule).[52]

Plummer concurs. "The food was superb and plentiful." Ditto from geophysicist Donald Plouff. "The food was excellent, thanks to two motivated cooks, who even provided little appetizer snacks on crackers." Somewhat to his surprise, Plouff was appointed (by teletype, in June) station scientific leader. "The title was a formality because everyone had their own job and performed what each understood their function to be." A medical corpsman was assigned, but it was doubtful that anyone could have been evacuated during melt season for a serious illness or accident.[53] (See pages 187 to 190.)

Cultural boundaries endured. Despite the civilian-military mix, most residents fared well together, living simultaneously and comfortably in two cultures amid professional disparities and very close living. The camp commander was in charge over all and represented a link in the military chain of command. The scientific leader, for his part, reported on overall progress to the appropriate co-ordinating centers (in 1957–60, the IGY Committee). "The responsibility of command on the drifting ice stations is diverse," Arnold Hanson, a veteran of eleven tours, recounts, "both because of the environment and of the scientific mission. Largely because the different investigators are responsible for their own work, the over-all station command cannot be authoritarian; rather, it must be diplomatic."[54] Any station leader must be a good communicator, Russians agree, and a good manager, and, certainly, the team must respect him. The *polyarnik* in charge, they add, should participate in the work of the camp, such as repairing the runway—odious labor. On ALPHA, for example, its air force and civilian leaders confined together, the question of leadership was solved by a kind of duality in command: a station commander and a scientific leader working together for the common goal.

What else constitutes good leadership adrift? In the Russian view, experience is the "main feature." And knowledge of the environment, the tasks attending the mission, the scientific questions being probed—each of these is vital. There is an art to exerting authority: knowing when to be fair and when to be firm, being alert to swinging moods. Obliged to lead, to manage, and to keep a supervisory eye on everything, station heads are an amalgam of strategist, psychologist, planner, parliamentarian. Their decisions are vested with grim significance because the lives of everyone on the ice depend on their character and judgment. They lead by example. Yet reciprocity is crucial if the camp is to function productively. On the ice, the character of each man is vital to the common interest.

Not by chance, the caliber of U.S. commanders was high. "They were experienced and responsible men," Hansen opined, "considerate of their troops, capable of planning ahead and not bashful of seeking advice of specialized personnel, either military or civilian."[55] Though there were exceptions, certainly, the two populations did not mix easily. "This was," Gerry Cabaniss explains, "because of the usual personality conflicts, socio-economic and educational level differences, civilian vs. military, supported vs. supporter, scientific and social snobbery, etc. etc. Conflicts, including those among scientists and among the military, were often exacerbated by isolation and propinquity."[56]

The onset of fall's half dark is prelude to winter (October–March). With its deep frost and clearer skies, the failing of the summer season was awaited by all hands. Below-freezing temperatures harden runway surfaces, ending the infernal melt and thus permitting landings. The air force—with eager civilian help—began logging long hours with the 'dozers, renovating and grooming a melt-abused ice surface.

The chill deepened to its petulant worst. Temperatures seldom rose above zero, and they could plunge to 40 below. During the latter half of December, air temperature at ALPHA averaged –51°F. Summer calm had yielded to racing winds that often made it impossible—save for snatches—to work safely beyond shelter.

ALPHA THREATENED

One definition of adventure is an uncertain outcome. All floe camps fracture; the arbitrary violence of deformation and then annihilation are sure. *When* this will happen is unknowable. The choice of ALPHA floe had proved fortunate; for nearly a year, there had been no cracking. But inevitably, the encircling white became restive. The prelude to breakup commenced in March, when a crack appeared. A second breach was noted five days later, a third on the 16th. By month's end, foundation's frailty was obvious. Continued pressure caused heavy ridges to erupt near one end of the runway and along the scientific area (near camp), the convulsions cleaving the floe by roughly half. More active fractures appeared. Early in April, the runway broke free; it now could veer relative to camp. Surveillance heightened, communications and power lines were spliced and repaired. The threat subsided, but did not cease altogether; as fractures, ridges, and leads multiplied.

The floe increasingly infirm, its campsite untenable, ALPHA was shrinking, if not failing.

A new site was chosen. Camp shift entailed moving an entire bivouac about two kilometers over active, ridging ice. Abandonment was deliberated. Bilotta and Untersteiner had caucused; by radio, the pair pleaded to maintain the drift using a skeleton crew retrievable by one C-47. If degradation persisted, necessitating a pullout, the plane, they argued, could effect final evacuation. (Icebreaker support is an uncertain business, even when penetration is practicable.) Handling disaster was, and remains, a military mission near the poles. The air force, understandably, preferred not to invite a search and rescue; headquarters was inclined to lift everyone off. The pair managed to force their will; concluding anguished exchanges, the Alaskan Air Command relented, agreeing (reluctantly) to furnish additional support.

ALPHA was allowed to continue. After nonessential personnel were evacuated, the camp was shifted to a replacement floe, a weeklong process. But the Jamesways were not designed for rough handling. "You know," Hunkins explains, "the mess hall is *really* the focus of a camp like that. If you don't have that, it's pretty dismal. There's nothing else for you. . . . You can have a pretty simple living quarters . . . but the mess hall, that's warm and cheery, you get together with everybody, food . . . that's very important." One evening, with the move still unfinished, a dog-tired camp paused, as if to collect itself. Most of ALPHA turned in. The damaged hut was set down in mid-move. This was the lowest moment. As the others slept, a sergeant and one airman somehow managed to complete the relocation (including stove) over ridged ice. "The morale really shot up again," Hunkins smiled, recalling that night's exertion. "That's one of those cases where a couple of people really stuck to it."[57]

The move to camp two was completed on 5 May. Programs resumed. But so did cracking and ridging. Runway Two split not long after it was finished. A third runway was leveled; it withstood a single landing before splitting lengthwise. Although difficult to reach, runway One floated undeformed and unbreached from pack pressure.

Danger concentrates the mind. During this fraught time, battle-weary hands strove mightily to sustain a viable facility atop dwindling level ice. Moving some of the equipment proved unavoidable. And some gear was lost. Still, investigators kept their data sets largely free of gaps. But the mood had changed: although operations persisted relatively undisturbed throughout that summer, "After that initial breakup, there was never any really easy time," Ken Hunkins continues. Battered yet resolute, ALPHA endured. (See Appendix 2.)

The transition to a summer energy budget now fed ice decay; by August, shelters crowned icy pillars. Huts had to be moved before they pedestaled too high, so another camp shift was ordered. At T-3, ablation "elevated" BRAVO camp up off a dirty, water-saturated white. Precarious structures were slid off their perches and the ice leveled, whereupon each was returned (skid-mounted) to its site.

ALPHA, ACOUSTICS, *SKATE*

The ALPHA party this field season included an acoustics specialist, Elton Kelley, from the navy's Underwater Sound Laboratory. Why was USL on ice? For the oceanographic science of underwater acoustics—in this instance, to collect data on Arctic Ocean parameters to apply to the prediction of sound propagation in the under-ice environment. Why underwater *sound?*

Light, radar, microwaves, and other electromagnetic waves [in air, radar] attenuate very rapidly and do not propagate any significant distance through salt water. Because sound suffers very much less attenuation than electromagnetics, it has become the preeminent tool for sensing, identifying, and communicating under the ocean surface. And yet, for decades, inadequate oceanographic information about the extraordinary spatial and temporal variability of this medium has hindered underwater acousticians in their desire to predict sound propagation.[58]

The navy's war-fighting priorities had focused elsewhere. For disciples like Waldo Lyon, the Arctic's strategic significance stood in alarming contrast to programs of oceanography, naval patronage, and arctic operations. What about submarines? Mastery of the under-ice medium depended upon under-ice acoustics. How did the physical properties of ice and water column relate to acoustic factors to allow prediction of underwater propagation? Acousticians now know, for example, that various properties of the canopy are integral to understanding reflection and scattering. In 1958, the piloting, navigating, and detecting sensors for sailing under ice cried out for improvements. Piloting sonar is crucial to safety and tactics—to reduce the danger of ice collision in shallow water and to allow tactical use of the canopy. "Early systems were relatively crude in nature," Robert Francois notes, "limited by the environmental information and state of the art (1960) at that time."[59] So experts were riding ice, working to understand the physics and acoustics of subsurface sound—the parameters affecting propagation, backscatter, reverberation, target strength, other phenomena. Data on temperature, salinity, density, currents, layering, and so on would yield design information to sensor researchers and engineers. Also needed for under-ice piloting were a reconnaissance survey of the deepwater Arctic along with bathymetric and environmental data for its approaches, particularly for the northern Atlantic. In shallow waters rich in ice, the growth and drift of sea ice is basic to navigation and communications, let alone undersea warfare. A 1950–54 joint U.S.-Canadian effort (by icebreakers) had produced bathymetric/oceanographic charts of the Beaufort Sea and channels of the archipelago. Still, uncharted waters included vast areas for which data would be collected by submarine or by drift-ice station.

> Submarine innovations provided the major thrust and emphasis behind navy funding for oceanography in the postwar world.... Without a more intimate knowledge of the ocean, the U.S. Navy might find itself at the mercy of faster and more lethal submarines with little hope of responding effectively.... In the minds of submarine veterans, the ferocity of the naval war beneath the surface between 1940 and 1945 forever bound together sonar development,

improved echo ranging, precise target identification, effective weapons, and a better understanding of the ocean environment.[60]

Regarding (belated) naval-research priorities for the deep-ocean basins of the Arctic and its marginal zones:

> During the three decades of the '60s–'80s the navy provided most of the funding for ice stations in the Arctic [Ocean]. And the navy always gave highest support priority to underwater acoustics of all the scientific disciplines. The reason of course being that because of the ice cover, the Arctic Ocean was a unique ocean from the standpoint of sonar efficiency and we knew very little about the two environmental factors that effect long-range detection of submarines: background or ambient noise, and transmission loss.[61]

The propagation of low-frequency underwater sound compares favorably with the air-propagation of radio waves. Although attenuation–the loss per unit distance of travel due to absorption and scattering–is far less than that for underwater light and radio waves, underwater sound presents two serious limitations: low velocity of propagation and a rapid increase in attenuation with increase in frequency. Moreover, background noise in the ocean is much higher at low frequencies. And noise adds echoes.[62]

Sound speed is a function largely of temperature and salinity (which defines conductivity). But propagation depends also on the ice cover, water depth, and seafloor. All submariners know the operational importance of temperature/salinity variability on changes in the sound speed profile. Water tends to stratify with density. And density increases with depth. Density, temperature, and salinity are stratified over most of the world ocean. Unlike the upper layers of the open oceans, however, the temperature and salinity profile–hence the sound profile–for the Arctic Ocean is almost constant. Here the speed of sound is generally an increasing function of depth, that is, speed increases with depth. As a result, sound waves are continuously *refracted* upward, causing repeated *reflection* at the ice-covered surface. The propagating sonic energy is trapped between the surface and refracting "turning depth," focusing energy into a "surface channel." The reflectors are far from perfect, however: signals are partly scattered by the random roughness of the seabed and water surface, by leads (open water), and by the bottomside morphology of the pack. This backscatter attenuates sonic energy–the higher the frequency the greater the bounce loss.[63] The basic determinants of sonar design are transmission loss and ambient noise. Working his USL counterpart on T-3, Kelly was taking propagation measurement and assessing how transmission conditions differ from those in ice-free seas.[64] (From 1958 through 1962, USL was the most active single organization measuring Arctic transmission loss.)

Nautilus had pierced the pole on 2 August 1958. Traversing the basin submerged, she proved it to be just another operating ocean to the nuclear submarine. The Arctic's strategic importance had been abruptly underscored. As she steamed, *Nautilus* had recorded almost continuous acoustic profiles of the ceiling overhead. How? Using up-looking "echosounders" (ice-profiling sonar), the pulses of which reflected off the pack's underside. The return signals were translated, electronically, into an ice-draft profile–draft is proportional to thickness.[65] In her pioneering wake, *Skate* logged 90° N latitude nine days after *Nautilus*' subsurface operation. Forty or so miles beyond the pole, Cdr. James Calvert, USN, had surfaced his command, to radio word; under-ice explorations then resumed, including vertical ascents calculated not to exceed the boat's design strength. To help solve the all-season surfacing problem, *Skate* would revisit the deepwater Arctic, to test the *winter* operational environment. "The art of routine surfacing through sea ice was devised on board the *Skate*," Lyon writes of this missionary work, "in the winter of 1959, by intuitive engineering, rudimentary knowledge of ice mechanics, and trial."[66] *Boarfish* had been first U.S. boat (in 1947) to log extended dives under ice. Now *Nautilus* and *Skate* were demonstrating unlimited movement beneath the canopy. Unknowns remained, certainly, among them a paucity of bathymetric, oceanographic, and acoustic environmental data. And better sensors awaited design, development, and testing. Still, piloting the open sea that underlies the ice, the transarctic submarine was demonstrable fact.

Skate's patrol orders had permission to rendezvous with Station "A" if the pole was logged by 14 August. At this particular moment, Kelley and (perhaps) Bilotta excepted, ALPHA knew nothing of a rendezvous. Confident of his executive officer and navigator, Cdr. John Nicholson, USN, and feeling ambitious, Calvert ordered course set for ALPHA. When near enough to each other, boat and base passed messages. Calvert's requests to the officer presiding included the exact position of the camp, as well as information on the number and size of polynyas near it. There were many, Major Bilotta replied, "But best only fifty yards from our main buildings." Then an inspiration: "Will run outboard motorboat continuously in polynya nearest camp." To augment the major's undersea beacon, Calvert arranged for ALPHA to drop detonation caps every ten minutes. Course was taken at nine knots. "I was listening for other things with my acoustic gear," Hunkins recalls. "I could also listen, in a sense, the way the navy did, but with not exactly their kind of gear. But I was aware of some kind of a transponder, some kind of signals going off." What he was hearing was a beacon dunked by Kelley, to home in the approaching submarine.

Bilotta had skimmed off. As the officer ran his boat-cum-beacon in circles,

Skate (SSN 578) at Alpha camp, 14 August 1958, following her run out from the pole. The all-season capability of nuclear submarines opened up a "new engineering domain" for Arctic science. Subsequent cruises continued to collect data, most of it relevant to under-ice operations and military concerns—and most of it classified. (Dr. Kenneth L. Hunkins)

word of this bizarre antic spread. "We had a little outboard motor boat," Hunkins remembers, "and he had that in the lead, . . . going up and down. . . . Then we were really aware of it."[67]

Calvert was to credit Nicholson with "one of the finest jobs of navigating I have ever seen." At 0835, 14 August, a periscope peeped forth. Its viewfinder revealed an unforgettable scene. From his boat, Bilotta was waving. "Small brown huts dotted the ice. A high radio antenna rose over them. The squat silo shape of a radar dome lay further astern. Near it stood a tall pole with the American flag."[68] Icemen were waving and taking photographs. Hardly had *Skate* surfaced when Bilotta maneuvered alongside. Crewmen clambered topside with stakes and mooring lines; by 2015, *Skate* was lying-to less than fifty yards from the main encampment.

On walk-through, sailors found a top-notch science station fueled by a powerful curiosity. The senior scientist was Dr. Untersteiner, a civilian glaciologist/meteorologist from the University of Washington. Charming and tactful, his personality is ideally suited for getting the most done in confined isolation. ("Of course, everybody thinks his program's the most important," one alumnus confides.) Untersteiner's energy and enthusiasm and engagement were plain. Stimulated and challenged, disciplined and tightly efficient, the drifters were teasing out secrets–interrelated components of a research organism labeled ALPHA. A near swaggering engagement offered counterweight to the drear hummocked plain enclosing the ice party. Untersteiner's colleagues, Hanson and English, were there, along with Ken Hunkins and others.[69] The normal workday ran from about 0600 to 2200–or round-the-clock. Busy times saw the men looking as if they had not slept, showered, or shaved. Cold attacks equipment, mandating incessant maintenance and repair. Keeping the runway to minimum specifications presented its own (chronic) headaches. But the schedule kept time from dragging. On Sunday, save for essential duties, the men were allowed to sleep in.

"Mounted on pedestals of ice." Calvert writes, "the huts resembled sadly misshapen toadstools. It had been necessary for Bilotta's men to build wooden stairways to each door. As the summer progressed and the ice pedestals grew higher, the stairways gradually became longer and more rickety. Furthermore, a large melt pond seemed to have grown under each ladder so that quite a leap was required to clear it. Getting in and out of the huts was quite an adventure." Arthur D. Malloy was a researcher from the hydrographic office aboard *Skate*. "They have a nice set up," he wrote to his log, "but the camp is pretty messy."[70]

Shuttles brought an exchange of goodies. *Skate* rewarded the base's hospitality with a cake inscribed "Greetings to Alpha from SKATE." As recompense, her sailors were allowed to exhaust ALPHA's entire supply of beer. (Also, a large timber, which later was exhibited in New York as the North Pole, was carried off.)[71] USL's Francis Weigle, an electronics scientist assigned to this cruise, visited with Kelley who, there since early summer, tarried sixteen weeks in all.[72] "Gus looks fine," Weigle reported, "and is anticipating his return to New London. The IGY station and the submarine swapped ice cream and polar bear meat, although naturally the station got the best of the bargain. Gus and I took pictures of each other and finally one of the crewman photographed the two of us together."[73] When ice movement began closing the polynya, an emergency recall brought its men back to the boat. The respite had delighted everyone, yet "[*Skate*] didn't change things for us greatly," Hunkins says. "It was obviously something very new; we just took it in stride. It came, and went." Malloy, for his part, took his leave with no regret. "The stop was nice," he records, "but we all felt we would not like to stay on the island [*sic*] to collect data. Just a little too primitive."[74]

END OF ALPHA

ALPHA survived the drop-in less than four months. September and October brought strong winds, which intensified cracking and pressure ridging. Gradually, deformation began to interfere with science. A ridge again approached the main area. Conditions degraded to dangerous. Every pound of logistic and scientific gear and all personnel floated hostage to airlift. Unbridgeable leads had to be crossed by boats loaded with men, gear, and data slated for retrieval. The camp itself floated hopelessly damaged. "The lack of food other than C-rations, the shortage of cigarettes, and the necessity for coffee rationing added to the discomfort." Tension mounted; some personnel, Major Bilotta reported, had almost reached the breaking point.[75] Beginning about mid-October, a succession of storm cells resulted in almost continuous ice movements. The destruction proved harrowing. For months, active ice had encircled the camp; unruly convulsions gradually gnawed away at the livable-workable area with awful ease. Deformation obliterated runway Two. Embattled and wounded, ALPHA was foundering. Thoroughly challenged, the camp again was shifted. Soon, though, the floe lay in extremis, hopelessly splintered. Now more than a mile of shattered ice separated camp from runway, like a wall. (A grader that had been left there had kept the airstrip usable.) Compounding matters, the first day of November saw rifts snaking into camp Three. The next day, a runway inspection signaled the order to evacuate.

A C-123J lifted clear on 6 November–the final abandonment. "That was the final exodus," Untersteiner recalls. Having endured a rough passage, "We had to leave everything but the people and our data. No one was hurt but we just barely got out."[76]

ALPHA had logged a net drift of 1829 miles (2945 kilometers) over nineteen months–an average 3.6 miles per day. Its drift track and that of *Severnyy Polyus-7* are strikingly similar. Indeed, a mere 150 miles had interposed between the pair. The overall results were summed up by Cabaniss, Hunkins, and Untersteiner in 1965:

> Station Alpha in the Arctic Ocean was the first floe-station established and maintained by the Unites States in the IGY. Its scientific program was thoroughly planned, and it lived fully up to expectations, providing much new information on arctic meteorology, geophysics, and biology. The cost, however, was considerable: Robert Johns, U.S. Weather Bureau technician, and A1C M. Fendley, U.S. Air Force, lost their lives while performing their duties at the station. . . . The success of the Station Alpha project was a direct result of the efforts and cooperation given by the officers and men at the station as well as at Alaskan Air Command Headquarters.[77]

Segments *had* been discovered by the Soviets. Still, the subaqueous Alpha Rise was first recognized as a distinct highland feature of the Arctic Ocean basin courtesy of drift station ALPHA. Indeed, the Alpha-Mendeleev Ridge is the largest mountain complex in the deep-ocean Arctic; in extent, it exceeds the European Alps. As well as providing the first photographs of the rise, ALPHA had taken deep-ocean cores.[78] Bathymetric, depositional, and structural (seismic) data from thousands of meters down revealed a "normal" oceanic basin similar to that cupping the Atlantic. ALPHA, in short, had lived up to expectations (as was widely recognized), providing much new data on biology, geophysics, meteorology, oceanography. "It was a model station, really, for all its problems," Hunkins says. "Looking back from my own standpoint, I can't think of a better one."[79]

ALPHA provided further information, as Max Britton explained. "ALPHA suffered the fate of most pack-ice stations. It just simply broke up. But what did it do? It broke ... across the runway. Well, big airplane[s] need big airstrips–half of it gone is *no* airstrip. It illustrates the fallacy of trying to work on pack ice with big airplanes."[80]

CHARLIE

The U.S. IGY contribution had stipulated a *pair* of ice-based stations sharing the neighborhood. Prior to the abandonment, the Department of Defense had authorized the continuation of scientific programs beyond the IGY at *both* ALPHA and BRAVO, the air force continuing logistical support. The year's close brought no announcement of a stand-in. If IGY programs were to be carried to completion, another floe had to be scouted, another camp installed before the melt curtailed airlift operations. In 1959, Louis Quam was Max Britton's boss. Quam knew Washington. His interest and push realized $750,000 in emergency DOD funding to meet the demise. Further, he elected to join forces with the air force in a common effort. Operationally, the Alaskan Air Command and its ICE SKATE staff were agreeable–and (of course) equipped. So the newfound pot would further support the IGY as well as air force and navy contract science–thus granting the navy a foot in the door. The balance, transferred to the air force, would support the logistical requirements of a replacement camp.

In sum, Station CHARLIE was "go." With the AAC directed to establish CHARLIE camp, a search was initiated using aircraft from both services. At this time–March 1959–a call was made throughout the command for ice-duty volunteers. Capt. James F. Smith, officer in command when ALPHA was lost, was named commander of the base-to-be. His staff comprised about eight. Not just any air soldier would do:

> In addition to specialized skills, personnel selected for duty were expected to have the proper temperament, maturity, and stability to undergo weather and isolation extremes. Duty tour at the drifting station was to be six months, following which volunteers would be given every consideration in the selection of a stateside base of assignment.[81]

Navy ice patrol P2Vs flew searches for nearly three weeks. On 13 April, two hundred miles due north of Point Barrow, a landing party was set down on prospective ice. It included Max Brewer, Arctic Research Laboratory (ARL) director, Captain Smith, and pilots Robert Fischer and Robert Main, the aircrew of a Cessna 180 light plane assigned to the laboratory. A temporary strip was marked out on a frozen-over adjacent lead, after which fuel was paradropped. The four then returned to Barrow. Next day, approximately 260 miles to sea, a ski-equipped C-47 returned with the initial construction party—fifteen men—as a pair of circling *Globemasters* began drops. Among the essentials: two D-4 Caterpillar tractors, two weasels (all-purpose track vehicles), Jamesway huts, generators, fuel, and food. At 75° N, 158°30′ W, CHARLIE camp floated about 900 miles (1440 kilometers) from the pole.

A fifty-nine-hundred-foot runway, capable of handling large cargo aircraft, would be leveled. Inasmuch as all transportation to and from the floe was to be by air, the strip would have to be fastidiously maintained and groomed. On the 25th, meantime, it received the station's inaugural airlift; five days later, an official dedication was logged when the American flag (forty-nine stars) was raised over CHARLIE camp.

> The camp contained 20 Jamesway shelters and helicopter huts ["Helohuts"] for laboratory, messing, sleeping, generator, shop, and storage space. The staff averaged one dozen Air Force personnel and an equal number of civilian scientists. An 1800-m runway, capable of handling large cargo aircraft was maintained, all transportation to and from the station being by air.[82]

Throughout May, some five hundred tons of equipment and supplies were shunted onto CHARLIE ice; by mid-June, the facility stood virtually complete. Shifted west-northwest by surface circulation, its path contorted by wind forcing, CHARLIE was to log a straight-line course of 132 miles (213 kilometers), yet cover a total distance of 1375 miles (2200 kilometers).

Jointly supported by the Office of Naval Research and the air force's Geophysics Research Directorate, the work of this latest station included meteorology, oceanography, underwater sound propagation, seismology, and marine biology. Its sponsors were the U.S. Weather Bureau, the University of Washington, USL, and Columbia's Lamont Geological Observatory. Daily seismic "shots" provided bottom and sub-bottom reflections. Bathymetric mapping was carried out using a precision depth recorder developed at Lamont. And, seeking keys to basin climatology and paleocirculation, scientists grabbed bottom samples for identification and later analyses.

Why sample deep-ocean ooze? The basin is roofed-over today; in the geologic past its ice was absent. The variability of its cover is studied by means of cores of the bottom. Glacial cycles alter sedimentary processes and microorganisms, which leave fossils. The abundance of planktonic foraminifera (one-cell protozoa common to all oceans) is a measure of productivity, which in turn is related to the amount of open water. When the sea surface is ice-covered, conditions are comparatively quiescent: on the bottom, silt and clay predominate with only sparse foraminifera. Overall circulation is constricted, and the upper layer receives scant sunlight. (Ice reflects and absorbs much of the total solar radiation, especially when snow-covered.) In contrast, open-water conditions of partial or complete melting permit wind-induced surface circulation. Drifting bergs release sands, gravel, and boulders that drop to the bottom. As well, planktonic and benthic foraminifera are abundant, courtesy of phytoplankton—microscopic, photosynthetic organisms that are the base the food web. In sum, sediments (and rock) on subsea ridges, slopes, and plateaus are clues to reconstructing paleo-oceanography and tectonic history.[83]

The pan hosting CHARLIE measured about four by six miles when colonized. Instead of drifting north, fixes showed the floe wandering to the northwest, signs of nibbling at its edges. Late in May, the strength of the forces controlling drift was made manifest when a one-mile-wide piece broke free. Slabs as large as one hundred feet sheered off or were ridged up as summer advanced. As well, ablation proved severe. Cores revealed a worrisome thinning beneath the runway: from more than six feet of ice to less than five.[84] Ponds overspread the floe. Heavy-equipment hazards multiplied, in part because of undercutting of its edges. In mid-July, all vehicle operation outside the immediate camp area (save for garbage disposal, which was halted in early August) was ordered stopped. Based on the ALPHA model, emergency procedures were prepared, ready for application.

Elsewhere, U.S. researchers were no less busy, if less harassed. In outline, the 1959 air force program for T-3 BRAVO included
1. Station position and orientation
2. Gravity observations
3. Magnetic observations
4. Seismic observations
    a. High-resolution studies of ice
    b. Bathymetry
    c. Sub-bottom structure
    d. Arctic seismicity
5. Long-line electrical resistivity program (supplementary information on the nature of ocean-bottom rock)
6. Oceanography
    a. Ice-movement studies

      b. Physical and dynamic character
      c. Chemical properties
      d. Biological properties
7. Micrometeorology

This field season found two dozen or so contractor representatives on the ice-island roster. The navy, for its part, had dispatched a five-man team from its USL, some of whom were slated for CHARLIE. The U.S. Geological Survey sent four field scientists, the Weather Bureau two. As well, five researchers represented the Arctic Institute of North America (including two Japanese oceanographers), three from the Cambridge Research Center. The summer season saw ten scientists. Support staff was made up of fifteen personnel under the command of Lt. Col. Edward Feathers, USAF.

These months did not lack for drama. That December, fire destroyed T-3's power shack and two generators. Within weeks, another blaze doomed a Jamesway, BRAVO's seismograph, and other equipment. A third fire (in April) ravaged the replacement shelter with its contents. More happily, light aircraft from ARL deployed explosives that April–May onto the pack north and south of BRAVO camp. The refracted seismic returns were recorded, for study of the Canada-Alaska continental shelf. T-3 now lost acreage: the slab had touched bottom, in about 140 feet of water. T-3 held stationary until strong winds pushed it free (July). On 2 September, a navy icebreaker called briefly at BRAVO. "Air traffic to the island was still suspended owing to melt deterioration of the ice surface, so data, mail and equipment were exchanged and several scientists left the station."[85]

The Beaufort Gyre is a clockwise motion of ice with a mean center in the Canada Basin. From April to October, as T-3 inscribed its halting coastwise orbit, the USL research team carried out the laboratory's second Arctic program. The camp staff took measurements in the vicinity of the slab as colleagues prepared sister equipment at CHARLIE. Thereafter, sound propagation tests were logged; the sound-velocity structure studied; reverberation and ambient noise (as functions of season and location) investigated; and electromagnetic (radio) measurements recorded.[86] In September, when a revitalized runway met its first aircraft, weatherworn scientists, technicians, and support personnel decamped. For a few, replacements would be dispatched to begin their own winter tours on ice.

—•—•—•—•—•—

Well beyond U.S. horizons, the 1959 Soviet springtime relief and resupply pounded to seaward. *Severnyy Polyus-6* was logging its fourth year. The Transpolar Drift Stream is a large-scale motion of ice away from the Siberian coast, across the pole area and through Fram Strait. Velocities increase toward the exit. So propelled, the ice island was heading inexorably for the Greenland Sea. The bivouac traveled some 5800 miles in all, 1600 miles as the crow flies. That September, its party was evacuated, *Sever-12* logging nine sorties to complete the operation. A hut with food and fuel was left at the site to serve anyone in want, but also to allow the ice mass to be identified as it shifted southward.[87] On 15 April, meanwhile, to satisfy IGY commitments, a new station was deployed about midway between Bering Strait and the pole. SP-8 was the first *komsomol* (youth organization) station. With about seventeen men riding, SP-8 would cover the sector north of the Beaufort Sea, heretofore drifted by SP-2. As usual, cracking would dictate relocations before a final evacuation (in 1962), when the floe fragmented entirely.

Four days before SP-8 raised the Red flag, *Severnyy Polyus-7* had been airlifted to the mainland. Moscow announced an evacuation from an unspecified floe that had spilt, separating huts and stores from the main camp. SP-7–for such it was–had logged a troubled career. Cracks and breakage had led to repeated disruptions and displacements, requiring *eleven* shifts of stores and equipment. For some reason, the broadcast omitted the evacuees' date of leave-taking as well as the identity of the camp to be evacuated.[88]

October returned the T-3 strip to usability, courtesy of the dozers, explosives, and grader. The runway was then exploited until the onset of thawing season.[89] As June 1959 blended into summer, melt-related labors intensified. (At CHARLIE, strip maintenance was further retarded when both caterpillars melted through the floe.) Ice dynamics control late-summer ice coverage. The slab conveying BRAVO would rotate noticeably this season. Its annual release had commenced with spring breakup, when loosening of the pack transformed an almost continuous sheet into a broken, relaxed cover. For the melt of 1959, temperatures were to average 37.5° (August) and, during September, 26°F. T-3 was still tracking off the Canadian Archipelago, about one hundred miles out. By late September, the slab floated at 71°14′ N, longitude 137°44′ W–a location off the Mackenzie River delta.

Hitherto, bears had kept their distance. Emboldened by the transition to dark, these great creatures padded in to maraud. That September, as Colonel Feathers started for his quarters, he heard a noise behind him, near the mess-hall garbage collector. Turning, he faced a huge bear, which spotted the officer that moment. A chase ensued, the animal in furious pursuit. One of the dogs distracted the brute just long enough for Feathers to scramble into his trailer. The racket, meanwhile, had aroused the camp. USL's Bert Reynolds rushed out and dropped the creature with a close-range shot. The animal was then dispatched. "Needless to say, the fellows will be even more cautious about traveling around at night from now on."[90]

Frederick Williams, a USL electronic specialist, had arrived in May. The benevolent "day" temperature, he reported, was usually 32° to 35° above zero degrees Farenheit, with 40° the highest recorded. The men slept in forty-foot trailers, each equipped with three double bunks. Food was stocked on a six-month basis and proved satisfactory, if at times monotonous. Occasionally, conditions permitting, airdrops of steaks and fresh vegetables supplemented the basics. As for a daily schedule, there wasn't any. The routine varied according to the tests under way or in planning. Facilities were available in the mess for card games, chess, and checkers. And for those desiring to watch a film, the recreation trailer hosted a different one each day.[91]

On 2 September, Williams was retrieved by helicopter and plunked down on *Staten Island*, hove to about two miles off. After participating in acoustic experiments held in conjunction with T-3 and CHARLIE, the vessel departed, and it lay dockside Point Barrow on the 6th.

T-3, ACOUSTICS, END OF CHARLIE, *SARGO*

Rather than ending in autumn, USL would stay active for a "cold-weather" acoustics program for the winter of 1959–60. What inspired this? The need to determine the changes in physical parameters under deep-winter conditions and to conduct cooperative work in connection with the arctic patrol of *Sargo* (SSN 583), ordered for February.[92] Among the field experts slated to the ice was Carl T. Milner, an electronic scientist.

Sea ice is unreliable; riders must be alive to its random cruelty. CHARLIE soon was in peril. Arriving at Ladd mere days into January, Milner found the ICE SKATE office pressed and preoccupied, on high alert. About 390 miles (628 kilometers) off Barrow, the CHARLIE floe had fractured. A storm had cracked free roughly 40 percent of the runway, hijacking nearly a third of the floe. The situation was dire. In an air of heightened readiness, imperiled routines continued offshore—in a fashion. Milner, for his part, scrapped plans for CHARLIE and initiated an all-out effort for T-3, instead. "This event was probably equally traumatic to the headquarters staff, to the men on the station, and to the support force at Ladd Field. . . . Not only was there real danger that the whole effort of drifting station research would be scrubbed, because of the short duration of both ALPHA and CHARLIE, but also the Navy had planned an extensive and significant research exercise to be conducted between the drifting station CHARLIE and . . . *Sargo*, scheduled for Arctic maneuvers in February."[93]

In a chaos of back and forth, Max Britton, Louis Quam (Director of the Cambridge Laboratory), and station skipper Smith (among others) pleaded against full evacuation, calling instead for withdrawal of personnel and all costly gear not required for the *Sargo* experiment—then reoccupation for the rendezvous. The AAC offered to accept this plan only if the navy would assume responsibility for quitting the floe by submarine or earlier, should breakup commence. Ice attrition might well take CHARLIE's power supply, however, or a storm further slice its airstrip. As for *Sargo*, who could predict her arrival? Access from the Pacific interposed one thousand miles of shallow (less than 180 feet) winter water roofed with ice keels draping perhaps ninety feet or more.[94] The chief of research bent himself to the inevitable task of ordering the AAC to proceed as planned, deploying all available C-47s to get the men off as quickly as possible. (The runways at both U.S. camps were too short to handle C-124s.) Delaying evacuation once breakup had begun would have been pointless, as ALPHA has shown.

On 7 January, transports were dispatched to assist the rescue, followed by heavy cargo aircraft. New cracks came, the nearest within five hundred feet of base camp. The commander of the 11th Air Force and project officer for CHARLIE, Captain Smith, flew out for a firsthand look. The result was increased anxiety, along with new orders to hurry the evacuation. In icy fog and great cold (Ladd AFB recorded temperatures of minus 45° to minus 50°), five of sixteen men were retrieved. A complete withdrawal was effected two days later, on 14 January. All high-value gear had been recovered as well, leaving the station stripped. By retirement, the floe had been bludgeoned to one-fifth its former size.

The air force had successfully executed its second aerial evacuation in less than fifteen months.

> The sudden termination of Station Charlie in January, 1960, due to extremely hazardous ice conditions, reduced the scientific effort in the Arctic Ocean to a single station. Consideration has been given to the establishment of a replacement station, but this possibility is not likely immediately, if at all. Meanwhile the scope of the research at Fletcher's Ice Island, T-3 will be increased, with a larger number of scientists participating and with the use of light aircraft for extending investigations to a larger area in the vicinity of the island.[95]

Trapped in a circuit to nowhere, CHARLIE camp (unlike "neighbor" SP-8) had shifted but little before its torture and demise. From June to January, it had ambled a straight course of 132 miles (213 kilometers) in a west-northwest direction. This drift track totaled nearly fourteen-hundred highly contorted miles, averaging a "remarkably high" 4.7 miles per day. This was attributable, in part, to loose summer/autumn ice conditions, the relatively low latitude of the station, and several stormy periods.[96]

Occupied less than nine months, CHARLIE produced a harvest that included full-scale studies of the heat budget, particularly of the heat-exchange factors and interactions between air and ice and ice and ocean—interfaces or couplings crucial to (later) numerical models. Continuous recordings of the magnetic field had been measured (as at ALPHA), twice-daily upper-air readings recorded. As well, oceanographic and other programs stood logged. "Instruments of considerable size and complexity," Hunkins writes, "were used for the first time in the Arctic Ocean.... All these instruments functioned and added greatly to our knowledge of the Arctic Ocean environment—crust, sea, ice, and atmosphere." Altogether, "We look back on [the IGY] as one of the landmarks, really, in geophysics." He might have named a dozen fields of research. "Those [IGY stations] were *rich* sources of information because almost anything you discovered up there was new at that time."[97]

CHARLIE (not notably successful) had been no replay of ALPHA. Unpleasantness and plain bad luck had fueled discord. Personalities had erupted, compromising camp performance. Domestic irritations had ranged from a colleague's personal hygiene or a pipe-smoking bunkmate to the recommendation (by a contractor) that further services not be renewed because of broken air force promises. "Delayed shipments, items damaged during air drop, lack of fresh food, tobacco, movies and mail were major vexations," one man catalogued. "To avoid irritations, no heavy, noisy equipment was operated during normal sleeping hours, and activity that would interfere with delicate instruments used in research was eliminated."[98]

Field programs at T-3 intensified. As for the sound propagation tests working *Sargo*, these were hurriedly shifted to the island (the sole alternative). BRAVO floated in shoal water, close in, and so was hardly ideal for the experiment. As CHARLIE's last evacuees were decamping for shore, an acoustics team was arriving on T-3. Robert H. Mellen, Ph.D., head of the research group at USL, was in charge. His associates were Elton P. Kelley (a four-month veteran of ALPHA) and Milner. The men linked up with an electronic technician from the Marine Physical Lab (Scripps Institute), with Eddie Hopson and Frankie Akpik, both native Alaskans with ARL, and with Lt. Cdr. Beaumont M. Buck, USN, special projects officer of ONR.

This team found circumstances miserable: a strong blow aggrieving minus-43° air. Following the safe off-loading of its equipment (including an ice drill), the men were met by lead scientist Roger Lewis, whereupon the party was ushered to billet spaces. "The trailer," in Milner's account, "appeared to be very precariously perched atop a ten-foot ice pinnacle—cold evidence of the large melt which had occurred the previous summer." Compared to those of Fletcher's era, though, the lodgings were quite grand: six bunks at one end and "a sort of catch-all lounge" at the other. USL would eat in the mess hall and wash and shower in another nearby. The next weeks were marked by acute anguish. The physical conditions would have been daunting under any conditions; now they skirted the edge of survival: frightful cold, wind, and hard-driven snow. A feeble sun teased the horizon. Going all-out to meet *Sargo*'s schedule, in the most extreme conditions, the scientific team (with precious assist from Hopson and Akpik) erected a low-frequency radio camp on T-3 proper, near an edge, plus an acoustic thin-ice camp on sea ice welded to the slab.

> In establishing the [advance] camp, it was necessary for the men to set the huts in position, get the heaters going as soon as possible, and put gasoline-driven generators in operation so that they could have electric power. In addition, the men had to set up their radio communications, including two antenna masts, in order to carry out the technical program and to communicate with the Underwater Sound Laboratory and with the base camp on T-3.[99]

Advance camp set, hydroholes for the acoustic installations were tackled. Off-island, atop the pack, a five-man team power-drilled and hand-cored to seawater through ten to eighteen feet of floe ice. Ice islands are colder because of their great thickness, which insulates them from "warm" seawater. Air temperatures sagged to 52 below, with punishing winds. In these conditions, exposed human flesh will freeze within sixty seconds.

> There were basically two separate acoustics experiments going on [Buck writes]. USNUSL [U.S. Navy Underwater Sound Laboratory] had a straight-line array they tape recorded for later analysis at their lab in New London. My experiment was the trial of the first digital sonar (called "DIMUS" for Digital Multibeam Steering) designed by Vic Anderson of MPL [Marine Physical Lab] and constructed by MPL under ONR contract. Our array of hydrophones was a circle of 48 elements on a 1000-ft diameter circle installed in the sea ice of "Colby Bay" adjacent to the ice island. *Sargo* was our target...We typically worked in −40 F and 20 to 40 kts [knots] of wind.[100]

As the last days of January faded, the most artful defenses tested the daily miracle of continuing.

> The oil in the stoves [Milner continues] was so thick and dirty that it wouldn't flow through the regulators. The storage batteries when filled with electrolyte didn't put out enough current to start the gasoline heaters either. The oil for the generators was so thick we couldn't fill the crankcases until we could get it warmed by blowtorches. The antifreeze was like jelly in its cans, and it had to be warmed before we could fill the radiators."[101]

By 15 February, at BRAVO, round-the-clock listening was the order of the day. That evening, word was received from *Staten Island* (via ham radio) that

Sargo was en route. Having surfaced at the North Pole on the 9th, she was still a day or more away. Commanded by Cdr. John Nicholson, of *Skate* fame, the boat had run out from the pole, then probed the western arm of the Northwest Passage. She cleared the mouth of McClure Strait on the 15th, outbound, and steamed toward T-3, course 273° T (true). Her under-ice track made a dogleg to give its USL team greater range and deeper water for measurements (and for *Sargo* herself to gather better bathymetry for the Beaufort). A flat stretch detected, Nicholson elected to break through the canopy, only his sail protruding. The ice was twenty-two inches thick.

> It was a wonderful day. The sun was up, about a degree above the horizon, and most of us went out on the ice to enjoy the scene. There was a light wind and the temperature was −13 F. The polynya was enormous, 700 by 1500 yards, and was surrounded by impressive ridges and hummocks. It looked like a large frozen lagoon. Fresh tracks of two polar bears were discovered out among the ridges and a safari was quickly organized. . . . Having some light while on the surface greatly increases the interests of the crew in the cruise and in the Arctic.[102]

At 0415, 16 February, the course was changed to 222° T, to head directly for the island's last reported position. *Sargo* continued to close—uncertainly. The island had been traveling up to three miles per day: three different messages to the boat (received via her floating wire antenna) held three different positions. And *Sargo* had yet to hear its underwater phone. Unable to contact the restless, shifting outpost, staff established a routine of slowing, calling on UQC [underwater communications], and listening every half hour. Early on the 17th, her sounds reached BRAVO earphones. "We could hear her sonar pinging away and the whirring sounds of her machinery," Milner reported. "In a short time we were able to communicate with her by underwater telephone."

BRAVO was close, Nicholson knew, inside 10 miles (16 kilometers). Yet *Sargo* had difficulty hearing its low-powered telephone transducer. Nor could range be estimated accurately. The icemen resorted to code on their sonar homing beacon. A bearing having thus been obtained, at 0136 the helm was ordered turned. *Sargo* would sail under T-3, to measure the underside profile and shape of the slab—an imposing ten by five miles, with a draft of 160 feet (49 meters). Eight minutes later,

> After establishing UQC communications with T-3 at 0133 we changed depth from 200 to 390 feet. Suddenly the depth of ice on the BQN [sonar] and NK-VF [variable frequency topside echo-sounder] changed abruptly from five

feet to 120 feet and we were under the ice island. A fantastic sight... This ice island is unique in comparison to the thousands of miles of sea ice we had passed under.... To us at this moment, however, it was just a huge piece of ice to be carefully avoided.[103]

Communication knots untangled, the experiment was finalized. *Sargo* informed BRAVO that it would make runs for the acoustics team and, next morning, attempt a surfacing. Cessna 180s, came the reply, would scout for floes thin enough for an underwater beacon to be inserted under them, to bring in the boat. At USL direction, then, the visitor logged runs for calibration, the first at a 5000-yard (5-kilometer) radius. This took her under the slab. "It was disconcerting the first trip around making 15 knots at 390 feet (nec[essary] to get good BQN trace). Our best dope was that it was a max of 220 feet thick but pinnacles?" As test-target *Sargo* performed for the array, her acoustic output was recorded under ideal conditions of known course, speed, and range. But the wind increased; by 1102, BRAVO was reporting wind flutter, making it impossible to record any longer.

Assisted by T-3, Nicholson was keen to surface close aboard the island, "to establish personal contact with the various groups stationed there." But instead of thin ice, reconnaissance had found a halo of compressed pack and twisted ridges— "by far the most rugged, cluttered ice I had ever seen."[104] Clockwise surface circulation (researchers now know) tends to advect ice toward the Canadian Archipelago and North Greenland; this push causes the canopy to thicken, realizing the greatest mean drafts in the basin. Beaufort Sea pack, *Sargo* discovered, held conditions as severe as those found anywhere in the central Arctic.

With the exercise concluded, her proposed track satisfactory for further acoustic work, *Sargo* departed the sector, to continue her epochal winter patrol.

—•—•—•—•—•—

The spring of 1960 saw two Soviet stations operational. *Severnyy Polyus-8*, established in April 1959, would persist to March 1962—1057 drift days in the service of science. Its foundation would fail repeatedly, however, obliging staff to relocate camp and airstrip several times because of fragmentation. U.S. visitors would inspect the last hastily abandoned SP-8 campsite–clandestinely.[105] That April, Moscow Radio reported that SP-8 had been relieved by the next shift, under N. I. Blinov. Basing at Tiksi, on the Russian Arctic coast off the Laptev Sea, the spring resupply expedition *Sever-12* had effected the transfer during March as *Sever-8* floated some 450 miles (725 kilometers) off Wrangel Island. A turbo-prop AN-10 had delivered the freight–the first time the model had landed on sea ice.[106] In mid-April, a protracted air search deployed an advance party at 77°23′ N, 163° E–the northern fringe of the East Siberian Sea. Thus was the drift of *Severnyy Polyus-9* initiated. On 26 April, regular observations commenced. All ice camps break up, but *polyarniks* say, SP-9 was "the most cracked up station." Its encasement would compel eleven relocations, from fragment to fragment.

These long-term encampments had become largely prefabricated affairs. Once off-loaded, the huts were promptly erected. "The typical [sixties] station," P. A. Gordienko[107] outlined, "consists of a cluster of sturdy houses not unlike small railroad cars. The houses are set up on sled runners so that they can be dragged by tractor or pushed by the personnel of the station should a crack threaten or split the campsite." Each was foam insulated and sheltered three men. Separate huts and hemispherical tents served as a wardroom, kitchen, and laboratories. Electrical power was supplied by large diesel generating plants, plus portable units, and distributed by networks of cables suspended on poles. (These also carried telephone communication lines.) Tents had been heated with propane, but with the advent of more permanent huts, coal-burning stoves were introduced.[108] The radio had its own wind-powered generator. The typical complement was ten

ARCTIC RESEARCH LABORATORY (LATER NAVAL ARCTIC RESEARCH LABORATORY) CESSNA 180S EN ROUTE TO ICE SELECTED FOR A SEISMIC REFRACTION SHOT, SPRING 1960. DURING AND AFTER THE INTERNATIONAL GEOPHYSICAL YEAR, RESEARCH PERSONNEL EMPHASIZED THE NEED FOR STATIONING LIGHT AIRCRAFT ON U.S. DRIFT STATIONS. THE U.S. NAVY PIONEERED THE USE OF LIGHT AIRCRAFT TO DEPLOY OCEANOGRAPHIC AND GRAVITY STATIONS ON PACK ICE. THE LABORATORY'S 180S OPERATED IN COMPANY, SO THAT NO MAN'S LIFE DEPENDED ON ONE ENGINE. (DR. GERRY H. CABANISS)

to fourteen scientists supported by a physician, a radio operator, a mechanic, a cook, and a four-man helicopter crew. Groups of five to ten might augment permanent staff for a month or two, conducting "special missions." If the ice was surviving well and drifting favorably, staff was replaced during April.

> Since the station operates all day every day [Gordienko continues] and supply problems limit the size of the crew, a good deal of versatility is expected from each man. The magnetologist usually serves also as astronomer, the physician doubles as housekeeper, the meteorologist takes readings of the direct-heating power of the rays of the sun and the fliers assist the hydrologists and aerologists. All hands take turns at kitchen duty. . . . And all lend a hand in hacking a runway out of the ice.[109]

Shifting along the rim of Arctic Canada, then the "top" of Alaska, T-3 passed within nineteen miles of Point Barrow that spring. Ice islands are deep-draft features. T-3 grounded northwest of the spit, about eighty miles out. Its utility was thus so compromised that the air force would elect finally to evacuate the island camp.

ALPHA and CHARLIE had been large bivouacs—the first attempts at floe-ice stations by the United States. The CHARLIE experience had proved expensive. As planners maneuvered for (shrinking) post-IGY funds, programs had been shifted onto T-3 (its camp no longer BRAVO). Ashore, the team of Britton and Brewer (ARL director) deliberated: How does one log ice-based research when surface ships and aircraft suitable to the task are lacking?

> The obvious answer was to utilize the ice itself as the Soviet Union had long been doing, and as the U.S. Air Force was doing. . . . We in ONR were determined to get into the drifting-station business. . . . Brewer and I had been discussing a different concept in drifting-station operations. It seemed apparent that we would never have funding for the type of station then in operation and that the realistic thing to do was to use smaller aircraft, landing on unprepared surfaces, thus becoming independent of airstrip maintenance and of the necessity of abandoning a station when such strips were shortened by ice breakage. . . . At the end of CHARLIE phase of operation Max Brewer and I sat in Fairbanks considering our next move.[110]

Barrow had had no air capability until 1952, after which it had leased aircraft. Eager to operate its own air fleet so as to extend its reach, ARL had pushed to acquire big-plane capability—platforms capable of exploiting (among other chores) sea ice innocent of prepared airstrips. "We had researchers/investigators coming along desperately needing to get elsewhere than at Barrow," Britton recalled. "So our problem, as always: to provide facilities. So airplanes were one of my *prime* objectives when I came to ONR in 1955." Instead of R4Ds, ski-equipped Cessna 180s had been secured. Within a radius appropriate to light planes (about 115 miles out), winter oceanographic and other types of research were being thrown to seaward, onto an ocean crowded with drift ice—where the navy had obvious research requirements. A new logistical tool was operational off the coast of Arctic Alaska, and on his return from T-3 via Barrow, ONR's Beau Buck was invited to observe it. This technique entailed finding the right ice and then establishing temporary oceanographic stations. But would longer, overnight loiters be possible?

> To Brewer, who was then trying hard to justify the [Cessna] 180s cost to ONR, *anything* was possible with his new birds. Buck was fresh from a noisy, overstaffed ice island and wondered what could be done in acoustics research from small, quiet ice camp set out by the 180s. Correspondence started the planning for the first attempt.[111]

The small ARLIS III, IV, V, and VI camps would result from this planning. Intended primarily for low-frequency (long-range surveillance) research, ARLIS V and VI would be dedicated *totally* to acoustics.

Chapter Five

ARLIS, ACOUSTICS, AIDJEX, AND SPS

> The climactic world is one world even if politically we are not.
> –Reid Bryson

THE SCIENCES OF THE EARTH ARE HEAVILY observational, that is, they rely on fieldwork and extensive (and improving) data sets. Interactions between sea ice, ocean, and atmosphere in the polar region strongly affect climate; although daunting, logistically, the poles are central to understanding world climate and thus global change. The best high-resolution records of climate in the recent geological past are from Greenland and Antarctic ice cores. (Scientific data, indeed, is Antarctica's greatest export.) A major component of the maintenance of climate and amelioration of climate variability is, we now know, Earth's oceans.[1] Shifts in circulation have triggered climate upheavals. In the 1960s, however, climate-change issues had yet to bloom. Study of the ocean climate was in its infancy and its models simplistic, driven by a few parameters. Protracted multidisciplinary, multinational studies require long-term funding. Still, sensitivity to a raft of environmental concerns was rising–and, so too, the fortunes of climate science.

Adhering to a coherent plan, the Soviet Union (bureaucracy notwithstanding) was far ahead of her rivals in Arctic science, and her second-to-none standing would persist.

On the other side of the pole, the United States pursued parity with the Soviet Union in Arctic studies. But the United States had no national policy with regard to that huge part of the world. Historically, the responsibility for planning and funding Arctic research has been apportioned among U.S. federal agencies, Alaska, private industry, academia, and national laboratories. Despite attempts to coordinate, not until 1984 did Congress pass the Arctic Research and Policy Act (since amended) "to provide for a comprehensive national policy dealing with national research needs and objectives in the Arctic." As previous chapters have described, the air force had scored most of the early successes, leaving the navy to steam hard behind. How does one pursue field science in ice-rich seas with no aircraft or ships fully suitable to the mission? Programs sponsored by the Office of Naval Research (ONR) helped to realize missions and assets. Given the responsibility of providing logistical support for the U.S. Antarctic Research Program of the National Science Foundation (NSF), moreover, the U.S. Navy shouldered that mission into 1999.[2]

The Soviets' first post-IGY expedition was airlifted onto ice in the spring of 1960. This was *Severnyy Polyus-9* (SP-9). Slowly at first, its floe raft gathered speed, tracing a course north and northeastward—roughly paralleling SP-8, whose drift path had cut that same piece of water. This season, "8" floated about 300 kilometers (186 miles) from its sister station. And both lay remote from the mainland. As it happened, the relief party for SP-9, led by I. Romanov, was not fielded. Suffering severe, repeated fracturing, "The ice [rafting SP-9] broke up 11 times, and 11 times we moved from fragment to fragment."[3] At the last, "9" gave way to the next in the series, the evacuation harassed by pressure and much cracking. March 1961 saw its staff and equipment air-transferred to SP-8 and thence back to civilization. Relieved by *Sever-13*, this eighth long-haul camp of Soviets worked on— the sole functioning station. SP-10 was brought in by ship that October—the first time aircraft were not used for that purpose. Its luck was to prove better than the abortive SP-8: 925 drift days (to April 1964). Resupply of "10"and its final abandonment would be by airlift only.

Probing north of the Beaufort, *Severnyy Polyus-8* worked light aircraft. Station-based that spring and summer, the AN-22 (with its limited range) was used for ice reconnaissance, scientific work, and ferrying stores from the main airstrip to the main camp.[4] At Barrow, the Arctic Research Laboratory (ARL) was dispatching its birds for reconnaissance and for setting oceanographic stations, exploiting thin lead ice. Lead ice is formed when relative ice motion causes cracks to open. Seawater fills the openings and quickly freezes; this creates undeformed, table-smooth ice. Throughout the IGY period, large camps had called for long fixed airstrips and the equipment needed to construct, maintain, and (in the fall season) rebuild them. "Work on the runways went on continuously. Almost the entire camp turned out whenever a plane came in to clear the snow, which frequently piled up in huge drifts all the way across the strip, sometimes in only a few hours time."[5] ALPHA had had tons of machines: two D-4 crawlers, one D-6, one road grader, two weasels, a snow compactor, and a thirty-foot runway drag. Yet when pressure sliced its runways, a near crippling had gradually overwhelmed the station until the final foil: emergency evacuation.

ARLIS I

Post-IGY camps demanded a reengineering to a simpler, stripped-down, more budget-friendly design. The shops at ARL got busy. The resulting ARLIS (Arctic Research Laboratory ice station) series would be operated for ONR under contract by the University of Alaska.

In August 1960, off the beach at Barrow, USS *Burton Island* took aboard prefabricated panels (enough for ten huts), fuel, generators, a weasel, 22,000 gallons of petroleum and oil, and other supplies, along with personnel for a new at-sea research station. The icebreaker sortied on 1 September, initiating a new approach to establishing on-ice bivouacs.[6]

Meeting unusually heavy ice, *Burton Island* steered mostly northeastward, around the edge of the multiyear pack, then north, along the western coast of Banks Island—the most westerly of the Canadian Arctic Archipelago. She then moved off due westward 210 miles (336 kilometers)—muscling deep into the pack. On 10 September, after ship's helicopter had made several reconnaissance flights, the icebreaker inched its way to a suitable ice floe and off-loaded. (Thirteen months later, the nuclear icebreaker *Lenin* would help to establish SP-10.) Her crew assisting, *Burton Island* soon had ARLIS I operational—the first time ARL had installed an at-sea station independent of the air force. Cdr. Griffith C. Evans Jr., U.S. Navy, ship's skipper, boasted that scientific work had commenced a mere six days after drop-off at coordinates 75°12' N, 136° W. A point about 420 miles (672 kilometers) northeast of Point Barrow, this was a new high latitude for a ship under power.

The expected duration of the demonstration model? No one could know. "It might last two days," Max Brewer, ARL director speculated, "and then it might last six months." However long, ARLIS hosted programs in physical oceanography, micrometeorology, sea-ice physics, marine biology and geology, and geophysics. Under K. Bennington, Ph.D., an ice physicist, six staff were encamped under conditions that might be described as comfortably bleak. No aircraft, as it happened, met the floe for eight weeks—not until an airdrop (3200 pounds or 1440 kilograms) on 21 November, the first of three that month made by Barrow's Lockheed Lodestar. The sole transport in ARL inventory (a recent gift), the big-bellied Lodestar proved unfit for low-level, precision operations.

> The paradropping called for expert piloting and infallible coordination between the cockpit and the fuselage, where a couple of men pushed the bundles with their parachutes out an open hatch. Since ARLIS had no vehicles for

hauling cargo, the drop zone had to be as close as possible without endangering the buildings. The pilot would ease back on the throttle and glide down.... The men [aft] would wait, unable to see where the plane was in relation to the target. The plane's velocity and elevation would decrease until [it] had almost lost flying speed and was just a few score feet from the ground. It was necessary that the plane be over the drop zone at that point because the pilot would have to give the engines full throttle to keep the power-starved plane from plummeting.... Otherwise, the plane would have to circle and begin another pass.[7]

The first aircraft to *land* at ARLIS I were ski-rigged Cessna 180s. Flight rules for this pair were to operate always in company and never to carry more passengers between them than one plane could return ashore. Resupply shuttles persisted from November to March 1961–twenty-six altogether, most completed in darkness. Indeed, ARLIS was sustained almost entirely by this light-plane pair. "The ARLIS-I effort," Britton recalls, "was the beginning of ARL's capability to carry Navy programs far to sea in the Arctic Basin. As the Navy Arctic Research Laboratory's (ARL renamed) fleet of aircraft grew, its capability was to be utilized repeatedly to establish and maintain a series of both floe- and ice-island research stations."[8]

ARLIS I investigators would secure over one hundred oceanographic stations along a contorted 615-mile (984-kilometer) track. No polar venture unfolds easily. Ever peripatetic, by March 1961 the bivouac was so far to the west as to be near the practical limit of air support. Deteriorating ice sealed the issue. At midmonth, the *Anchorage Times* ran the headline "Ice Floe Research Camp Being Evacuated." On the 18th, the Cessna 180s departed ARL. Fortuitously, the first of two reconstituted R4Ds reached Point Barrow the very next day and were pressed into immediate (evacuation) service. Landing on a large floe near ARLIS, the R4D was "fed" by light-plane shuttles to and from the camp. In temperatures of −10° to −25° and winds of up to thirty knots, the R4D logged seven return runs to Alaska from its runway of opportunity, retrieving 24,000 pounds (10,800 kilograms) of ARLIS equipment. A week was needed to effect the evacuation, during which the Cessnas flew eighty-two landings onto unprepared ice, remaining in camp for four nights. Its own floe lay nibbled to near uselessness–from an eight-square-mile pan to a quarter square mile. "Once more," Britton records, "ice breakage had taken its toll but this time the cost was small." Exclusive of the services of *Burton Island*, the total estimated outlay for slightly less than seven months' field research was about $75,000.

All scientific gear and data and all other equipment save for the shelters themselves were salvaged, and although the short life of the station was a disappointment, the great economy of its operation, calculated as about one-twentieth the cost of previous stations and providing equivalent results, fully justified the hopes and expectations of its sponsors.[9]

ARLIS II

The constraints of economy had made ARLIS I into a template. Its shelters were a genuine bargain at $10,000 cash invested. "You know it can't be done for that. The thing is we had so many assets at Barrow we didn't have to buy things. We had lumber, we had plywood, we had insulation . . ." Brewer and his native assistants had designed a safe, dry, warm, and well-insulated hut. ARLIS was soon again offshore, on ice.[10] (See Apendix 3.)

Both East and West sought target ice. In Washington, the Department of Defense authorized a continuation of the Arctic Ocean Drift Station Program; T-3 would carry on. Off North Alaska, delayed sweeps were initiated for new ice for ARLIS. Candidate floes were located in the area of 76° N, 158° W. On 23 May 1961, Barrow's R4D and the Cessna pair lifted off, carrying generators, radios, and survival gear, as well as three personnel: station leader John Beck, a radio operator, and ARL director Brewer. En route to the chosen sector, quite suddenly, an unfamiliar surface pattern was spotted at 73° N, 156°05′ W—much closer in. Brewer sought a closer look. The slab occupied a roughly rectangular area of about three by two miles. "Looks like land," chief pilot Robert Fischer remarked. "Ice land," suggested Brewer. "Maybe ice island." The value of this find was instantly recognized.

> Although its location was farther south than desired, the relative permanence and stability of an ice island far outweighed the southerly locale. [In some excitement] the aircraft circled and inspected the island. The first Cessna [Fischer] landed on the island and the second landed on the sea ice, on a refrozen lead that cut deeply into the island, to investigate possible landing sites for the R4D.[11]

John F. Schindler, a longtime assistant director who would assume the NARL directorship in 1970, describes the technique of aerial search, to locate and install an ARLIS-type floe station.

> We were looking for multiyear floe ice . . . 6–7 or more feet thick . . . a mile plus in all directions . . . AND with a relatively recently refrozen lead next door, to serve as a runway. There were always lots of suitable floes, but the

lead-runway requirement was a bit limiting. We wanted ice that was thick enough to support the R4D (or a C-130 on some later stations) and with a minimum of snow cover so that we could land the R4D without skis. We used the skis once (ARLIS III I think) but found they weren't really that much help. We would land the Cessnas, drill the ice to determine thickness and then land the R4D if everything was OK. The R4D contained a complete building, radio, stoves, food, sleeping gear, etc., for three men that would be left on station with a radio beacon to bring in the supply flights. The Cessnas would not leave the station until everything was set up and operating. Often we would have a second R4D there by that time and the supply phase had begun.[12]

Upon landing, a possible camp-runway site was scouted; after three trial landings, the big plane was waved in. (The R4D required lead ice recent enough to offer little snow yet thick enough to offer ample support.) A look around satisfied Brewer. "Let's get started," he said. Prefabricated sections for two huts were unloaded, generators and radio gear set up, a homer beacon connected. "Five weary men slept in the two huts the first night, and in the morning Beck and his men began a week that consisted primarily of hut building."[13]

ARL had crafted a prefabricated marvel. The huts' wall panels were plywood sheets with glued and nailed studs mounted with plywood to the inside; over the outside were three inches of fiberglass covered with aluminum foil. Thick translucent polyethylene had been added to the outsides in lath-like strips. (These "wanigans" were temporary, so moisture condensation was not a problem. Beginning with ARLIS II, a layer of polyethylene was added between the studs and the interior plywood, to exclude condensing moisture. Later, an outside layer of plywood was added as well.) The panels could be toenailed together in a few hours to provide a flat-roofed twelve-by-sixteen-foot hut seven feet high. Three windows were built in; the single weatherstripped door was inserted in the field. Designed to be used as sleeping space for two or for laboratory space, each box was heated by a diesel-burning stove in one corner, near the door. A tiny fan circulated the stove's output, thereby preventing stratification. The cost of materials and prefabrication for each had been $1,500.

Why didn't the camp use tents? The interior space of fast-to-erect wood structures can be better exploited than even the largest tents because the walls lend themselves to horizontal surfaces—shelves and bunks—thus getting men and gear off the floor. Extra room also means stand-up comfort. And wooden walls offer (some) defense against marauding bears.[14]

A telegram reached ONR headquarters, near Washington:

AN ICE ISLAND–REPEAT ICE ISLAND–THREE AND ONE QUARTER BY TWO MILES . . . WAS SPOTTED AT 73 DEG 10 MIN NORTH AND 156 DEG 05 MIN WEST.
ARLIS II WAS OCCUPIED AT 1600 ALASKA STANDARD TIME THIS DATE.

A "massive airlift" rushed materials to seaward. All ARL personnel were pressed into duty as "round-the-clock freight handlers." Inside of two weeks, twenty-eight flights by R4D brought 156,300 pounds (70,335 kilograms) of freight to the station. Shelter is vital; a wanigan was up by the evening of day One. ("We never left the ice," Schindler notes, "until the first hut was up.") The shelters were nailed up in speedy succession; so quickly did they rise that the pilots reported new ones between successive arrivals–thirteen wanigans within five and a half days. The first investigator was William McComas, from the Navy Hydrographic Office, who deplaned on 25 May. Oceanographer Leonard A. LeShack arrived next day. On the 28th, the R4D delivered the nineteen-kilowatt generator, the heaviest single freightage; it would serve as the main power source. ARLIS II recorded its inaugural measurement–a gravity reading–by 1 June. Within days, all research programs were under way.

But frustration set in. The ARLIS raft was ablating, softening, melting. Its surface was blotched by puddling and, near the camp, a lake–an invasion by water. Soon, the summer sun closed the airstrip. A crosswind pushed the inbound R4D into a series of ridges hidden by snow, skidding it into a channel filled with wet snow. The damage proved minor, but the plane was out of action. No powered vehicle was on hand, so every pound of freight now had to be dragged by sled into camp–a quarter-mile haul. On 6 June, the airlift was completed using *three* light planes: two Cessna 180s and a Cessna 195. By the 14th, the ice ever mushier, this fleet-of-three had logged forty-one sorties. Altogether, 21,700 pounds (9765 kilograms) of freight and twenty-two personnel, including the full-size summer staff, had been ferried onto ARLIS ice. Courtesy of a repair crew, the deadweight R4D was manhandled (no mechanical aids) free of a snow bank and across a melt run, onto the runway. There the propellers and tail wheel were replaced; on 16 June, remarkably, the ship splashed off for Barrow–the season's final liftoff. Left adrift were eleven investigators plus four support staff. All faced, one resident said, "a long summer that would be cold, cloudy and lonely, but not in the least boring."

Fourteen huts sheltered, equipped, and then sustained ARLIS. The living huts served as both work and living quarters, with two men assigned to each. The station was laid out in the shape of a horseshoe and use was dictated by experience. There were four huts across the top–one each for storage, toilets, the generator, and support personnel. Four huts on each side completed the "U." The four huts nearest the storage hut sheltered the scientists and their gear; opposite stood a radio shack and living space for the station leader and camp cook, a kitchen and mess space (actually two huts end to end) plus two work huts for scientists. Placing the radio hut near the generator reduced the probability of separation should a crack interpose. The storage hut was across from the kitchen, the investigators well removed from the generator, which might affect electronic equipment. The open end of the "U" faced the runway, thus granting vehicles and sleds easy access to the courtyard area.

Men require reasonable comfort, so, while stressing economy, Barrow knew that on-ice personnel had few opportunities for recreation. Hence, it was attentive to certain fundamentals.

> The ARL theory is that, if a man is warm, has excellent food (steak and strawberries are more economical to use than poor-quality items because of the transportation costs for wasted food) and mail, he can do most anything—deprive him of these and frustrations will whip him. . . . Both scientists and support personnel should have sufficient work so as not to allow them idle time, which is differentiated from leisure time. The former breeds discontent for all personnel on the station. In addition, the scientists should not be required to help build the station but should be put to work on their projects immediately, certainly no later than four or five days after a new station is established. [15]

That summer of 1961 ARLIS-rafted science embraced underwater acoustics, marine biology, oceanography, geomorphology, geophysics, and sea-ice micrometeorology, as well as strain measurements of the ice.[16] To support its parties, ARLIS II received an upgrade: *two* camps, the main camp and a thin-ice camp. The latter stood at a floe edge from which various studies would be easier: the micrometeorology of sea ice (via thermocouples), depth sounding, ice deformation, and underwater acoustics. (In 1967, so as to continue in-situ acoustic studies, T-3's "quiet station"–four huts–would be moved about three miles from its main camp.)

Structurally, ARLIS II was an irregular shelf-ice block to which were welded broad areas of pack ice. Wedge-like salients penetrated the central block, nearly cleaving it at its midriff. Its surface was unusual, being marked by a series of rock-covered hills reaching heights of thirty-three to forty feet (ten to twelve meters). Summer sun "works" the ice, of course. During July alone, the melt totaled nearly two and a half feet, transforming the island surface into quagmire. More ominously, cracks appeared–and widened. Huts were shifted clear.

> Ice ablation and melt of the island were extensive that first summer. Huts

had to be moved off their pedestals as early as the first of July. Rivers and streams loaded with silt flowed off the edge of the island. A lake collected just off the opposite side of the runway from camp, and while providing a formidable barrier to moving supplies, it also provided recreation as some personnel enjoyed boating in free moments.[17]

Despite a somewhat unpromising beginning and (later) severely extended supply lines, ARL proved unwilling to quit the island, and ARLIS II persisted for three summers more. Meanwhile, pressure caused sudden shifts. In mid-July 1961, an extensive system of fractures developed along an edge. About two weeks later a fracture formed along the ice island's short axis, cleaving it in two, leaving the southern half to raft the station.[18]

Staten Island (AGB-5, fresh from Antarctica) overhauled ARLIS II that season with little difficulty; the icebreaker "docked" on 19 August. Among the ship's treasures was a D-4 Cat, two weasels, three new buildings, lumber, 360 drums of POL (petroleum, oil, lubricant), explosives, and food—125 tons altogether. The next day, while standing out, she detonated charges for seismic stations on the listen at ARLIS and T-3 and at Barrow. Homeward bound, *Staten Island* stopped off at T-3 to backhaul some equipment in need of repair.

The usual warm-season irritants—water, pedestaling, a pervasive damp—arrived and then persisted. "The monotony of the ever-present overcast was broken only by the paradrops from Barrow and the welcome mail from home." John Beck, a retired navy officer (whose experience included duty aboard *Burton Island*), declared 1 July 1961 "hut-moving day." All hands helped move shelters off sagging foundations and onto adjacent flat ice.

As the source of support and services, ARL had the responsibility of airlifting—on time and intact—the stores, food, and odd and sundry items requested by researchers and staff. The melt had scrubbed the chances of any landings, and icebreaker sorties are chancy. For the summertime, then, the only option for resupply was airdrop. Five drops were made that season. The first of 1961 was made on 25 June, the first that fall on 18 September. Newly arrived staff were shown about (this didn't take long), after which the men settled in and set to work. (Not far away, on 17 October, the Soviets inaugurated SP-10.) In the long shadows of late boreal sunlight, *Staten Island* revisited—berthing at 76°20′ N. There, another one hundred tons was off-loaded—mostly POL—comfortably provisioning the winter station. November began with unusually heavy snow that prevented airlifts. Now came a rare moment of extreme stress. R4D 39078, having delivered supplies and personnel, was returning Beck to Barrow (his successor would be John Sater). The transport lifted clear at 1620, Max Brewer and John Schindler among its passengers, aft. Unknown to all, as the plane was being prepared in the dark, a drum of diesel oil had been pumped into the fuel tanks along with three of "AvGas." About twenty minutes out, the port engine rebelled. Brewer conferred with the flight deck then ordered a U-turn. (About 350 miles of open water interposed between Barrow and the pack-ice edge.) The angry engine was feathered, cargo was jettisoned. Still, the pilots could not maintain altitude on just one engine. Mayday-ing as they went in, they finessed a landing onto a refrozen lead.

The plane was known to be down, but its position was unknown. Rescue, its party hoped, was but a matter of time. Until retrieval, the fuselage could offer fuel, survival gear—and the best landmark in the area. The most urgent task for the stranded was to ascertain the thickness of the ice (two feet) and erect tents. Their second most urgent task was to stay uninjured, dry, and warm.[19]

> The landing had been smooth, with the wheels in the up position to prevent a possible noseover since it was after dark and visibility was poor. Visible damage to the aircraft was limited to the right engine cowling and the prop that was still turning when the plane landed. The plane was immediately unloaded and an emergency camp was established, using the two survival tents.... All hands turned in except for the man on watch and surprisingly everyone slept well. No personnel had suffered injury even to the extent of a bruise.[20]

The following afternoon, as an air force C-54 circled overhead, contact was established via the R4D's radios (operating on batteries). At about midday on 17 November, in maximum available light augmented by improvised flares, an Alaskan National Guard C-123J exploited the lead and retrieved the downed men.

Back on ARLIS II,

> The darkness of winter brought no further surprises, and the collection of knowledge and the drift of the island went on. Polar bears visited, pressure ridges built around the island, leads opened and closed. The only break in the routine was when a meteorite was spotted (it was later learned to be part of the Discoverer Satellite Capsule) and provided a pyrotechnic display when it struck the ice.[21]

—•—•—•—•—•—

That fall, in Siberian waters, a surface vessel—powerful and all business—steamed an easterly course, two ships trailing astern. The nuclear icebreaker *Lenin* had stood out from Murmansk on 29 September 1961. After assisting traffic off the Taymyr Peninsula, in central Russia, course was taken toward the eastern sector of the Soviet North. Its mission included depositing a winter scientific party on pack ice north of Wrangel Island—the first Soviet drifting station so installed.

On 14 October, *Lenin* approached but was unable to penetrate to the chosen floe. Huts, stores, and equipment had to be hauled (three days of cold, onerous work) over a two-kilometer-long trail to the campsite, then located at 75°27′ N, 177°10′ E. The ship's crew helped carve out an airstrip near the station-to-be. Returning in stages, SP-10 (1961–64) in its wake, *Lenin* followed the margin of permanent or multiyear ice; at intervals, she pushed in to deploy drifting automatic radio beacons (DARBs) and drifting automatic radio meteorological stations (DARMS). Curving north of the Laptev Sea in winter dark, she reached latitude 81°30′ N in the region of the 119° meridian. Ice reconnaissance characterized her next leg, after which *Lenin* rounded Novaya Zemlya and gained open water and the Barents. The icebreaker floated pierside, in Murmansk, on 22 November, having concluded a remarkable round-trip.[22]

Severnyy Polyus-10 had gone operational on 17 October 1961. It would function until 1964—925 days for science. Its first shift was fourteen men, under the leadership of N. A. Kornilov. Three station heads were to succeed him.

North of the Beaufort Sea, SP-8 was working waters remote from Soviet territory. The nearest mainland airfield was about 2000 kilometers (1250 miles) away.

The autumn airlift to the station therefore required an intermediate airstrip on the ice, and this was set up in the neighborhood of NP-10 in early November.

The nuclear icebreaker *Lenin* standing by station SP-10, 14–17 October 1961. A Kamov Ka-15 attends the autumn operation—the first time that an aircraft had not been used to deploy a drifting station. Launched in 1957, *Lenin* was the first surface vessel to employ nuclear energy for propulsion. Reactor: pressurized water type, the same as for *Nautilus* (SSN 571). (Boris Vdoveinko, courtesy R. E. G. Davies)

Between 30 and 40 flights were made to the station during the month. The position in the spring of 1962 was no easier from the re-supply point of view, but much importance was attached to getting more data from this area, which had never been traversed by a Soviet drifting station.[23]

As 1962 opened, Leningrad could boast more than 6000 days to seaward—a gallant, hard-fought choreography totaling 25,000 miles, according to news agencies. And more than 10,000 high-altitude radiosondes—balloon-borne instrument packages—had accrued observations. In the Soviet North, scheduled air services were flying, to be sure, along with as-needed operations. Still, air transport here concentrated on specialized missions: ice reconnaissance, meteorological and hydrological support work, and the establishment and logistical support of drift stations and high-latitude expeditions.[24]

T-3 REOCCUPIED

The U.S. Air Force had abandoned T-3 in October 1961. The raft had grounded; for about eighteen months it held to, stuck fast at about 71°45′ N, 160° W, in shoal water off Point Barrow.[25] When or whether the berg would bestir was unknowable, so everything movable had been lifted out—the first full evacuation. "The camp's immobility and nearness to the elaborate scientific station at Barrow," the

A PIONEER TEAM CORING TO CONFIRM FLOE THICKNESS PRIOR TO SETTING CAMP. EXPERIENCED EYES ENJOY AN INTUITIVE SENSE OF ICE THICKNESS; NONETHELESS, THE PROCEDURE IS A PRELUDE TO SETTING PACK-ICE STATIONS. OLD OR MULTIYEAR ICE IS PREFERRED, WITH REFROZEN LEADS NEARBY TO SERVE AS RUNWAYS. NOTE THE LISUNOV LI-2 "FLYING LABORATORY." (BORIS VDOVIENKO, COURTESY R. E. G. DAVIES)

Ski landing gear of a Lisunov Ll-2, SP-10. Note the tent camp. *Lenin* had deposited fourteen men and their stores at 75°27' N, 177°10' E, a position north of Wrangel Island. This tenth semipermanent camp of Soviets was evacuated in April 1964, ending two and one-half years' occupation. Its replacement, SP-13, was deployed by high-latitude air expedition *Sever-16*. (Boris Vdovienko, courtesy R. E. G. Davies)

Times noted, "reduced its value as an observatory." Now ARLIS II hosted the only active U.S. research station on the Northern Ocean.

Belying the apparent invulnerability of ice islands, the T-3 mass—grounded, rotating and moving but little—broke apart. Advised to keep a weather-eye for T-3, ARL shuttles to ARLIS conducted searches for the mass, but without result, obliging the Alaskan Air Command to concede it "lost." In February 1962, however, a routine supply run rediscovered the errant slab. A fragment had floated free: it lay about 100 miles (60 kilometers) north of the last known position. "Lost U.S. Ice Isle Is Found in Arctic," ran the *Times* headline (4 March). When first discovered, T-3 had exceeded forty square miles in area. Carved out by multiple fractures, this chunk proffered roughly one-fifth that area. With ARLIS II nine months young, should ONR reoccupy the long-serving berg? "Without sufficient funding to do an outstanding job of research and support on ARLIS II," Britton explains, "how could we possibly justify a second ice island?" Approval to reoccupy was nonetheless secured because as an ONR/ARL research station, T-3 offered an alternate airfield as well as a fueling way station. The wisdom of this decision was to be demonstrated on almost every ARLIS II-bound sortie during the next four years.

Thus, two arctic islands would sustain U.S. marine-based work. On 17 February 1962, three light planes skied onto T-3 tableland. Its surface had ablated fourteen feet: improbable mushrooms, camp structures sprouted on icy stems. Reoccupation meant a mass relocation onto flat ice. (The mess hall proved hopeless.) Meanwhile, to forestall another disappearance, a homer beacon was set out. Basic camp infrastructure was serviced, restoration of the runway begun. Forty-eight hundred feet of erstwhile airstrip floated ready by early March. That April, the Alaskan Air Command officially transferred custody of T-3 to the navy and ARL on an "as is–where is" basis. Also in April, the first science party arrived from the Underwater Sound Lab. Late spring saw a number of programs activated: acoustics, seismic studies, gravity, magnetics. Still, the newer camp hosted more science; throughout 1962, ARLIS II averaged fifteen tenants, T-3, nine.

—•—•—•—•—•—

Dormant during the IGY, international tension had revived. Polar science

THE T-3 ENCAMPMENT, SUMMER 1960. RESUPPLY OF THE T-3 BASE—FUEL, FREIGHT, PERSONNEL—WAS ALMOST ENTIRELY BY AIRLIFT. THE RUNWAY SUSTAINING T-3 WAS CONSTRUCTED, GROOMED, AND MAINTAINED USING BULLDOZERS PLUS A GRADER. VIRTUALLY A NATIONAL LABORATORY, THE OCCUPATION INAUGURATED BY THE AIR FORCE'S COL. JOSEPH FLETCHER PERSEVERED—WITH INTERRUPTION—FOR MORE THAN TWO DECADES. NOTE ICEBREAKER ON HORIZON. (DR. GERRY H. CABANISS)

lay rooted in a protracted cold war shaped, steered, and dominated by military concerns. The Eurasian superpower continued to cause discomfort. "Understanding the polar sea is crucial to our security; the barren waste separates North America from Siberia and provides an avenue of approach for Russian attack. The bombers and missiles of any future war would be propelled over the Arctic's desolation. The country with ice island or ice-floe missile bases would have an enormous advantage."[26] Scientists tend not to share the enmity of their governments; reason prevails over pious, patriotic rhetoric. The planting of flags aside, investigators refused to glare at each other across the white. For most, it is not about opposing ideologies; the mission is *science*—which tends to trump ideology. "The Cold War is a temperate region phenomenon ignored by polar scientists the world over." "Even in Cold War," an official told me, in St. Petersburg, "we managed to keep good relations in Antarctica."[27]

Science cultivates debate; researchers converse through conferences and peer-reviewed professional journals. Papers based on SP-derived data had reached Soviet journals. And the observations themselves were being published *in extenso*, in a series of volumes called (in English) *Scientific Results of Drifting Stations*—twenty volumes by 1961. (Only nine, however, had been received outside the USSR.)[28] More than seven hundred research parties and nine semipermanent stations had so far been fielded by the Soviets. Best placed, perhaps, to judge this continuum were U.S. counterparts. This systematic reconnaissance from Leningrad was well conceived, massive in scale and scope, realistic, and vigorously pursued.

U.S. and Soviet aircraft again hunted candidate ice, and this season, units from those forces would cross paths. The two sovereignties share this ocean. And surface circulation is transboundary, so bruised sector claims are inevitable. ARLIS II crossed the International Date Line (180° meridian) in November 1961–an interdiction of Soviet "sovereign" ocean. This resulted in fatuous headlines ("Island Defects to the Reds"). The U.S. camp would float in "tomorrow" for nineteen months, geopolitics notwithstanding. "We just don't get much cold war here, either between amateur radio operators or between scientists," Milner advised the *Daily News-Miner*. "For example, we send all weather information into an international exchange, and we broadcast our position from time to time. We are all scientists, and we enjoy being able to communicate and exchange data." On the 28th, a Soviet plane flew three passes over ARLIS. "As the beacon was not on," officials noted, "the contact was either very lucky or reveals that they keep a close check of our position."[29]

Personnel at U.S. camps noted numerous overflights. For instance, APLIS 78 and APLIS 88 both "pushed the sector." Acoustics scientist Robert Francois recalls working Beaufort ice in 1988, *inside* Soviet declared waters:

I remember I estimated the altitude of the Bear [TU-95] at about 1,000 feet. I could clearly make out the dual, counter-rotating propellers on the front of the four engine nacelles. They made an enormous racket as their four blades crossed each other . . . At a slightly higher altitude came the U.S. escort [and tanker], flying a zigzag course to keep its net forward progress at the same rate as the Bear . . . We were told later that the Bear had been picked up and trailed as it entered the Alaskan ADIZ, turned around at Banks Island and headed directly toward APLIS . . . Only one pass was made. . . . Two days later, we had a second visit . . .[30]

No camp can avoid detection. Indeed, to assert (or challenge) sector rights, explicit, premeditated violations were more or less obligatory; in those waters, sectors were not so much guarded as surveyed. "It should be pointed out [Francois adds] that APLIS always used the same mid-frequency radio band for communication with its shore support base. This was intentional, to allow monitoring by any nation interested in the region we were 'occupying.' We had many post operation reports of Soviet patrol flights whose track passed over our camps."

Leningrad had planned to resupply and to re-man *Severnyy Polyus-8* despite its tortuous drift, so as to continue operations a further year. Nature intervened. Storms during February and March 1962 brought the ice alive, effectively dooming the bivouac. Cracking and hummocking made prospects so uncertain that the decision was made to vacate the camp. Using intermediate (located *between*) strips and light aircraft (because of a shortened runway), the Soviets did not effect a full retrieval from the embattled station until 19 March. SP-8 had logged 1057 days, a total drift of 6090 kilometers.

Severnyy Polyus-11 was established by the air expedition *Sever-14* at 77°10′ N, 165°58′ W—well north of the Russian littoral.[31] It commenced operating on 17 April under station head N. N. Bryazgin. The spring of 1962, then, found two stations serving the Soviet state. Within the circumference of the Arctic Ocean, a quartet dotted the central basin. *Severnyy Polyus-10* floated nearest the geographic pole. Its sister in science, SP-11 (1962–63), would speed to the north-northeast as 1962 faded, nearing the so-called Pole of Inaccessibility. In American waters, ARLIS II floated well away from the Russian outpost, at latitude 78° N. By December, its new coordinates were 81°12′ N, 161°45′ W—far to seaward of Alaska, about halfway to the pole.

The shift manning SP-11 was encamped mere weeks before being found. En route to ARLIS II, R4D pilots Fischer and Quates came upon it quite by chance about halfway between Barrow and ARLIS. Banking close, they spotted a village spread below. A report sped shoreward. Neither side had yet visited the other, but, despite "touchy" diplomatic relations with the folks across the water, Brewer radioed ARL for permission to land.

> Upon receiving permission from Barrow to land, the two pilots paid a brief visit to the personnel of NORTH POLE 11, then the newest of Soviet drifting stations. Because of the language barrier, the amenities were limited to smiles, handshakes, and taking photographs of all present, but this in no way impaired the cordiality of both parties. After [a somewhat awkward] 15 min. on the station the plane continued on its way to ARLIS II.[32]

That same day, a Soviet plane circled ARLIS II for twenty minutes. "Although the plane lowered its landing gear during the course of four passes over the camp and both pilots waved, no actual attempt was made to land."[33]

A bond of respect and camaraderie unites those who study polar worlds. Not long after his crew's drop-in, Director Brewer set forth a protocol for visits. "Now that we have landed on North Pole XI and been received with great friendliness," he advised, "it is entirely expected that we will experience a return courtesy from our Soviet counterparts.... Cooperation and courtesy are a way of life in the outer stations of the Arctic and we wish to perpetuate this as much as possible." As well as offering tactical opportunities, visits promote cooperative research. Accordingly, Soviet personnel were to receive "every courtesy" and be granted the "complete hospitality of the station"—including a refueling and a look around, if requested. "It is magnificent opportunity to contribute immeasurably to the advancement of research in the Arctic and to promote the establishment of working stations similar to those we now enjoy in the Antarctic."[34]

For whatever reason in this new kind of war by other means, the Russians did not come.[35]

—·—·—·—·—·—

ACOUSTICS PRIMER (QUIET CAMPS)

The farther removed one is, the more attenuated will be the noise a camp generates. Given background interferences, T-3 had proved less than satisfactory for undersea acoustics work. Ice used as acoustic stations needed smaller, noise-free camps (at least at certain frequencies), less self-noise (man-made interference). Beaumont Buck explains:

> One requirement for the in situ study of [background or ambient noise and transmission loss] was quietness of the measurement platform (both acoustic and electromagnetic). The only way to achieve this was keeping the stations

small (the bigger the camp the more people, the more food and fuel, the more air flights, mess halls, longer airstrips etc.) QUIET STATIONS required special consideration be given to electrical power generation and could not live with hydrohole winches for oceanographic and/or biological casts, or graders and cats trying to keep the runway smooth. Many times more than a single camp was needed in a season for the propagation work. . . .

While permanent and safe from breakup, experience at T-3 indicated that [big camps] were hardly ideal sites for low-frequency acoustics research. Four KW generators going full time, road graders on the runway at all hours, uncleared personnel, electromagnetic radiations from strong HF [high-frequency] radios and aircraft beacons—made an acoustician's life a hard one. Small, quiet camps were obviously needed.[36]

And so, small, short-term (weeks) bivouacs were designed expressly for acoustics research. Their primary mission: to study some aspect of the underwater environment and/or to establish experimental technology such as signal processing or satellite telemetry. Sundry camps—though not all—were integrated into the SUBICEX (Submarine Ice Exercise) program. Initiated in 1959, these operations averaged one per year and included, typically, an icebreaker, aircraft, or drift camp. They were intended to create an advantage in sonar technology, and they produced revolutionary instrumentation. Naval research and development reflected the strategic situation: military technology was transforming maritime warfare. One particular system implemented in 1982 would reenergize interest in Arctic operations: the ballistic missile-firing submarine. Armed with long-range missiles and deployed into northern frontier waters—a new battlespace—the Soviet "boomer" threatened targets throughout most of North America. Indeed, a lone fleet ballistic missile submarine, or SSBN, could hold the entire Free World hostage to nuclear attack. The U.S. Navy wanted these hard-to-target enemy boats (deemed the most dangerous maritime threat to the United States) detected at long-range and surveilled.

The Arctic Ocean and its marginal zones (the interface between open sea and ice cover) are acoustically unique. A complex dynamic of solid plates atop a refracting medium, the ice cover necessarily influences the sound-propagation properties of the water column beneath. Water density is a function of temperature and salinity, and increases with depth. Distinct water masses occur having stratified, sloping (because of the Earth's rotation) density surfaces. This thermal-salinity structure provides unique protection by rendering antisubmarine ships and antisubmarine aircraft blind. The U.S. Navy knew little about how the effects of density-temperature-salinity on sonar efficiency and under-ice piloting would, in turn, affect the detection and trailing of subice submarines.

> The acoustic regime under the Arctic Ocean ice is *extremely* complex because of the *boundary* [media interface] between the ice and the water, which is slushy—and the density changes all the time. If you ping and listen with sonar (or any sonic device in the Arctic Ocean) it's confused by reflection and refraction from the underside of the ice which confuses the ping-listen system. There are multiple paths. Then when you put it over the continental shelf you've got the bottom—which we understood pretty well. But the ice above we did not understand very well.[37]

Further, the pack itself is an acoustic source: ultimately, near-surface stresses deform then fracture the canopy.

For attack nuclear submarines (SSN), these behaviors confound operating and listening *and* effectiveness as an under-ice platform for firing torpedoes. Could the arctic submarine become an effective combat weapon—fight it out if necessary?[38] Improved piloting sonars were crucial, to reduce the danger of ice collision in shallow water and allow tactical use of the canopy. Research on these questions as well as surveillance concerns secured top-priority status for environmental acoustic parameters; this subsumed man-in-the-arctic research to the naval mission and threat. Thus certain fleet priorities came to be resented in academia. In Ron McGregor's words, "Instead of being science driven, ONR became navy-driven."

Low-frequency sound, investigators found, propagated with good efficiency, with ambient noise low much of the time. So, it looked as if long-range surveillance would be practicable if large hydrophone arrays were used—and as long as the hydrophone results could be taken back to a warm lab for processing. As for antisubmarine *weapons* performance, understanding the performance of *high*-frequency systems—both active and passive—in that singular environment was fundamental to acoustic sensors related to torpedoes and mines.

The first "quiet camp" was deployed in 1962. GMIS I (General Motors Ice Station No. 1) was put out 100 miles (160 kilometers) northeast of Barrow, in south Beaufort. For deployment, the project would rely on Barrow's ski-equipped Cessna 180s. GMIS was small and rudimentary, intended only for a few days occupancy by two men. (However, power-supply glitches would cause a three-week sojourn.) Low-frequency surveillance acoustics was the mission. Its final objective was to develop a DIMUS that would allow all-season, long-range detection and tracking of nuclear submarines. But GMIS I had a different mission, as well: to test over-the-horizon telemetry en route to developing an autonomous-buoy replacement for manned camps.

> So the cart would not get too far out in front of the horse, however, it was realized that a considerable amount of work must be done in researching the

acoustic environmental effects that would determine how well such a sonar would perform. What were the statistics of ambient noise levels? Heretofore only spot measurements had been made, primarily from ice islands. Under conditions of rapid ice drift, cable strum had prevented measurements altogether (no form of cable fairing had been tried). Propagation loss measurements had a plus or minus 10 or more dB [decibel] spread. And, no measurements of noise or transmission loss had been made in summer ice conditions. Under the upper-refraction conditions that were known to prevail in the Arctic, what effect would surface reflections have on signal coherence? Would a large aperture array work at all? How well? Was the noise anisotropic, and, if so, to what extent? What were the effects of source and receiver depth? All of these questions are still with us today, but to a much lesser extent than in 1960—thanks to small-floe ice research stations.[39]

GMIS II followed in April 1963, about 230 miles (370 kilometers) out, staffed by five, who would undertake propagation and ambient noise studies in cooperation with the Naval Ordnance Laboratory. ARL-based Cessnas helpfully deployed explosives out to 132 miles—55-pound block TNT charges pushed into an open lead after landing.

A partial roster of U.S. quiet camps is a long one (Appendix 3), incorporating the first summer pack-ice station (1971), the first fall-season camp (1978), and the first U.S. Eurasian Basin stations (1977).

—•—•—•—•—•—•—

The first anniversary of ARLIS II, on 23 May 1962, "passed almost unnoticed, with the press of the resupply activity"; this included last-minute requests—all of which the last plane of the season cannot carry. The meandering path it had been following now totaled about 600 miles (960 kilometers). Anniversary day was marked with "due ceremony and full allowances for propitiating all applicable gods."[40] That December, a short experiment would be conducted at ARLIS II, then floating a few hundred miles off the pole. A colleague assisted Buck in using correlation techniques to study noise anisotropy. "The most serious problem then," he writes, "and as it remained throughout the sixties and seventies, was the lack of a suitable field tape recorder. In those days FM tape recorders were made for lab, not field use in such a rough environment."[41]

ARLIS II shifted generally to the northeast during 1963. Its second anniversary had the island station 14° north of its initial coordinates. T-3, for its part, floated over the Canadian Abyssal Plain and the Alpha Rise.

Sever meant survival as well as sustenance for ice-based assets. *Sever-15* from Leningrad, the 1963 resupply and re-manning expedition, took flight at February's end. The distance to SP-11 called for an intermediate airstrip, onto which *fifty* sorties ferried supplies. From its ice also, late in April, SP-11 was evacuated from inside permanent ice north of the New Siberian Islands, outcroppings that separate the shelf of the Laptev from the East Siberian Sea. (The campsite was left stocked with food and supplies.) SP-10, then floating 1100 kilometers (683 miles) from its resources, was restocked with one hundred tons of cargo.

> The airstrips at both stations were so badly damaged by ice movement that new sites had to be found for them before the relief could take place. A new station, SP-12, leader L. Belyakov, was established at lat 76°30′ N, long 165° W and commenced regular reporting on 1 May [1963]. The number of operating Soviet stations thus remained two.[42]

This twelfth station was positioned well north of the Bering Strait, within the limit of multiyear ice. Deploy coordinates put it somewhat south from where SP-11 had begun its odyssey. The transpolar current (the Arctic and Antarctic Research Institute had calculated) might convey "12" poleward, and, with luck, grant a lengthy drift. Twelve, though, was to log an untimely breakup after two years' service. Its drift ended at 81°06′ N, 145°47′ W–in the Canada Basin. Its latitude was the same SP-11's had been when it had been quit two years before, its longitude only seven degrees east of that coordinate.

The 1963 field season saw ARL press fully to the pole. Bob Fischer (a superb pilot) flew one of the Cessnas, with researcher Arne Hanson in the right-hand seat. Cliff Alderfer had the controls in the other, with Jan Black as his passenger. The squadron of two advanced in stages, exploiting T-3, then the ARLIS runway. (Without these intermediates, the operation would have been impracticable). Approaching Ninety North, they saw a bit of luck floating on offer–a refrozen lead. Fischer touched down safely; Alderfer followed him in. "It was a little after midnight, Alaska time, on May 24, 1963," Fischer commented. "Arne Hanson remembered it was his birthday. How about that? The North Pole for a birthday present!" This was the first recorded landing by single-engine light aircraft at the northern axis.[43] A gravity reading was taken, an ice-hole drilled, an explosive charge detonated. The returning echo required six seconds–two and a half miles (four kilometers) of water. This reading was identical to that taken by *Nautilus*. Thirty minutes after touchdown, the expedition again was airborne, pressing south.

T-3 wandered its useful way. That July (1963), T-3 radioed that one of its scientists was gravely ill. Within twelve hours a doctor was circling overhead, radioing a diagnosis, and medication was dropped. Plans for an aerial pickup– always risky–were cancelled, revived when complications in the man's illness erupted, then again dropped. A relapse ensued, after which the afflicted man collapsed and died. Unfortunately for Barrow and camp managers, the T-3 runway floated useless, almost completely submerged. The victim was retrieved in August by a B-17 mounting a special harness and cable arrangement (devised for project COLDFEET) designed for snatching a (living) body off the ice.

ARLIS II logged the northernmost point of its journey on 13 December: 88°39.7' N. Year-end coordinates implied two possible tracks the slab might take next: a reentry of the Beaufort Gyre, followed by a shift southward and westward (which would replicate the net track of T-3), or a shift to the east, and thus to extinction as part of the export into the far northern North Atlantic.

In 1964, T-3 and ARLIS II (in concert with a small, temporary camp) were used for long-range propagation experiments. ARLIS III had been established about 170 miles northeast of Barrow, on multiyear ice. A two-day search had turned up candidate ice, and on 10 February, a three-plane armada had set down on a floe at 73° N, 151° W. Following procedure, the Cessna pair had selected a refrozen lead then landed to drill the ice. When a foundation of three feet had been confirmed, a runway was marked with flags and the R4D called down—a first with wheels on unprepared ice. The light planes tarried (low air temperature: −48° F) as the transport hauled in further freight—a total of 32,000 pounds to complete the station.[44] Its "four tiny brown specs"—an advance over its GMIS predecessors—were intended for telluric-current and magnetic studies. The station staff comprised three scientists and Barrow staffer. ARLIS III was evacuated (16 May) by light aircraft exploiting runways of opportunity, that is, neighboring ice. During its useful life—101 days—the United States had boasted a trio of ice-based bivouacs.

Numerous factors influence morale on the ice. Weather is one, especially the change from constant daylight to continual dark. For ARLIS II and T-3 winterers, the dark of 1963–64 brought unique punishment.

> The year 1964 started with true arctic weather on both stations. Wind and cold hampered all operations, especially the continuing flights to supply fuel using the DC-3 aircraft. Storms, blowing snow and mechanical problems with the aircraft continued to plague the resupply in spite of tremendous efforts. At the end of January, the official temperature at ARLIS II was −55 degrees Fahrenheit and the temperature at T-3 dropped below the scale on all available thermometers. The temperature was estimated to be −72.5 degrees Fahrenheit. At the end of the month, the appearance of twilight to the personnel of T-3 was highly encouraging, though very dimly manifested on the southern edge of the horizon.[45]

This interval, then, was characterized by "dogged persistence well tried by wretched frustration." Novices "would find it hard to imagine how greatly the depressive combination of cold and darkness affects men's everyday lives. The simplest tasks become chores and difficult tasks become well-nigh impossible."[46]

—.—.—.—.—.—.—

Relief and resupply by Soviet airmen in 1964 were conducted by high latitude air expeditions. Led by V. M. Rogachev, *Sever-16* was in the field from March into May. That April, its aircraft relieved SP-12 (on the 8th) then retired SP-10 (on the 30th), the latter having logged 925 days at sea. At mid-month, moreover, a replacement for "10" was deployed. Station coordinates—73°50′ N, 166° W—put "13" northeast of Wrangel Island off eastern Siberia, a locus for long-haul drifts. On 1 May, under leader A. Buzuyev, *Severnyy Polyus-13* (1964–67) commenced operations.

> [*Sever-16*] was the biggest yet of these expeditions, employing turbo-prop and piston-engine aircraft and helicopters. An intermediate base was established in the northern part of the Laptev Sea to facilitate relief [evacuation] of distant NP-10. During the expedition 103 oceanographic stations were worked, and 20 automatic weather stations were set up on the drifting ice (bringing the total of such stations set up since their first use in the mid-1950's to about 400). . . . The scientific programme remained, and remains, substantially the same as before, with emphasis on meteorology, oceanography, terrestrial magnetism and sea ice studies.[47]

SP-13 would billet four station heads as it ambled to the "suburbs" about the pole, en route to destruction. It would log 1107 days of operations. That same April, the scientific station riding T-3 calculated its coordinates as 80°30′ N, 140°20′ W—a point close to the center of the Canada Basin.

—.—.—.—.—.—.—

At ARLIS II, outlanders had come. In April 1964, an R4D lost an engine prior to takeoff. This was the last flight out. One outcome was an unplanned layover for a Japanese Polar Group. Mere days later, the Staib North Pole Expedition (traveling on foot from Ellesmere) arrived. The camp roster ballooned—temporarily—to twenty-two plus thirty-one dogs. "By the time of ARLIS' third birthday [23 May] the population had returned to normal except for a few dogs left to become members of the station." Summer 1964 brought little melt. Runway construction commenced on 1 September at both U.S. drift stations. The fall season's first landing at T-3 stood logged on 17 September, the 19th at its sister camp.

ARLIS II was nearing Greenland. Oceanographically, the eastern Arctic Ocean is the outlet for ice and cold, low-salinity waters exported from the basin.[48] Located between Svalbard and Greenland's northeast corner, Fram Strait is a deep-water, high-energy connection between the Arctic and lower latitude oceans and commands exceptional scientific interest. Following Papanin's cold trail, ARLIS' research and support staff kept busy as they closed on the corridor. By late fall of 1964, ARLIS floated at 85° N. Long had ONR wanted an ice island to retrace the drift of SP-1 down the spine of the East Greenland Current. It also coveted the surface gyre off Arctic Canada-Alaska that propelled the T-3 platform.

> Now suddenly ONR was faced with the opportunity of conducting research programs from two ice stations each in one of these very favorable drift positions. The scientific activity on ARLIS II received the most emphasis; and currently active programs were intensified, other programs reactivated, and new ones added. Every possible effort was made to extract the greatest amount of information from the relatively unknown area that lay ahead of the station.[49]

From April on (the final segment), all flight ops had staged out of Iceland. "The plan," Schindler writes, "was to continue the station as long as possible to obtain as many scientific measurements as possible without compromising the safety of the station." Storms buried it in snow—"drifted right up to the roof level of the wanigans. Efforts to clear it were wasted as it would drift in again in a matter of 1 or 2 days. Winds with gusts up to 60 knots were often reported. Four loads of fuel were paradropped to the station from C-130s in January (1965), when the consistent winds kept the runway drifted."[50]

The guests on-station had aroused curiosity. Since to-and-from flight times were now short, interested media were accommodated—a public unveiling of sorts. On 1 April, seven journalists deplaned at ARLIS, among them the sole woman to visit the ice-island outpost.

ARLIS's position in the spring of 1964 mandated termination. At ONR, Max Britton's call to the Chief of Naval Operations (CNO) reached the desk of Cdr. Ron McGregor, Polar Operations Office—the CNO's representative for the Arctic Ocean. (Britton was a wizard at getting things, for example, icebreaker support, through the department. ARL worked for the navy; administratively, it answered to the university.) Under new orders, USS *Edisto* (AGB-2) stood out from Boston for Keflavik, Iceland. At Andrews Air Force Base (AFB), near Washington, D.C., Britton and McGregor shook hands for the first time. In Iceland, the pair met with Max Brewer and John Schindler,[51] with NATO officials and with the media. Off East Greenland, meanwhile, ice reconnaissance found ARLIS II beset by a halo of pressure ice and old floes. Wind-induced leads obliged, allowing Cdr. Norvel E. Nickerson, USN, commanding officer, USS *Edisto*, to steam to within seventeen miles (twenty-seven kilometers) of his objective by 6 May. Flown onto ARLIS via helicopter, Nickerson disembarked "amidst a very happy scientific party." The next three days had the icebreaker fighting all but impenetrable ice, logging mere miles each twenty-four-hour cycle. Having closed to a straight-line distance of just under seven miles, *Edisto* moored to a heavy-ice girdle.

A weasel train tugged about 3500 pounds (1575 kilograms) alongside. (Flight operations stood hampered by poor visibility and whiteout conditions.) In camp, the colors were lowered. The mood was ceremonial: "We had the privilege and honor, less than six months ago," Nickerson apprised the assembled, "of establishing the newest United States station in the Antarctic. And here we find ourselves at the other end of the earth at the evacuation of ARLIS II, an ice island that has been manned continuously since late May 1961." Then coordinates: 66°43.5′ N, 27°01′ W. "We all consider it a distinct honor," the officer continued, "to be here to evacuate the station, to welcome you aboard *Edisto*, and to take you back to civilization as fast as we possibly can."

ARLIS II had logged slightly more than forty-seven months as a science instrument (twelve days shy of four continuous years)—an astonishing run. The last man off, Carl Johnson, who had served as cook and station leader, had logged forty-one months on the ice. ARLIS' research harvest was huge. "We got a hell of a lot of good use out of that station," Max Britton recalled with satisfaction. "That was quite an experience, an innovative one. And it never cost us much money." The berg had rafted a total of fourteen separate projects. (See Appendix 4.) Altogether, 118 scientific personnel, 132 support staff and visitors, and 87 aircrewmen had come.[52] Lowell Thomas Jr., a two-time visitor and an eyewitness to the decampment, wrote in epitaph, "Arlis's buildings are at the bottom, but its great record is in the books."

ARLIS IV

On 25 February 1965, ARLIS IV was deployed 150 miles (240 kilometers) northeast of Barrow. Eleven hours' flying (five R4D sorties) conveyed nearly 30,000 pounds (13,500 kilograms) of cargo onto ice. A pair of light planes loitered two nights, for the erection phase. A temporary station (eighty-one days), its mission was to continue the acoustic work begun by ARLIS III. The bivouac itself consisted of a twelve-by-sixteen-foot mess hall, a radio and science hut, living quarters with a homer beacon, a main power plant, and a workroom, along with an auxiliary plant and storeroom. A small enclosure stored frozen food.[53]

The United States was again fielding three drift camps. The newest was shifting fastest, as much as twenty miles a day. So swift was the rate of drift for ARLIS IV that core samplers jabbed into the sea bottom reportedly were passed over and bent. The deploy positions requested for both III and IV offered unfavorable ice. Cracked and repeatedly broken, ARLIS IV was sliced to less than a quarter of a mile across. At the last, station materiel was sledged (over rough ice) to a new landing area about a mile off. By 16 May, the evacuation was complete. Thirteen sorties by light aircraft had shunted 8000 pounds (3600 kilograms) from the closed-out station back to Barrow.[54]

SP-13 — SP-17

In the spring of 1965, air expedition *Sever-17* took to the field. Deploying AN-2, LI-2, and IL-14 piston aircraft, an AN-12 turboprop and MI-4 helicopters, six hundred tons of cargo was shunted to seaward. A new shift, headed by V. F. Dubovtsev, took charge at *Severnyy Polyus-13* in early April. The station position was about six hundred kilometers northeast of Ostrov Novaya Sibir, the eastmost of the large islands of the New Siberian group. North of the Chukchi Sea, the next long-haul station was established, Yu. B. Konstantinov head. *Severnyy Polyus-14* (1965–66) would be shifted westward, its ice raft refusing to quit home waters. Northing less than three degrees, its ice field collided with Janetta Island and then Henrietta Island—lonely rocks inside the limit of multiyear ice. A disappointing

THIS PREFAB IS A TYPICAL HUT OF THE ARLIS (ARCTIC RESEARCH LABORATORY ICE STATION) SERIES — IN THIS INSTANCE, ARLIS V — A THREE-MONTH, LOW-NOISE CAMP FOR UNDERWATER ACOUSTIC STUDIES, FOR APPLICATION TO UNDERSEA WARFARE. THE GEOMETRICALLY COMPLEX, DYNAMIC PLATE (ICE) FLOATING ON A REFRACTING MEDIUM (WATER) MAKES ICE-ACOUSTIC INTERACTION A CENTRAL ISSUE FOR THE PROPAGATION OF ACOUSTIC ENERGY. AND THE ICE ITSELF IS AN ACTIVE ACOUSTIC SOURCE. (BEAUMONT M. BUCK)

284-day occupancy would end in February 1966, when its complement of twenty-five were retrieved. That first season (1965), the runway serving "14" served as a stepping-stone to SP-12. Before its abandonment near the end of April, "12" would glean a bit more data via an automatic weather station.

Sea ice is never a safe berth. In mid-January 1966, pack pressure fractured the "13" camp area, obliging a hurried shift to safer ice. (Observations were impossible for a week.) Late that spring, conveyed by *Sever-18*, a relief party was airlifted to "13" camp. Meanwhile, during December and January, SP-14 suffered violent cracking, obliging staff removal. Eight men gamely stayed, conducting a reduced program. But the winter night promised even further drubbing, and as prospects diminished, the station radioed for a full evacuation rather than await destruction—"and this was carried out between 7 and 12 February, with much difficulty and in almost total darkness."[55] Including its rescue, this fourteenth camp of Soviets had been sustained wholly by air.

The demise of SP-14 provoked a successor. Operation *Sever-18* was activated early, to install SP-15. Established at 78°50' N, 168°43' E (inside multiyear pack, near the East Siberian shelf break), *Severnyy Polyus-15* emphasized oceanographic work throughout its 1966–68 drift. The station leader for that first shift was V. V. Panov. (Ultimately, SP-15 would transect the basin, pushing closer to the pole than any previous station before its floe was accelerated toward Fram Strait, toward its demise.) And so the resupply expedition of 1966 departed SP-15 and the two-year-old SP-13, the latter enclosed by perennial ice off the New Siberian Islands. That fall, *Sever-19* would be activated, to resupply both.

On the North American flank of the Arctic Basin, the T-3 airstrip was declared closed on 14 June. Ablation and melting during the warm months of 1966 proved severe, causing facilities to pedestal four to six feet higher than the exposed, encircling white. Along a meandering path, the berg was slowly conveyed southward and east.[56]

Early in October that year—1966—cracking obliged SP-13 camp to shift to a new site thirty kilometers off. Fifty tons of stores were transferred by helicopter. Continuing the arc of its course "13" passed within 75 kilometers (47 miles) of the geographic pole late in January 1967, after which it promptly lost latitude. The final party took its leave under difficult ice conditions on 17 April at coordinates lat 87°52' N, long 2° W, north of the Greenland massif.

At SP-15, station staff were relieved by a party under the leadership of L. V. Bulatov. For whatever reason, "15" drifted on the next year—alone; no sister station was deployed. The fourth day of December found it lying just two kilometers off the geographic pole—closer than any prior SP station. *Sever-20* air-evacuated SP-15 (near the 86th parallel) on 22 March 1968, after 693 days of observation, information- and data-gathering. "Food was left at the site, with a note in English and Russian inviting callers to help themselves." *Severnyy Polyus-16* (1968–69) went operational on 9 April, less than three weeks after its predecessor had ceased to be. *Severnyy Polyus-17* (1968–69) commenced observations on 29 April. Staffs had encamped at 450 kilometers (280 miles) northeast of Wrangel and 430 kilometers northeast of Henrietta Island, respectively.[57]

CANADA AT THE POLE

Canada borders three oceans and boasts the longest coastline of any nation. In 1967, Ottawa dispatched an expedition to the geographic pole. The mobilization was fielded by the Polar Continental Shelf Project, part of Environment Canada, a federal government ministry based in Ottawa. "Polar Shelf" (as it is popularly called) is Canada's arctic logistics agency, supporting studies by various government departments and universities in all fields of science.

> The Canadians . . . qualify as one of the U.S. Navy's best and most competent scientific and operational allies. Their contributions, especially in the northern latitudes, in the polar regions, and in the Gulf Stream proved vital to oceanographic knowledge acquired for both scientific and national defense purposes.[58]

The prime objectives of the 1967 expedition (done in collaboration with several federal departments) were geophysical: to establish a line of gravity stations from Alert, on Ellesmere,[59] to the geographic pole and to obtain as many observations as possible near that sector, across the Lomonosov Ridge. As well, new techniques for navigation would be tested. An old Bristol freighter (Don Braun, pilot) provided transport. There was no fuel cached along its track, no backups, no checked-out airstrips. It logged Ninety North on 6 May, seven men on board. The Bristol thumped down hard, then lifted off—to circle round and note wet spots induced by impact. "When we landed," Neil Anderson said, "before we shut the engines off, we had to leap out and measure the depth of the ice. That was my job. So with the engines going full blast, ready to take off, I jumped out onto the ice with a one-inch auger and cranked it through."[60] The party labored commendably for eight days conducting experiments, many of which had to be curtailed. Still, "The knowledge gained about navigation in the polar area proved to be invaluable for planning the next trip," one researcher recalled. At that moment a mere supporting player, Ottawa had been outdistanced by Washington and Leningrad. But the 1967 venture was to seed the impressive Lomonosov Ridge Experiment (LOREX) investigations of 1979.

—•—•—•—•—•—

Elsewhere, the foundation of *Severnyy Polyus-15* yielded repeatedly to pressure; its floe was riven by cracks. "We spent all summer moving from place to place," a second-shift *polyarnik* recorded (1967).

That September (1967), a U.S. Coast Guard Cutter stood out from Seattle with orders to resupply T-3. *Northwind* found hard going. Pressing to 79°25.5′ N, 168° W—634 miles (1000 kilometers) off the pole, the farthest northing off Alaska-Canada of any U.S. surface vessel—she abandoned the effort because of extremely heavy ice and the advancing season. The cutter had steamed to within forty-two miles of its objective. Operating under a NARL contract with the University of Alaska,[61] the T-3 facility persisted in the service of science. At mid-year it floated about 600 miles (960 kilometers) northwest of Barrow.

Its floe nearing the exit, SP-15 was quit on 22 March 1968. A sixteenth station went operational in less than three weeks at 74°58′ N, 171°40′ W. And on 29 April, an air operation off the East Siberian coast established *Severnyy Polyus-17* well inside the permanent pack. Meanwhile, SP-16 continued to function; indeed, this bivouac was to persist longer than any before it in the series—approximately fifteen hundred days, or about four years. (Floes were changed twice.) Working the Canadian Arctic in 1971, SP-16 would attract overflights by maritime surveillance aircraft.

Washington's longest running ice asset, T-3, continued to host research teams from a variety of disciplines and agencies. In 1968, abandonment was six years in the future. In quiet darkness, representatives from universities in California, New York, and Washington State set about their icy work. There were meteorologists, seismologists, oceanographers, geologists, and biologists among their ranks. (Staffs at SP stations included two meteorologists, two or three oceanologists, four aerologists, three to four geophysicists, two electrical engineers, and one or two mechanics. A physician and a cook also were fielded. See for example Appendix 5.) Weather reports were radioed out daily; and letters and scientific reports describing results were dispatched. Among the subjects of study was how isolation affects individuals and the whole group.

Fred Roots was the first director at Polar Shelf. A superb field man, Roots understands well the pitfalls, perils, and frustrations that attend northern scientific endeavors.

> The tremendous expense involved in living, working and travelling in the Arctic, the uncertainty of all plans because of changes in weather and equipment breakdown, the physical difficulties in making measurements and taking samples in the severe northern environment, even in summer, all have a deleterious effect on research. . . . Researchers suffer psychologically from their total dependence on the decisions and services of those who manipulate their lifelines.[62]

FRAM—BACKGROUND

The continuity of support from ONR notwithstanding, funding reflected a "larger, central problem: the lack of a recognized and accepted national program."[63] At about mid-decade, the National Science Foundation had proposed a nuclear-powered version of the surface ship *Fram*—intended to facilitate ice-based investigation. Science does not operate in isolation from society. By 1968, this NSF project had stalled, hostage to the fiscal demands of Washington's war in Vietnam as well as to opposition to the proposed vessel.[64] (See Table 4.)

> Fifty years ago Nansen's *Fram* spent three years locked in the ice and now the NSF hopes to build a modern, nuclear-powered version of *Fram* as a permanent Arctic research station. . . . Right now though, the American project is at full stop; the gaping maw of the Vietnam war effort has seen to that. But even so, a great many US oceanographers will shed no tears if the money is never provided. The expense is not justified, they say: the ice islands are effective—and for free.[65]

Resenting the proposal as wasteful of limited navy-arctic funds, fleet-oriented researchers (quite unlike the academics) had worked to scuttle the proposal. Yet within a decade, segments of the polar-science community would return to the notion of an icebreaker-platform devoted to offshore investigation.

SP-18 AND SP-19

In April 1968, during a fly-over of the Chukotsk Sea attending SP-16, hydrologists chanced upon an ice island. Leningrad elected to exploit this gift—its second drift camp on shelf ice. The newfound fragment was huge, about eight by four and a half miles in area and ninety-eight feet thick. An ice-strengthened freighter in company, icebreaker *Leningrad* off-loaded nine hundred tons of stores to initiate and sustain an ice-based station. Dedicated to the one hundredth anniversary of Lenin's birth, SP-18 was installed on 9 September 1968. Its coordinates were 75°10′ N, 165°02′ W—an ice-infested nowhere well to the northeast of Wrangel Island. Although put in by ship, resupply and evacuation were to be entirely by air.

SP-18 (1968–71) floated active, monitoring, probing, compiling.

Sever-21 was the spring relief and resupply expedition of 1969 (called also *Sever-69*). Larger than usual, the operation was led by the highly regarded Gordienko. SP-18 was transferred in May onto floe ice, thus vacating its foundation to make way for a komsomol station. Reestablished near Jeanette island, "18" was to persevere until October 1971—a one-year home for four shifts. Supplied and ultimately evacuated at mid-pack (86°) by airlifts, "18" hosted extensive aqualung work and served as temporary base for at least one *Sever* expedition. Its

TABLE 4.

Rate of Annual Expenditure of U.S. Federal Funds for Arctic Activities, 1968.
Representing a broad spectrum, scientific programs were orientated toward the biological, atmospheric, and earth sciences. "These activities suffer from a lack of coordination, cooperation, and adequate funds, and can be considered a national effort in only the loosest sense."

Agriculture (USDA)	$1,000,000
Atomic Energy Commission (AEC)	201,135
Commerce	
Coast and Geodetic Survey	136,700
Weather Bureau	1,043,000
Institute for Telecommunications Sciences	497,900
Department of Defense (DOD)	
Army	5,000,000
Navy	2,076,000
Air Force	2,835,000
Advanced Research Projects Agency	538,334
Health, Education, and Welfare	
Public Health Service	1,113,000
National Institutes of Health, and others	1,975,567
Interior	
Office of Water Resources Research	185,974
Bureau of Commercial Fisheries	3,125,300
Bureau of Sport Fisheries and Wildlife	1,279,125
Bureau of Mines	492,000
Geological Survey (USGS)	3,768,200
Bureau of Reclamation	15,000
Federal Water Pollution Control Admin.	1,427,822
National Science Foundation (NSF)	2,720,032
Transportation	
Coast Guard	10,515,000
Bureau of Public Roads	15,000
Total Federal Expenditure	**$39,960,089**

Source: "Summary of Federal Arctic Research," reprinted in *Polar Record,* 14, (1969).

interim *Sever* duty saw the number of on-ice staff soar to more than a hundred individuals.

In October 1969, burdened with cargo, aircraft again set course north of the Soviet mainland for a jumpover airfield on the Indigirka River, which debouches into the East Siberian Sea. Conveyed by IL-4, more than seventy tons were deployed onto the ice island first spied two springs before. Cargo unloading operations proceeded smoothly, A. Chilingarov, the first station head (of three), noted with satisfaction. Ashore, the men worked twelve-hour shifts—two crews around the clock. Immediately, a small group (living in tents) commenced building an airstrip. Arriving over the place (74°34′ N, 161°48′ W) on 1 November. "I saw the island clearly in the dusk light. I could not help but see it. It was clearly distinguishable from the surrounding ice. In the middle of the white snow-covered desert was an enormous gray monolith, nearly rectangular in shape. Three of its faces were clearly visible and the fourth was too far away to be seen. Stream beds and hummocks were clearly visible."[66]

A weather report sped southward on 7 October. After speeches were delivered and a flag planted, and before dinner, a bottle of Armenian cognac was savored. The nineteenth manned drifting ice station in the *Severnyy Polyus* series (1969–73) was operational (though incomplete). Its mission included paying particular attention to abyssal currents in the region of the Lomonsov Ridge.

Organizing and construction were prime concerns in the setup work that consumed weeks more. Twelve-month homes, such stations proffer a bit of comfort.[67] "Chilingarov quoted A. Sarukhanyan writing of his first days aboard,

> They had already succeeded in erecting 10 buildings and, importantly, the mess hall from two of them. It was not yet in order and was half filled with boxes of canned vegetables, bags of potatoes, everything that had to be kept warm, but the tables were already arranged and we lived in rather comfortable conditions. . . . Gradually a street began to take form. On one side stood the diesel plant, machine shop and wardroom. On the other was the doctor's and cook's house, the chief's house and the hydrologists' house. At the end of the street was the radio and weather shacks and near them antennas and flagstaff bearing the state flag of the USSR. Off to the side . . . was the aerologists' house. . . . Stockpiles of fuel gas, food and equipment are scattered here and there behind the houses. We are not threatened by breakup of the island, but, as they say, God save us should it happen. Therefore the supplies are scattered, as is customary to do on drifting ice.

The physical demands of setting up are immense. Chilingarov concluded this grinding toil (the unloading especially), with the following dispatch.

> URGENT LENINGRAD KRUTSKIKH [the then director]
> CONSTRUCTION OF THE CAMP AND SCIENTIFIC FACILITIES COMPLETE.
> HYDROLOGY HOLE MADE IN EDGE ICE.
> 8-METER HOLE FOR OBSERVING TEMPERATURE OF ICE COVER BORED.
> FIRST RADIO SONDE LAUNCHED 1 JANUARY.
> CELEBRATING NEW YEAR IN GOOD SPIRIT . . .
> THANKS FOR SENDING NEW YEARS TREE. HOPE IT BRINGS GOOD LUCK
> TO ALL–CHILINGAROV[68]

A New Year's (1970) sortie to the nineteenth camp of Soviets delivered the hydrologists, who arrived "businesslike and with serious expressions on their faces, concerned about where the island is drifting." Their fretting was justified. Instead of deep water, the island had loitered near the DeLong Islands. In January dark, land was felt–strain energy transmitted from downstream through the wind-driven winter pack. Refrozen leads became ridges as young ice crushed and wrinkled. The raft for "19" grounded. "Cracks began to appear in the [disrupted island] ice," a *polyarnik* writes. "One passed just 15 meters from the radio shack, right next to the antenna. There was a roar in the air. The breaking up of the ice sounded like cannon fire."[69] Some equipment was lost. The campsite was relocated onto the main fragment and a new runway constructed. The drift soon increased; with one-fifth its volume broken free, the larger slab "sped up sharply. During spring and early summer we sailed like a flash to the northwest, leaving the islands behind." Science and survival were to last three years more as the drift arced fully across the basin. In the end, SP-19 and its staff would be moved ashore from a position north of Greenland in April 1973, the station having logged 1257 drift days.[70]

ARLIS V AND VI

In the south Beaufort, two pack-ice stations were set out in March 1970. Dedicated to (strictly classified) acoustics measurements, ARLIS V and ARLIS VI functioned for two months. With about 150 miles interposing, propagation work was conducted between the pair by exploiting sea ice and T-3 for the setting out of explosives. At the ARLIS pair, "special efforts" were made to prevent self-noise. The acoustic data so gathered represented "the only reasonably long-term and carefully controlled central Arctic ambient noise experiment on pack ice until 1975" and the Arctic Ice Dynamics Joint Experiment.[71] That May, ARLIS V and VI were evacuated by a trio of light aircraft, operating out of Barrow.

—•—•—•—•—•—

Canada, for its part, returned to the Central Basin in 1969, landing a Twin Otter at the pole on 2 April. The party loitered four weeks (until 3 May) exploiting grand weather and starlight—the sun was still low. Methods of working the ice were maturing. Satellite-related technology attended in the form of a Magnavox transit satellite. (Its raw data recorded onto punched paper, for processing back in Ottawa.) The Twin was used for spot gravity readings and for soundings, navigating with an experimental Omega receiver handling a very low-frequency signal. "Despite a four-day storm, which blew the [Canadian] ice station 40 kilometers [25 miles] to the southeast, the expedition was a success."[72]

AIDJEX—BACKGROUND

Interactions among atmosphere, ocean, and ice factor into a number of problems, ranging from ice forecasting and climate prediction to oil production and transport in high latitudes. If the Arctic Ocean were to become ice-free, researchers realized, the climate of much of the Northern Hemisphere might alter.[73] Why? The Arctic is one of two primary "sinks" for solar energy (the other is Antarctica). Moving, breaking, shifting sea ice thermally insulates the ocean. Heat flux through a changing cover might affect the surface heat budget and, hence, the global climate. The United States and the USSR, the *New York Times* reported in 1970, would initiate an "intensive investigation" to determine whether the Arctic climate was altering. Washington's plan, the Arctic Ice Dynamics Joint Experiment, or AIDJEX, had as its purpose gleaning *synoptic* data instead of single-station observations.

Databanks from participating nations enhance the body of knowledge. And the work is ever unfinished. Although certain special meetings, symposia, and scientific journals (e.g., "Underwater Acoustics") were restricted, the results of successive Soviet and American programs (happily) were proving complementary. Take basin morphology:

> The varied drifts of the Soviet drifting stations have enabled their scientists to conduct research in many sections of the Arctic Ocean that American stations have not visited, but both groups of scientists have made significant and related discoveries concerning the geomorphology of the ocean floor. One of the most important early findings was that the ocean floor was not a simply flat basin as had previously been thought. The findings of both groups of scientists have complimented each other and made this interpretation clearer.[74]

A coordinated program of multiple field experiments, AIDJEX sought to improve understanding of air-ice-water interaction by direct measurement of wind stress, water stress, and ice deformation. Led by Seattle's Applied Physics Laboratory (APL) at the University of Washington, the project involved measurements from *arrays* of manned camps *and* data buoys. With its robust application of speed-of-light technology, AIDJEX was the most ambitious international study yet of Arctic ice.[75]

Two laboratories contributed. The university-affiliated APL was one, its basic research carried out by the Polar Science Center. Its founding year was 1971. AIDJEX was the lab's first on-ice experience. APL's APLIS series of ice camps (Appendix 3) has since supported high-frequency acoustic research into the 1990s, its products (classified as well as public-release) distributed throughout the interested naval community. On board *Northwind* that long-ago summer of 1971, Francois and Buck were conveyed into the Chukchi, in which two research camps were installed. This experiment was MIZPAC–Marginal Ice Zone-Pacific. Planned, directed, and funded by the Arctic Submarine Laboratory, MIZPAC operations sought to understand the performance of high-frequency acoustic systems—sonars and sonobuoys, for example (or, if applied to weapons systems, mines and torpedoes). MIZPAC 71 was the first summer drift camp intentionally set up to investigate arctic water-column properties.[76]

The Polar Research Laboratory represented the civilian side. Formed in 1973, it conceived, installed, and operated all the "data buoys" supporting AIDJEX.[77] PRL would become the primary facility for almost all of the Free World's unmanned drifting stations. Why use autonomous instrumentation for real-time data? On moving ice, the dark months bring danger. Field operations are brutal, manned stations often inaccessible. Mounting a suite of sensors, unattended buoys can acquire, monitor, navigate, and relay, via satellite telemetry, environmental data sets in all seasons, something U.S. floe-based camps seldom did—unlike those of the Soviets. (Still, some observations cannot be made by automatic devices.) Air-deployed from springtime camps, fully automatic drifters have since compiled buoy-years of synoptic data over much of the Northern Ocean. AIDJEX, indeed, seeded the U.S. Interagency Arctic Ocean Buoy Program.

Waldo Lyon had encouraged the creation of PRL—one of three laboratories organized to extract the utmost from navy-arctic funds. ARL represented the government element of this troika, APL its university side. "He split the responsibilities between PRL and APL in a simple, unambiguous way: by sonic frequency. [Lyon] gave APL everything in excess of 10 kHz, and PRL everything below that frequency. So, [PRL] had the long-range surveillance problem, and [APL] had the under-ice torpedoes."[78]

T-3 INCIDENT

Later that July (1970), a macabre story hit the headlines: polar murder. U.S. Navy investigators, the media reported (erroneously), were being airlifted onto T-3 because the camp's civilian scientific leader had been fatally shot.

The last plane had lifted clear on 8 June, after which the island's runway was closed. Save for radio communications and airdrops of supplies and mail, T-3 floated free of the rest of the world. Its government-industry party of nineteen could expect no retrieval until late September or October; accordingly, any serious illness or accident meant genuine emergency. In remote field camps, the personalities of one's comrades can make for good fellowship—or nightmares. A written opinion for the Fourth Court of Appeals outlines certain of the hazards attending that unfortunate T-3 summer.

> Discipline and order on the island depend on the cooperation of all the men and the effectiveness of the group leader, particularly in the summer months when it is virtually impossible to remove any wrongdoer from the ice. There are no medical facilities on the island, no doctor and, indeed, during the summer of 1970, no person trained in any aspect of medical science.[79]

Bernie P. Lightsy, an employee of the U.S. Weather Bureau, had been designated station manager. On 16 July, the ownership of (and access to) home-brewed raisin wine became the subject of an increasingly heated discussion. Somehow a rifle was discharged, wounding Lightsy in the chest, near the heart. Weilding the gun was Mario Escamilla, an employee of the General Motors Defense Research Laboratories.[80]

None of the individuals directly involved were military personnel or even employees of the navy. Yet the camp was an ONR (University of Alaska contract) operation. Word was flashed to Washington. "I knew it was going to be a long day," the chief of naval research, Adm. C. O. Holmquist, recalls.

Arctic waters—and ice—are largely international. A ship is a particularly rule-bound precinct. Was T-3 a vessel and the reported killing a high-seas crime? The question of who had rights to U.S.-occupied ice inside the Canadian sector added further tangles. Most ice islands are large enough to permit semipermanent or permanent occupation, which raises questions of jurisdiction and control. The legal aspects of exploiting ice islands for at-sea research was (and remains) unclear in international law. So Ottawa could not simply wash its hands of this unpleasant T-3 business.

> The question of jurisdiction became apparent immediately. The laws of the sea are clear for jurisdiction over ships, boats, rafts, and debris, but are undefined for ice islands. If an ice island has the same status as a ship, clearly the United States had jurisdiction. From discussions with personnel from the

Office of the Navy Judge Advocate General, it appeared as though T-3 could be considered within the "special maritime and territorial jurisdiction of the United States. . . ."

In a situation such as this, it is considered wise to avoid possible disputes with foreign countries over jurisdiction. The U.S. government was uncertain of the potential legal implications of landing with the suspect on Canadian soil prior to returning to U.S. territory. Consequently, it was highly desirable that the suspect not touch down anywhere but in U.S. territory on the way back to the States.[81]

Aroused, the news media—bits of the story on the wind—began publishing sensational (but true) pieces on the "mystery killing" aboard a remote ice island. The exotic locale and the circumstances surrounding the slaying, as well as abstruse questions of international law, conferred an intense interest in the incident.

On T-3, the suspect could not escape the outpost. Nor could his colleagues escape him. (Locked doors were unknown because of the hazard of fire.) Notwithstanding, the need for instant action was plain. "We knew nothing about the suspect's state of mind," Admiral Holmquist observes, "and little about the feelings of other personnel on T-3. We could only speculate on the possibility of further trouble. Should the suspect be locked up or be allowed to continue his work routine until September? What would be the effect on other people on T-3 of having him free to roam about the camp?"[82]

Prompt fixed-wing retrieval of Escamilla was impracticable. And no helicopter possessed sufficient range to reach T-3 from either Alaska or Greenland and avoid touchdown for refueling on Canadian soil. Discussions with the air force helped bring about a solution: the H-3 type helicopter, air-refueled by a C-130 tanker, could gain the island. The defendant was retrieved, the incoming H-3 replenished over the Arctic Ocean—a unique operation. Arriving at Thule AFB, Escamilla was transferred in custody to a conventional aircraft. Two weeks after the slaying, the accused was deposited at Dulles Airport, in the Eastern District of Virginia—his first touchdown on U.S. territory.

The case was pressed to trial. After hearing arguments, the judge announced a tentative decision to take jurisdiction. (At no time during or at the conclusion of the subsequent trial did he expound on the grounds upon which he based this decision.) The U.S. attorney argued that T-3 was "high seas." And though it could not be navigated, the court agreed, and the island was deemed a frozen form of the high seas ("like a ship"). Quite sensibly, the question arose as to what authority, if any, a coastal state might exercise over the activities on such work platforms when they penetrated the territorial waters of adjacent states. The incident had occurred within the sector claimed by Ottawa; on 16 July T-3 had floated 213 miles (343 kilometers) off the Canadian Archipelago. (Before the decade was out, *Severnyy Polyus-22* was to raise again the issue of jurisdiction.)[83] The U.S. government had not hesitated to exercise criminal jurisdiction over the accused. Neither would the courts. If the United States did not assert criminal jurisdiction, hence enforcement, who would?

A preliminary hearing was held that August; the U.S. attorney's office presented an indictment to the grand jury charging second-degree murder. Nine months later, on 5 May 1971, the case was argued before the U.S. District Court of the Eastern District of Virginia. A jury convicted Escamilla of involuntary manslaughter. But the verdict was appealed, and a retrial, held in November 1972, resulted in a "not guilty" verdict.

To its relief, the court had found it unnecessary to deal specifically with the question of jurisdiction in international law. By formal diplomatic note, the Government of Canada had chosen to waive any objections to the U.S. exercise of jurisdiction in the case (though it reserved Canada's position in other respects). Donat Pharand is a professor of international law at the University of Ottawa. "Such a waiver," he writes, "was quite proper for at least two reasons. First, Canada could not have justified any territorial jurisdiction on the basis of the sector theory, and second, both the victim and the accused were of American nationality and the United States was quite justified in exercising personal jurisdiction regardless of whether the offense took place outside its territory."[84]

"This incident and its aftermath," Holmquist concludes, "point up some of the frustrating problems involved in conducting research in the Arctic Ocean, where communications are intermittent and transportation ranges from the difficult to the impossible."[85]

—•—•—•—•—•—

The long-serving T-3 station was nearing closeout. The berg had become "stuck" near Cape Stallworthy, the northernmost tip of Alex Heiberg Island. Still, scientific research persisted. In May 1972, APL arrived to deploy its unmanned Arctic research submersible for full-scale testing. (The slab floated about 150 miles off Ellesmere.) A mobile instrument carrier, UARS was a shallow-running low speed torpedo-like vehicle. The battery-powered, nine-hundred-pound submersible was launched by means of a hole through floe ice (hydrohole). Deploying an upward-looking, multi–narrow beam sonar, UARS imaged the under-ice topography along its trajectory as it recorded temperature and salinity–a mobile instrument carrier. Atop the canopy, hydrophones arrayed in a one-thousand-foot square arrangement tracked the device. All systems performed as designed.[86]

Lt. Cdr. Brian Shoemaker, USN, in command of NARL in 1974, was the man responsible for the operation at T-3.

It had not moved for three years [he writes] and consequently we were getting no useful scientific information. We needed generators and other equipment from T-3 to use for our ice stations that were part of AIDJEX. We left the huts behind and the camp pretty well intact with the intention of reoccupying T-3 if it got moving again sometime in the future. VXE-6 flew to T-3 with a ski-equipped C-130 and pulled out the gear that we needed. There were no ceremonies except lowering the flag.[87]

The pullout took place in April 1974.

AIDJEX PILOT STUDIES

Concern for the world environment was giving a new shape to thinking about Earth: the planet was seen as an integrated system of atmosphere, ocean, cryosphere (ice and snow), land surface, solid earth, and biosphere. To attack such complexities and thereby improve the basic understanding of how the climate system works, multidiscipline, multiagency, multinational initiatives would be key. AIDJEX, an exemplar, escalated ambition. The experiment itself had been conceived in 1969, when ONR signed a contract by which Dr. Kenneth L. Hunkins, of Columbia University's Lamont Geological Observatory, and Dr. Norbert Untersteiner, University of Washington, would design a multistation field research program in the Arctic Ocean. (The head of the AIDJEX science committee was NSF's Joe Fletcher.) Translating the final scientific plan into a full-scale field event took five years, during which field and laboratory efforts worked to develop special equipment and refine the applicable mathematics. The objective was to build a realistic dynamic ice model incorporating all the physics of sea ice—that is, to understand the interior ice and the forces that control its movement. Success, in turn, would help decode the air-ice-ocean system. As yet, global climate models tended to use elementary physics to represent sea ice. One outcome was poor representation of boreal processes in long-term climate predictions.[88] A pure research problem, AIDJEX tendered a quantum leap in scope, duration, complexity, data storage, and retrieval.

Its observing systems represented key innovations by AIDJEX planners. Indeed, the campaign would eliminate the need for multiple, long-term, manned ice stations by introducing a different logistical approach, that is, the use of remote-sensing devices. The project's observational requirements had inspired the rapid development of data buoys. AIDJEX and its corollary field programs would exploit the random access measuring system (RAMS) carried on *Nimbus*-F. A multiagency effort was under way to exploit satellite-borne remote-sensing methods for use in sea-ice monitoring and research.[89]

The field design called for a minimum of four camps about 60 miles (100 kilometers) apart, surrounded by a rosette of at least eight air-deployed buoys "reporting" to a central data-acquisition system. Laying out instruments over a roughly circular area 500 miles (800 kilometers) in diameter, researchers hoped to amass data sufficient to create a viable model.[90] The project's budget (despite cuts) was shouldered by several agencies. The research leading to AIDJEX had been underwritten by the Arctic Program of ONR (through NARL at Barrow), and, after 1969, by the National Science Foundation. The major participants of the main field effort–during 1975–76–would be the Lamont-Doherty Geological Observatory of Columbia University, the Cold Regions Research and Engineering Laboratory, the U.S. Geological Survey, the National Aeronautics and Space Administration, the National Data Buoy Office, and the University of Alaska. "Probably the only part of the U.S. government that didn't participate," McGregor smiles, "was the Civil Service Commission."

Ottawa was a major player. "We [at Polar Shelf] supplied and paid for the Twin Otter aircraft that supported that operation and there were at least three Canadian scientific parties as participants in that fourteen-month operation. Canadians were very much a part of the planning and execution of AIDJEX."[91] "We took advantage of the Continental Shelf program," McGregor agrees. "Hobson provided a *tremendous* amount of logistics support." Final full complement comprised about eighty Americans and Canadians and six Japanese.[92] "It was exciting, it was different, and there were a lot of young hard-chargers around who wanted to test their manhood," McGregor adds. "It turned out to be a significant change in how we did things." A spirit of cooperation is evident in the record. Canadian contributions notwithstanding, private sentiment in Ottawa was less idyllic than the public transcript:

> At present our relations with the United States in environmental and economic operations are delicate, and U.S. involvement in northern development displeases some Canadians. I can only say that in northern scientific affairs we cannot do without it, and the both parties need to learn how to collaborate more effectively and less emotionally.... A good example is the jointly sponsored and conceived AIDJEX project out in the Beaufort Sea.... The United States could have mounted this alone, and with some effort so could we. But by pooling resources we have much increased the potential usefulness of the project. In all northern scientific work the worst shortage is in human skills. By pooling these we make AIDJEX a stronger implement than either party could have achieved alone. And both of us will profit technologically by what we have learned.[93]

In preparation for the main multistation experiment, a series of pilot studies were conducted—one each spring during 1970–74—to resolve scientific and technical issues. The third field study, conducted in 1974, would be the most complex and sophisticated experiment yet. Meantime, "A base camp is being established on a lonely floe," the *New York Times* declared, "as prelude to the most ambitious effort yet undertaken by the United States and its allies to understand what forces control ice movement in the Arctic Ocean." Logistically, the center of the manned array had to be near enough to shore to be manageable yet far enough to seaward to be largely free of coastal effects. NARL had chartered an AIA C-130 for this phase. Spring 1972: Barrow logged a heady, frenetic pace as AIDJEX again deployed.

> POINT BARROW BASE CAMP—The big C130 emerges from the thin cloud layer, roars down onto the runway and then taxies quickly to the parking area. While it is still rolling, the pilot begins to lower the tail ramp, and a ground crewman with upraised arms directs him to the loading zone. Before the four turbine engines are shut down, an assortment of vehicles converges on the Hercules. Over the next hour, while the flight crew eats dinner in a nearby messhall, fuel in drums, fuel in rubber cylinders, and other materiel needed to keep alive a small community of some 55 persons 300 miles [480 kilometers] out on the Arctic Ocean will be put aboard the aircraft. Then pilot Bob Murphy will fly it to AIDJEX Ice Station and return for another load before finally heading back to Fairbanks. It will be 10:30 p.m. before parka-clad ground crewmen, most of them Eskimos, are through for the night. While loading the Hercules, they will be exposed to below-zero temperatures and a steady wind off the ice pack. The work will proceed swiftly. For weeks this activity has been going on here, in support of the largest research effort ever undertaken on Arctic Ocean ice by the United States. Canadian and Japanese scientists are also participating in the effort, designated AIDJEX (for Arctic Ice Dynamics Joint Experiment).[94]

That March, the C-130, or "Herc," made the first-ever landing on essentially unprepared sea ice—a frozen-over lead. Barrow's R4D logged thirty-three flights to the main floe during a two-month study (late February to late April 1972). Leader of the team from Lamont-Doherty: Dr. Ken Hunkins, a geophysicist. "We hope to develop a dynamic ice model which can be used in forecasting ice conditions," he told the *Times*. "We expect that this model will be an important step towards understanding—and eventually towards predicting—long-term climate problems."

Nascent camp for AIDJEX—Arctic Ice Dynamics Joint Experiment. Here is its first C-130, on the ice off North Alaska. The C-130 was first used in Antarctica in 1960, replacing the C-124 Globemaster. Following a series of pilot studies, the main field experiment was launched in March 1975. AIDJEX was the first to fully exploit earth-orbiting satellite platforms, thus ushering Arctic science into the space age. Drift programs for the IGY laid the scientific and logistic foundation for AIDJEX. (John F. Schindler)

The costs of deployment then sustaining polar outposts make on-ice research stations prime users of remotely sensed data. Focused yet wide-ranging, AIDJEX was emblematic of change: direct environmental observations coupled to remote sensors, exploiting advances in sensing and related emerging technologies. Ice drift would be monitored with satellite navigation systems, predecessors to the now-prevalent GPS (geographic positioning system). As well, autonomous stations (data buoys) would augment the manned stations. The AIDJEX 1972 pilot study saw six buoys arrayed about the triangle of manned camps. Developed by University of Washington's APL under contract with the National Data Buoy Office of the National Oceanic and Atmospheric Administration, each was equipped with air-pressure and temperature sensors at two levels above the sea-ice surface. Exploiting advanced technology, the principal buoy for the 1975–76 main experiment would hold radio data-link to a central station at one of the on-ice camps. Indeed, autonomous drifters have proven to be an answer over the wide central Arctic and its marginal zones. Instrumented, self-navigating on-ice buoys remotely measure, monitor, and transmit a suite of environmental parameters.

In all fields of science, the digital flow of information has become staggering. In AIDJEX time, at the University of Washington, a databank was established for information handling and retrieval. Processed and archived into formats useful to researchers, AIDJEX results stood computer-accessible. Historically, in contrast,

> the most important equipment needed to do good science in the arctic was a keen eye, a strong back and legs, and perhaps a perverse ability to enjoy physical discomfort in glorious natural surroundings. These are still very important assets, but they are no longer sufficient to ensure a scientific contribution.... To be in the forefront of arctic science today, a country has to have the ability and the will to develop and maintain the expertise and facilities that will enable it to take part in world science.[95]

In sum, AIDJEX represented a striking departure from previous scientific methods. Exploiting a network of both rude and manned camps, modern technology had been integrated deep into the experiment: unmanned submersibles, computers, lasers, navigation satellites, even satellite telephones—a multilayered assault. AIDJEX had ushered arctic science into the space age.[96] Not surprisingly, succeeding polar studies have relied heavily on remotely sensed data and attendant technologies.

This same April 1972, Yevgeny Fedorov was flown out to SP-19. Its boreal foundation—an ice island—was novel to this comrade of Shmidt and Ivan Papanin. "Only once, in 1954, had I flown above such an island examining the remains of an abandoned US drifting station," the explorer writes. His comments shed light on the instrumentation, equipment, and comfort factors at "19."

> In general, living conditions had improved considerably not only as compared with North Pole 1, but also compared with North Pole 3 and 4 which I had visited. The standard houses were quickly bolted together out of panels. Each had a small entrance compartment and over 10-square-metre-room. There were bunks, a table and a stove. Formerly gas stoves had been used, but now they were switching over to liquid fuel, petroleum. Electricity and phones were everywhere. There were two or three diesels in the power plant. One was running, the other in reserve. The scientific instruments were much more modern, of course, than at North Pole 1, but had not changed much in the 20 years since I visited North Pole 3 and 4.... Fine motor-powered winches made it possible to quickly lower and lift bathometers, current meters and other instruments into the ocean, but the instruments themselves had not changed significantly. The meteorological devices were also old, of the "classical" type.... The magnetic observatory was very good. In effect, it was a big crate with an effective thermal insulation and a thermostat maintaining a

constant temperature. There were very small magnetic variometers inside: they registered the magnetic field variations on slowly moving film.[97]

SP-20–SP-22

As dry runs prepared for the main AIDJEX event and SP-19 dutifully recorded observations, *Severnyy Polyus-20* was installed–11 April 1970–on old ice at a position 640 kilometers (400 miles) north of Wrangel Island. Its multiyear floe was to sustain a two-year colonization, under station heads Y. P. Tikhonov and N. Shevchenko. Its work completed, abandonment was logged on 10 May 1972– 750 days in combat with nature. And so "20" floated concurrent with "19" camp in 1970–72. Mere days before SP-20 decamped (on 1 May 1972), SP-21 was airdeployed at coordinates 74°42′ N, 175°58′ E. Supplied entirely by aircraft (like its predecessor), the station persevered for two years, also–approximately 760 days, from April 1972 to May 1974. When its ice became unreliable, the second shift was airlifted off, investigations done.

The drift of *Severnyy Polyus-22* was to prove unusual in two features: an autumn deployment by surface ship and longevity (1973–82). Altogether, it would float *nine* shifts of personnel and base two *Sever* air expeditions on its ice-island foundation. Hostage to the Beaufort Gyre early on (March 1976 to May 1978), "22" camp would shift westward across the gyre before arcing toward the geographic pole astride the transpolar current. SP-22 would transect the Arctic Basin. The camp leaders were V. Morozov, N. V. Makurin, Nikolai D. Vinogradov, and I. M. Simonov.[98] Hugely productive, scientifically, the bivouac would incite a diplomatic dustup when, in 1977, after no formal request, it penetrated Canada's two hundred-mile zone–Canadian territorial waters. (See Preface.)

AIDJEX-MAIN EXPERIMENT

Off Arctic Alaska, staff representing three universities mounted the AIDJEX lead experiment north of Point Barrow, recording data on the modification of the air and ocean boundary layers near an open lead. "The rapidly shifting ice, the quickly freezing water during the arctic spring, and the sophistication of the instruments used in the experiment required painstaking preparations, high-precision logistics, and luck." The main field experiment began in March 1975, under promising conditions. This drift year, "some 200 scientists, technicians, and support personnel from nearly a dozen university and government organizations will man 50 buildings and tents at four camps and monitor instruments at ten additional stations."[99] But poor weather, along with a paucity of good sites, delayed deployment until the 12th. Sixteen missions totaling 125 hours of search by Twin Otter and R4D were logged prior to set-down northeast of Point Barrow at about 76° N, 145° W–250 miles (400 kilometers) to seaward. The weeks thereafter saw about nine hundred metric tons of supplies shunted out, primarily via L-100 aircraft under contract to NARL. Three satellite camps were set about 37 miles (60 kilometers) from the main camp; by the end of the first week of April the trio was operational.[100] June saw four camps installed, an eight-buoy rosette surrounding the quartet. With some variation, this constellation functioned until early in May 1976, when the experiment ended. From aloft, the main camp resembled "a small orange speck amidst the stark expanse of ice that stretches as far as the eye can see." On ice, AIDJEX camp "looks like a cluster of orange box-like structures connected with wires. The runway [checked daily for cracks] is marked by black oil drums, and caches of fuel drums can be seen nearby. It is difficult to imagine a more barren place to conduct research."[101]

AIDJEX (as noted) was satellite-assisted. Launched in mid-1975, Nimbus 6 made its first flight. The random access measurement system was designed to relay data and position information automatically from relatively inexpensive surface drifters (buoys). Late April witnessed initiation of data retrieval through SATNAV–the Navy Navigation Satellite System. Positioned at eight locations, the system's stations were set in a 250-mile (400-kilometer) radius about the main (Beaufort) camp. The first of July saw buoys deployed. Ice dynamics are the most important influence on low-frequency ambient noise. Instincts aroused, PRL felt it "logical and highly cost effective" to conduct an ambient-noise project in concert with AIDJEX, thus exploiting "the logistic support and later ancillary oceanographic and meteorological data that would be taken as part of that experiment for supportive correlations with acoustic data." Ten buoys (called SYNRAMS) had been developed, eight of which were deployed over the sea sector 72° to 79° north and 130° to 170° west. Recording noise (via chain-suspended hydrophone), barometric pressure, and air-temperature data, the drifters operated for the full AIDJEX year 1975–76 "with few failures."

Incorporating AIDJEX data, "synoptic measurements of ambient noise made during all seasons provide[d] one of the most extensive ambient noise data banks ever measured in any ocean." Later, predictive models of Arctic ambient noise were derived.[102]

Pack ice is an ensemble of multiple floes, not a solid plate. Internal stresses within this jumble are large during winter and spring–hence the potential to erupt.[103] For those riding the ice, losses to pressure are inevitable. Within months, the main "Big Bear" camp was threatened. On 1 October 1975, a crack was discovered to be running under the main mess hall. Within an hour, the breach had widened into a lead. Electric lines were hurriedly disconnected and plans made to reestablish power, so as to keep radio and scientific gear operating. Additional leads developed at the runway and around the now-compromised bivouac. The science buildings were relocated away from the rupture onto the larger floe, the

Fracturing threatens the main station floe, AIDJEX, 1975. On 1 October, a crack split the "Big Bear" camp, widening into a lead within one hour. A series of ensuing cracks led to abandonment. The satellite ice camps continued their regular observations and data acquisitions activities. (Dr. Kenneth L. Hunkins)

mess hall itself pulled clear. Cracking did not cause immediate destruction because the leads promptly refroze. The main camp would now ride it out. Still, matters were calamitous enough to warrant a (reluctant) order to evacuate. Ten of two dozen residents were airlifted to "Snowbird," a satellite camp nearly 100 miles (160 kilometers) off. Staff began dismantling equipment. Later that day, ten more personnel were retrieved. But four experienced men were left aboard, to keep the base running and to observe ice conditions; distressed ice or no, the core AIDJEX program continued to its predetermined termination.

Boreal summer, many assumed, would be uncongenial to rotary-wing aircraft. AIDJEX proved otherwise. The Bell 204 and 205 were deployed throughout the field year, hauling external and internal cargoes as well as granting close support to research parties. (One helicopter was stationed at Big Bear.) In April, satellite telephones had been installed at the main camp and at the AIDJEX office in faraway Seattle. As Untersteiner reported, "The high degree of reliability of the system and the routine daily contacts for one hour proved to be most beneficial."

Continuing, "We believe that this large and costly effort produced a quantum jump in our knowledge and understanding not only of ice dynamics but also of many other aspects" of the polar environment. And in terms of logistics, the AIDJEX main experiment was, he said (not without satisfaction), on "an order of magnitude more ambitious than any previous experiment in the Arctic."[104]

Among its achievements, the AIDJEX program developed the most detailed predictive model to that date for ice—its motions, stress states, distribution of thickness—by relating strains to the distribution of thickness. As it happens, thickness is important not only to computation of stress-strain behavior but, as well, for determination of heat transfer between sea and atmosphere.

The period known as the Cold War devoured forty-five years, during which the push-pull of crisis defined the singular danger of nuclear confrontation. Security dominated everything. The mid-seventies found the Kremlin wary of Pentagon plans in the (overstated) "Arctic operational zone." In the world according to the central Soviet government

> The military leadership of the U.S. assigns great strategic importance to the Arctic. In accordance with the development by the Pentagon of a "polar strategy," U.S. Arctic research work mainly serves military purposes. The influence of climatic and hydrographic conditions on the operation of forces and on the possibility of using weapons and military equipment in the Arctic is being studied. . . . The main efforts of the USSR in the Arctic are directed toward continued research and economic utilization for peaceful purposes.[105]

Severnyy Polyus-23 (1975–78) had been shunted to sea on 5 December onto an ice island floating north of Wrangel Island. This sector of sea is a focal area for extended drift expeditions; most camps so deployed are conveyed poleward.

FRAM SERIES

Research is constrained by capability. Drift ice poses hefty limitations: high cost, seasonal ablation and melt, cracking, the hazards of airlift resupply, and a quarter-century ago, lean computer backup. Unlike the Amerasia Basin, the Eurasia sector of the Arctic Ocean was little known, due in part to its relative inaccessibility to U.S. aircraft. Moreover, ice dynamics implied a canopy ill-suited for manned stations; survival prospects, some said, were highly dubious.[106] Still, exploration of the western central Arctic shifted to the much more dynamic (and interesting) eastern Arctic and Fram Strait area—a major refocus. The strait is the passage through which the East Greenland Current exports cold water and multiyear ice—the main gateway between the deep-ocean Arctic and high-latitude North Atlantic, where most water, salt, and heat exchange take place. The Transpolar Drift Current debouches there, interlayering Atlantic water with that of the Arctic Ocean.

Another feature beckoned investigators: the Arctic Midoceanic Ridge (also called the Nansen or Gakkel Ridge). A transarctic segment of the world-girdling midoceanic ridge system, it extends northward between Svalbard and Greenland into the polar basin. A number of geophysical, oceanographic, and climatic questions related to that sector. Nansen had been first to conduct oceanographic work in the Eurasian Basin region, after which its waters were left unstudied.[107] "We already had the basic idea about the Arctic–the big features were known. And now we needed some detailed studies."[108]

When the notion of a ship-based station was reintroduced, the National Academy of Sciences was invited to prepare a blueprint for working the Transpolar Drift Stream–pressing as far upstream as possible, then letting a ship drift itself out. The chairman of the Fridtjof Nansen Drift Station (NDS) Committee was Norbert Untersteiner. Drawing on experts, the Polar Research Board produced a plan. "The desirability of repeating the historic journeys of the *Fram* (1893–1896) and *Sedov* (1937–1939) across the Eurasian Basin of the Arctic Ocean," its report opens, "has been discussed in the scientific community on various occasions over the past 20 years." A platform lay to hand. "The scheduled decommissioning of the WIND-class icebreakers by the U.S. Coast Guard offers, for the first time, a definite opportunity to acquire a suitable [frozen-in] platform to carry out a wide-ranging research program in a little known area of the Arctic." Reconfigured, the World War II vintage *Burton Island* might float a wide-ranging program[109] augmented by surface traverses and auxiliary ice-rafted camps.

Certain logistical hurdles had seeded NDS. The eastern Arctic lies beyond easy range of the U.S. logistics support base at Barrow; in consequence, a pack-ice camp would float unacceptably remote from potential rescue. Further, an icebreaker is immune to breakup. And serious geological oceanography had been proposed. The sedimentary record of climate is provided in part by cores. But no one had yet recovered *long* cores, that is, cores deep enough to penetrate to layers deposited before the last glacial cycle. (The Soviets, for their part, had taken hundreds of Arctic Ocean cores by 1976.) "You wanted to do really big coring," Hunkins explains, "good long cores, and all that kind of thing is very hard to set up on an ice station. Taking long cores is not easy."[110] As well, a ship proffered a below-decks lab and processing facilities unthinkable for the largest shelter. Computers had become indispensable, granting far more detailed, complex, and voluminous calculations. Further, a steel hull would largely eliminate the costs, inefficiencies, and hazards attending an air-dependent, ice-based station.

The bill for NDS amounted to $14,365,000 for science and another $17,672,000 for logistics and operations, including reconfiguration of the vessel. Compared to present-day costs, these figures may seem quaint, but such things are relative. The dollars would have to come from programs more attuned to the fleet submarine. Inside *that* world, the proposal was deemed far too expensive, hence wasteful of limited navy funds earmarked for the Arctic. Haggling was to frustrate the plan, and by mid-1977, it was apparent that NDS was not going to garner sufficient interagency support. However, the proposal would bear fruit, of sorts. That August, meantime, a small group met to discuss options—lesser dreams. "It was the unanimous and enthusiastic conclusion that some action was needed to spur scientific research in the eastern Arctic. The concept of *Fram I* was initiated."[111]

That May–July, APLIS 77 deployed off Arctic Alaska. Installed atop multiyear ice, this summer camp investigated the change in acoustic response of the under-ice surface with progression of the melt season.

> We started on fast ice in about 80 feet of water on 20 May about 30 miles east of Barrow, broke loose at the end of June and enjoyed a looping ride to about

THE APLIS 86 ICE CAMP—ARCTIC POLAR LABORATORY ICE STATION, MARCH–MAY 1986, OFF NORTH ALASKA. THE LABORATORY'S APLIS SERIES (1971–93) WAS INTENDED TO FURTHER THE UNDERSTANDING OF HIGH-FREQUENCY ACOUSTICS SYSTEMS, BOTH ACTIVE AND PASSIVE, IN THE MARITIME ARCTIC ENVIRONMENT. (ROBERT E. FRANCOIS)

100 miles northwest of Barrow, alone on a shrinking floe. Our camp buildings (4) had pedestaled badly as the 24-hour sun ate away. Bears and walruses moved in on us. We had 14 consecutive days of bear intrusion into camp.[112]

That summer of '77, the Red flag flew triumphant over Arctic ice. By every standard, the Soviets had explored more of the boreal canopy than anyone. Now a dramatic seaborne lunge added luster to a splendid record when, on 17 August, the nuclear icebreaker *Arktika* logged the geographic pole, thus becoming the first surface vessel to gain the extremity—an ironic counterpoint to the scuttled NDS. This was brilliant stagecraft. "The eternal dream of generations of sailors and polar explorers has been realized," Soviet television exulted. Timed to mark the sixtieth anniversary of the Bolshevik Revolution, the arrival of the 23,460-ton *Arktika* at 90° north latitude came at 0900, on 17 August. "The *Arktika*'s accomplishment," the *Times* reported, "may enable the Russians to supply their floating ice research stations by sea instead of by air and it may now become possible to expand Russia's existing system of [automated] weather stations" set up on the ice.[113] (At that moment, *Severnyy Polyus-22* and *23* floated operational.) The vessel had determined her precise position assisted by satellite. "The old method is the sun," her captain informed reporters. "Unfortunately, when we approached the Pole there was a thick fog. We had to spend several hours establishing whether we were at the Pole or not." Having tarried fifteen hours, *Arktika* resumed way.

Why had the Soviets done it? The minister of merchant marine denied that the voyage was a stunt. "The Arctic interests us as a national transport route that will insure our country's economic development through shipping," he explained at a news conference. "You know that a distinguishing feature of our country is that a significant expanse is in the Arctic." The immediate problem was to build ships capable of following the icebreakers through the paths they had cleared. Not long after the *Arktika*'s feat, the journal *Nature* compared SP stations to the icebreaker's sortie—and pronounced it wanting. *Arktika*'s role in gathering meteorological, geophysical, and oceanographic data, in its view, had probably been "limited." Though it had been a prestigious expedition, certainly, "only 15 hours were spent at the Pole," one editor sniffed, "which included time spent in ceremonial flag-hoisting, rocket-firing and holding hands round the pole."

Russia deploys the largest fleet of icebreakers and icebreaking bulk carriers. The USSR could boast nearly two decades of nuclear operations, longer with conventional types and ice-reporting services. Nuclear icebreakers were conceived in 1956, during the Twentieth Congress of the Communist Party of the Soviet Union. In approving the sixth five-year plan (1956–60), construction of a prototype had been directed as well as "the further development" of the Northern Sea Route. The sixteen-thousand-ton *Lenin*, launched in 1959 and as revolutionary as

Nautilus, was the world's first nuclear-powered surface ship. And the first vessel intended solely for polar use. Highly experimental, *Lenin* was a research and training platform.[114] *Arktika* was launched late in 1975; two years thereafter, *Sibir* floated complete.[115] Key platforms, they helped extend the commercial shipping season along the Sea Route, thereby aiding in the exploitation of metals, natural gas, and other resources vital to foreign exchange.

No other nation yet had fielded such platforms. In the United States, the ownership and operation of icebreakers is the responsibility of the Coast Guard. In 1977, the Guard was operating eight conventionally powered ice-breaking vessels for Arctic and Antarctic service.[116]

Circumstances now conspired to squeeze the U.S. Navy from Barrow. Because the Eurasian oceanic sector lies well east of North Alaska, a new logistics train was called for. The threat of closure hung over NARL, anyway. An agreement was therefore negotiated with the Danish Defense Command to use its facilities in northeastern Greenland. Nord (82° N) is a well-placed jump-off point able to accommodate Ski-130s. And it sits well within range of Thule AFB (76°N). And so a new logistical approach emerged in the form of short-duration, camps with small staffs, set up in strategic but understudied, undersampled locations and targeting specific features, phenomena, and geophysical problems. An experiment was mounted preliminary to a major program. RUBY and PEARL (1977) were pioneer camps for the *Fram* series (1979–82), shakedown bivouacs deployed to assess the feasibility of exploiting the pack in a dynamic, high-energy sector. No polar expedition can succeed without thoughtful logistic support. For this East Arctic Program, the Polar Science Center, University of Washington, was under contract to the Arctic Branch of ONR.[117]

And so 1977 realized

> A major shift from the western central Arctic to the much more dynamic eastern Arctic and Fram Strait area. . . . Moreover, PRL data buoys were becoming more and more sophisticated, could be installed all over the Arctic by airdrop and aircraft landings, were economical, and nicely filled the off-season data requirements (i.e., outside of the manned station spring months.)[118]

ONR had initiated the program using Nord as the prime support base, given its maximum reach over the eastern Arctic. Canadian, Danish, and Norwegian scientists and organizations participated in organizing *Fram I* as an U.S. ice-based station. Ken Hunkins was lead scientist for *Fram I* and *III*. "The *Fram* series," he recalls, ". . . moved everything over to the east for the first time." Staging out of Nord, transports would air-drop to the ice; landings, though, would rely on short-takeoff-and-landing aircraft, to eliminate the need for long airstrips. Ice-based stations are limited instruments, probing strips of ocean. However, by exploiting camp-based helicopters as adjunct platforms, regional geophysical surveys could be flown. "As [each station] drifted, instead of just having one line of data, you had a *swath* of data," Hunkins reminds. *April–July* data. These bivouacs were seasonal: get ready, deploy onto springtime ice, escape by the end of July.

The basic plan envisaged an extended drift for *Fram I*, southbound, into the little-studied sector upstream of Fram Strait. Deployed into the abyssal-plain region of the pole, its party would be conveyed (conditions obliging) over the southern axial valley of the Arctic Midoceanic Ridge—an active spreading center and part of the ocean-ridge system—thence across the Nansen Fracture Zone and up the continental slope of the Greenland massif.

As planning matured for the *Fram* series, APL and PRL jointly deployed to seaward (from Nome), though into separate sectors. "It was logical that we operate apart," Buck explains, "because man-made background noise limitations were different below and above 10,000 Hz. Almost nothing bothered [Bob Francois's] measurements and everything screwed us up." On 9 October, *Northwind* off-loaded APLIS 78 onto the sea-ice cover of the Chukchi—one element of the navy's MIZPAC 78 exercise. APLIS would work BIRDSEYE, a P-3 with special instrumentation for hunting microwave signatures on or beneath neighbor ice. In the distant Eastern Arctic, PRL deployed onto the same waters. BOXER represented the first U.S. autumn pack-ice station.[119]

The year 1978 proved especially active. That May, on an ice island adrift near coordinates 75°40′ N, 166°30′ E, north of the East Siberian Sea and off the DeLong Islands, *Severnyy Polyus-24* (1978–80) commenced scientific work. (Construction had begun in March, sustained by airlifts.) The ice (like that rafting SP-22) would sustain a productive scientific run, hosting three shifts and basing a trio of *Sever* air expeditions (1979, 1980, 1981). Riding the transpolar drift, SP-24 would log a transect en route to the marine dynamics off West Greenland.

On 12 February 1979, all field equipment plus two *Fram* personnel sortied from Anchorage, Greenland bound. Approximately 30,000 pounds (13,608 kilograms) of equipment reached Nord, one man accompanying. A second remained at Thule, to coordinate efforts there. In late February and early March, project aircraft, equipment, and personnel began to reach Thule's flight line. *Fram I* was established on the 11th at approximately 84°24′ N, 6° W. All personnel were on station within nine days, all primary equipment and supplies in two days more. Twin Otter and Tri-Turbo 3 aircraft shunted about 80,000 pounds (36,320 kilograms) of cargo and personnel to the nascent station.[120] Military Airlift Command C-130s out of Thule followed up, air-dropping palleted explosives and drummed fuel.

Chance interfered with plans. But *Fram I* would demonstrate that a failing floe need not stifle the determined advance of science.

During the last week of March, as the science program had just started into full operation, the camp was split in two by a crack running through the Bedford Institute hydrohole. With minor relocation away from crack, the sampling program continued almost uninterrupted from what had now become two camps separated by approximately 1 km [0.62 mile]. The lead separating the camp eventually refroze and became a Twin Otter runway. With some ongoing ridging and cracking, the ice held together until the end of the program."[121]

Save for an initial shift from southerly winds, the *Fram* camp followed the expected trajectory (albeit too slowly). As it moved, personnel monitored depth and gravity continuously, sampled the upper ocean for temperature, salinity, and velocity, and recorded deep as well as shallow seismic profiles.[122] Via helicopter, regional geophysical surveys were flown–landing at three-mile (five-kilometer) intervals for gravity and depth observations. Ice camps tend to drift in company. While Canada's LOREX explored the Lomonosov Ridge (see chapter seven), *Fram I* floated over the Nansen Ridge. As in Underwater Sound Laboratory's acoustic work between ALPHA and T-3, explosives were detonated at both *Fram* and "Iceman," a LOREX satellite camp, for long-range sound propagation experiments. In addition, the effect of water and ice properties on low-frequency propagation was studied, as was ambient under-ice noise. "It was considered important," Dr. Hunkins writes, "that the scientific program be as broad as possible in this region which is so little explored in nearly all scientific areas. Timing was coordinated to take advantage of the Lomonosov Ridge Experiment (LOREX), so that cooperative programs could be arranged."[123] The focus of individual projects ranged from the sub-bottom and water column to the ice canopy and beyond, into the upper atmosphere. The Canadian contribution to *Fram I* was the seismic methods of studying the structure of the oceanic crust. The deeper structure was assessed by seismic refraction, the explosives deployed up to eighty kilometers from camp. Sediment thickness was measured with an airgun reflection system.

> The subsea structure was studied with gravity, magnetics, and seismics; the seafloor with dredges and corers. In the water, data on acoustics, chemistry, currents, biology and physical properties were gathered. The ice platform the camp was on was itself studied, as was life on the ice in the form of polar bears, and above the ice in meteorology, air pollution and ionospheric radio propagation. Twenty scientific and support staff had only two months on the ice to cover this broad spectrum of science.[124]

Hunkins himself conducted much of the off-station hydrographic work:

helicopter-supported conductivity/salinity, temperature, and depth (CTD) casts. First, he scouted out oceanographic stations; alighting by a refrozen lead (easily dis-tinguished from older ice), he either chiseled through to the water column or set up on an edge of the lead. On a small, sled-mounted A-frame, a CTD profiler was lowered to nearly 2300 feet (700 meters).[125] "It was set up very fast, so we didn't waste helicopter time, and would maximize our number of stations in a day."[126] Once spot data was in hand, everything was collapsed and shoved back aboard, after which Hunkins was airlifted to the next station.

Early May saw a gradual reduction in data collection. The ice in and around the *Fram I* camp was becoming "very active." With the approach of summer, moreover, fog was curtailing flight operations. Happily, "science objectives for the most part were met." Data collection ceased on 13 May, with all personnel off ice by mid-month.

Poleward of the New Siberian Islands (at almost 82°), meantime, the second shift for *Severnyy Polyus-22* relieved the first, ending twelve months for weary comrades.

Kenneth Hunkins remembers the comforts of ALPHA and the later *Fram*

HYDROGRAPHIC STATION AIRLIFTED FROM *FRAM I*, APRIL 1979. THE EXCHANGE OF FRESHWATER, SALT, AND HEAT THROUGH FRAM STRAIT WAS POORLY UNDERSTOOD AS THE 1970s ENDED. THE *FRAM* EXPEDITION SERIES OF SHORT-DURATION CAMPS (1979–82) WAS ESTABLISHED FOR OCEANOGRAPHIC AND GEOPHYSICAL RESEARCH IN THE EURASIAN BASIN, THE MAIN THRUST COMMENCING IN 1979; AFTER 1982, IT CONTINUED WITH (PRIMARILY) ACOUSTICS CAMPS AS PROJECT AREA—ARCTIC RESEARCH IN ENVIRONMENTAL ACOUSTICS. (DR. KENNETH L. HUNKINS)

series—1958 to 1979—as not so very different. "I don't think the actual level of comfort changed that much," the geophysicist remarks—a testament to the U.S. military. He says he would probably prefer the *Fram* stations—"they were smaller, you had to do a lot outdoors, but we didn't have to maintain a really big, long runway." This was a contrast with the *Globemaster* of IGY vintage; the C-124 had required a long, well-groomed surface. Still, when the strip serving *Fram I* split, concern arose over whether even the Twin Otter could get in. "So we had to get out there and work"—to build a new runway by hand. "So, from fifty-eight to seventy-nine, there wasn't [chuckling] that much change, really. You always got to anticipate that its going to break up—that's of course one of the hazards of those stations."[127]

CHAPTER SIX

GLOBAL CHANGE: ADVANCING THE CASE

> On our watery planet the climate is governed largely by the oceans.
> SPENCER WEART, PH.D

CLIMATE RESEARCH HAD ATTAINED CACHET. Throughout the 1980s and '90s, however, consensus eluded the community of atmosphere-ocean professionals. The so-called greenhouse warming lay hidden amid natural fluctuations over geologic time–climatic noise. An unambiguous anthropogenic (human-caused) signal did not emerge from the geologic and other records until 2001. Still, interdisciplinary collaborations across fields of research helped to advance a broad understanding the whole earth system.

> The need to understand and predict global change is forcing scientists to redefine the way they look at the world. In the past, science carved the Earth into separate chunks. To geologists went the planet's crust. To biologists went the living kingdom. Oceanographers got the great basins of water. Yet environmental change cuts across traditional academic borders. Scientists dealing with the Earth as a whole will have to broaden their vision and collaborate more than they have in the past.[1]

In the realm of climate variability, the introduction of coupled models of the atmosphere-ocean system represented a major step. The science of climate change continues full ahead–more tests of ever-more-refined models against the real climate. And the policy debate persists, with certain well-placed interests preferring to emphasize the uncertainties in understanding the causes and consequences of warming.

Scientific judgment is an exercise in the assessment, reduction, and quantification of uncertainty. Uncertainty creates its own issues. Understanding the greenhouse effect–human-induced "forcing" (causes of change)–cried out for proxy (paleoclimactic) records and long-term observational evidence (instrumental data), in order to identify and quantify the natural forcings, feedbacks, causes, and probable impacts of change. A standardized, more integrated approach to world-data gathering (and processing) had become an imperative–long-term actions requiring commitment and continuity. Afloat, ashore, and in space, high-latitude programs of observation and measurement yield important findings that become subsets of huge databases for climate-model simulations.

LOREX

North America and Europe were once joined together with no Atlantic Ocean between. When did the Arctic Ocean open up to become an ocean? The Arctic Islands probably are continental. But were they once part of mainland Canada or Alaska or Russia? AIDJEX had granted Ottawa experience with short-term camps. Was there a problem Canada could tackle similarly? The subaqueous Lomonosov Ridge bifurcates the central Arctic into two major basins. Energy, Mines, and Resources (EMR) Canada undertook a full-scale multidisciplinary project to study the nature and origins of the highland. This was the Lomonosov Ridge Experiment of 1979, whose name was abbreviated as "LOREX." Three years in the planning, devoted purely to scientific discovery, the $1.2 million project (George Hobson is proud to note) "was mainly Canadian with a couple of U.S. capabilities in there that we didn't have. Hans Weber, a geophysicist in EMR's Earth Physics Branch, was chief scientist. Logistical support came courtesy of Polar Shelf.

On 21 March 1979, the main LOREX station and two satellite camps were deployed near the geographic pole. About thirty men were fielded, the tents and huts dotting a large floe. As motion shifted these dots of occupation across the ridge, their position was tracked and computed by satellite. Meanwhile, hundreds of echo and shallow seismic soundings were recorded by instruments on the station; ashore, these bottom-surface mappings would translate into a detailed contour map of the seafloor enclosing the pole. More than 250 depth soundings and gravity measurements were taken. As well, forty-two sediment cores were recovered, pulled from the adjoining basin floors and from the ridge itself, and heat-flow measurements taken.

One fundamental question attending continental drift is whether ocean basins are passively pulled apart at mid-ocean ridges or, instead, driven apart by convection-heat forces originating deep within the mantle. One possible model applied to the north polar basin had extensional rifts forming during the opening

Geologists examine a bottom core for Canada's Lomonosov Ridge Experiment, LOREX, April 1979. The climate history of the maritime Arctic is preserved in its ocean-floor sediments. The long-term variability and stability of the ice cover are deduced from the study of such cores. (George D. Hobson)

of the proto-Atlantic. Geologically, this is a tough puzzle to sort out. As one LOREX member said, "The last area in the world where we don't know much about plate tectonics is in the Arctic." "While we're up here," geophysicist Jack Sweeney continued, "we'll be testing one of these ideas—whether the Lomonosov is a fragment of a continent."[2]

LOREX defined a remarkably uniform highland—a flat-topped, steep-sided rise overlain by a kilometer or more of flat-lying sedimentary layers.[3] This supine bathymetry—so distinct from, say, the Alpha Rise complex—belies volcanic action. Instead, the ridge is a block-faulted fragment from the fringing continental massif—strong evidence that powerful forces from within the mantle are in fact shaping that seafloor. LOREX, in short, had bestowed valuable insight into the formation of the Canadian continental-shelf margin.

The following August (1979), *Severnyy Polyus-22* welcomed its relief—the station's *seventh* rotation of staff. Two more shifts would come aboard before, at

Divers and photographic equipment beneath LOREX ice, April 1979. (George D. Hobson)

the last, the shelf-ice slab was air-evacuated, thence exported out into the far North Atlantic. The last aircraft supporting the operation would lift clear on 8 April 1982.

Fram II was installed in March 1980, with the mission of again scrutinizing the Fram Strait sector, specifically the sedimentary basin between the Nansen and Lomonosov Ridges. "Our objective was to compare it to the younger, 10–20 million year-old crust studied by *Fram I* on the Nansen Ridge, to see if crustal properties change with time," the Geological Survey of Canada reported. The program was more restricted than that floated in 1979, comprising studies of the subsea crust, long-range/low-frequency acoustics, and "some" physical oceanography. Marine geophysics (seismic refraction) assessed crustal properties as well as structure. Two manned satellite encampments augmented the primary station. The drift rate proved fast, abbreviating the March–May geophysical program conducted by the Canadian party riding *Fram II*. "Before the crustal studies could be fully completed, the camp was out of the deep water and up onto the Greenland margin. However, we did get enough data in the basin to contribute to understanding the crustal evolution."[4]

Fram III is illustrative of advances in logistics. Late in February 1981, specialists were deployed into the Fram Strait sector. Two military units—both from North Carolina—accompanied the advance team: parachute riggers from the army's Fort Bragg plus air force support for three *Hercules* assigned to Pope AFB. Departing Thule AB, the transports airlifted all scientific and logistic gear to the jump-off point (Nord) while riggers back at Thule prepared the necessary lumber, fuel, and explosives for paradrops via C-130. A Twin Otter and the Tri-Turbo-3 then sortied to locate, establish, and support the nascent camp. This pair would be used exclusively in 1979–88 for the *Fram* series of research camps and for PRL's quiet stations—the latter under Project URSA (Under-Ice Research in Surveillance Acoustics).

> Those of us who have done extensive field work in the Arctic and especially those who have worked in the field of ice dynamics . . . knew that there is statistical basis for predicting open water conditions suitable for missile launch at a given time. Many of us work with satellite imagery of the Arctic and have preached this gospel for many years. But by late 1982, it was observed by our intelligence people that the new Soviet missile submarine Typhoon remained in the Arctic on its maiden strategic deployment. This created a potentially destabilizing situation if we could not, at least, demonstrate our ability to successfully counter that threat.[5]

Acoustic test and research work by the Defense Department escalated in the Arctic Basin. One measure: the navy's weapons-related APLIS drift-ice camps (see Appendix 3) jumped in both size and scientific ambition.

Mid-March saw *Fram III* deployed onto a multiyear floe measuring 1.9 by 3.1 miles (3 by 5 kilometers). Average thickness: 13 feet (4 meters). By mid-April, 203,000 pounds (92,162 kilograms) had been paradropped to the bivouac via *Hercules*. An additional 75,000 pounds (34,000 kilograms) of scientific and logistic equipment gained this newest camp courtesy of two-dozen missions logged by Twin Otter augmented by five TT-3 sorties.[6]

Self-assured and assertive, the Arctic Institute this April had *Severnyy Polyus-25* (1981–84) manned and functioning. Installed to the north of Pevek, in far eastern Russia, the station's starting coordinates were 75°01′ N, 168°35′ E—within multiyear pack northwest of Wrangel Island. In the "North Pole" series of deployment cycles, Leningrad often deployed into that sector—the wastes straddling the Date Line (longitude 180°), bounded west and east to about the 160° meridian, north-south by the 75° to about the 80° line. Seemingly impervious, the decade's inaugural camp of Soviets would survive the vagaries of weather and ice pressure for four years. Northing ten degrees, "25" tracked across forty-six degrees of longitude before its abandonment deep within the interior pack, far poleward of Canada's Arctic Islands.

Less than four weeks following commencement of this Soviet odyssey, *Fram III* decamped off northeastern Greenland, its scientific goals accomplished.

> The resulting net drift of 361 km [224 miles] proved to be much longer than that of *Fram 1* (163 km) and *Fram 2* (83 km) stations during the previous years. This not only allowed experiments to be carried out over a larger geographic area but also over a range of ocean depths, from a maximum of 4088 m [13,409 feet] in the Nansen Basin to a minimum of 727 m [2,385 feet] above the Yermak Plateau.[7]

Logistical assistance from Barrow for the APLIS series (1971–93) of high-frequency camps continued. An early objective had been to investigate the change in acoustic response of the under-ice surface during melt. Now, in October 1982, the icebreaker *Polar Sea* off-loaded APLIS 82 onto Beaufort ice for a three-week drift—one element of SUBICEX-1-82.

CESAR 83

Unlike the Eurasia Basin, the geologic origins of the Amerasia Basin were poorly known, particularly the nature and origin of the Alpha and Mendeleev ridges. Was the Alpha Rise a remnant of the Siberian continental shelf or in situ? Was the ridge part of North America? Few features of the deep waters of the world ocean generated more hypotheses as to their origins than the Alpha-Mendeleev Ridge complex. In 1983, the Department of Energy, Mines, and Resources deployed a large force to pursue the answer to this question. Planning for CESAR 83—Canadian Expedition to Study the Alpha Ridge—had in fact commenced while LOREX was still in preparation.

Put together by Hans Weber of EMR, the system of logistics came (again) courtesy of Polar Shelf. All its experience and expertise was called upon: the logistics for CESAR were "enormous," its equipment requirements "gigantic." That February, for instance, the ice patrol of the Atmospheric Environment Service, Environment Canada, conducted a side-looking airborne radar survey for the proposed site area as part of its synoptic ice reconnaissance. (The darker the radar reflection, the smoother the ice. Thickness, though, cannot be estimated.) Until the actual search for a campsite, target-area position was estimated by plotting the drift path of SP-25 and two satellite-tracked data buoys of the Arctic Ocean Buoy Program.

The air hunt began on 12 March. A suitable lead was occupied within two days, at 85°56′ N, 112°35′ W. The next weeks are an example of aggressive northern logistics. On ice, the search party erected camp and made radio contact with Alert as the deploying Twin Otters returned to a Eureka runway, on Ellesmere. Mere hours thereafter, four military *Hercules* standing by at Trenton, Ontario, took off for Thule Air Base. The Twins, meantime, returned to CESAR ice on the 14th conveying fuel, equipment, and supplies after which each made for Alert. The Twins retrieved a six-man military team responsible for flagging the drop zone and controlling a paradrop. On the 15th, a trio of *Hercules* hung by the CESAR site.

> A few minutes later men and equipment parachuted from 400 m [1,300 feet] altitude, making a perfect landing with no injuries and no equipment damage. The air temperature at the time was –42 C. . . . Within minutes a bulldozer, grader weighing over 12 tons and three snowmobiles were running.

> This field team–highly experienced engineer talent–set up camp and radio communications; over the ensuing eight (punishing) days, a 1600-meter (5248-foot) long runway 30 meters (98 feet) wide emerged.

> Using explosives and the bulldozer they demolished pressure ridges that crossed the runway, removed the snow with the grader and smoothed the runway by flooding the ice with seawater pumped from below the ice. During that time the wind chill dropped as low as –90° C.[8]

The base at Resolute, in the central archipelago, helped support fixed-wing operations. Gear and supplies were lashed to pallets by military personnel and made ready. Sundry freight was heavy–equipment for collecting piston-core samples and gravity cores from the ridge itself and flanking basins. As many as six sorties per day were flown during 24–30 March, freighting some 500,000 kilograms (550 tons) of cargo plus personnel onto ice. Among the loads were twelve-hundred drums of fuel.[9] As the CESAR airlift (thirty round-trips) shuttled to seaward then back, men worked to enlarge facilities to house forty-four scientists and support personnel, including lab huts and hydroholes set with winches.

Science à la CESAR kicked off on 3 April. The longest sediment cores yet extracted from the subsea Arctic would be retrieved, to study lithology (composition) and age—useful to deciphering basin evolution as well as paleoclimate. (The minimum depth recorded over the ridge was 1087 meters [3566 feet].) As well, twenty samples were dredged from the walls of a major graben on the highland itself—the first rocks recovered from Alpha Ridge and, therefore, the first direct evidence for its composition and origin. Spot soundings and gravity measurements were taken by helicopter and recorded also at base camp. (The outpost was equipped with a continuous gravimeter and echo sounder.) Adding spice, an in-country visit to ice-based neighbors was flown—the Russian Embassy and External Affairs having cleared the matter.

> During CESAR, one day we nipped over to see SP-25. We had wonderful treatment.... We were there just as they were changing personnel. About a week later we were on television in Leningrad. [The utilitarian AN-2s], huge things, on wooden skis, [had a] bloody big padlock on the doors. And the station manager had the keys. Pilots had to *turn in* the keys. Because with the range in the tanks that they had, they could very well escape to North America.[10]

WELCOME TO *SEVERNYY POLYUS-25*, MAY 1983. A TEAM FROM CANADA'S POLAR CONTINENTAL SHELF PROJECT HAS DROPPED IN FOR A MAY DAY VISIT. ESTABLISHED IN MAY 1981, THIS SOVIET STATION HAD SHIFTED INTO THE MARITIME SECTOR CLAIMED BY OTTAWA. (GEORGE D. HOBSON)

That May, a major storm breached the pack girdling the Canadian base camp. Station and skiway floated on intact. The 23rd brought another vigorous storm cell—and another beating. Personnel commenced teardown. As the seventh retrieval mission lifted clear, and with it the last of the equipment and staff, the airstrip began to buckle. Mere days later, a reconnaissance "revealed a jumble of broken ice with no trace of either campsite or runway."

—·—·—·—·—·—

As CESAR wound down, SP-26 (1983–86) was airlifted to sea at 78°30′ N, 174°46′ E–vacant seascape northwest of Wrangel, in permanent pack ice. The bivouac floated almost four degrees north and six degrees east of the deployment point for "25." This pair would operate concurrently during 1983–84, until the abandonment of "25" in April 1984.

Problems in the fields of oceanography and geophysics awaited investigators in the very far north of Canada. And, increasingly, pollution of the Arctic environment warranted a series of campaigns.[11]

> The Arctic is a natural storage reservoir for atmospheric and water pollution. Industrial effluents, aerosols and gases from lower latitudes in Europe, USSR and North America appear in the form of "arctic haze" over large regions of the Arctic. The absolute concentrations of pollutants are comparatively small and may not pose immediate health hazards to arctic residents. However, their long-term accumulation in the food chain and subsequent release . . . as well as the magnitude of their effects on the radiation balance and the chemistry of the atmosphere, while still difficult to appraise quantitatively, are still of clear potential concern.
>
> Concerning water pollution, the Arctic Ocean receives as much as 10% of all the world's riverine input. . . . Since it is surrounded by populated land and has very limited outflows into the world oceans, it is much more vulnerable to industrial, urban and agricultural pollutants discharged into rivers than any other ocean.[12]

Stratospheric ozone shields ultraviolet radiation. Since the latter seventies, startling reports of an ozone hole over Antarctica have attended each polar spring.[13] Man-made chemicals drive this phenomenon. President George H. Bush's proposed research and development budget would, by 1991, earmark $193 million for the NSF-coordinated U.S. Antarctica Program, as well as $1 million to begin monitoring Arctic pollution. To fund research on global change, the Bush administration requested $1.18 billion, distributed among several federal agencies. Thus began the U.S. Global Change Research Program (GCRP), whose expenditures to 2003 equaled $20 billion.

HOBSON'S CHOICE—BACKGROUND

Knowledge of the western Arctic Ocean had come largely from researchers riding the Beaufort Gyre, on ice islands. Notoriety belies the rarity of these tabular bergs. (In 1984, about six hundred had been identified, many mere fragments.) Edging to seaward of Disraeli Fiord, the Ward Hunt Ice Shelf is the largest of five that front Ellesmere Island. During September and October 1982, Ward Hunt calved an ice island. Canadian researchers on hydrographic and gravity survey chanced upon these oddments.[14] The discovery realized considerable excitement at Polar Shelf. The agency had yet to occupy an ice island; still, it "*had* done a hell of a lot of logistics before." Now a vehicle to support long-term, multidisciplinary research floated on offer.

Director George D. Hobson (supported by senior management) scheduled a landing onto the largest of the newfound bergs for 11 August 1983. A buoy was left that would transmit temperature, barometric pressure, and, most importantly, position via satellite, thereby allowing Ottawa to track the mass. Equipment, supplies, fuel, staff support—planning commenced from that day. When the device ceased its transmissions, in February 1984, a search (to replace its batteries) proved futile. That May, happily, the device resumed its useful transmissions. And so a second buoy was set.

Researchers, meantime, were solicited for proposals for the Ice Island Project. Prepared by EMR, a field program called for geological, geophysical, oceanographic, meteorological, environmental, and related studies. "Anybody interested in the Arctic Ocean basin," Hobson said of his platform-of-opportunity, "will have a wonderful base from which to work." The estimated cost of establishing and operating off Canada's northern coast that first year, including the capital cost of EMR's work plan, was $4.3 million (Canadian). Amortized over successive years, the annual expenditure would be reduced to $1.9 million.[15] "The Arctic Ocean is one of the least surveyed oceans in the world," a sister agency noted; "little is known of its physical, chemical and biological characteristics." Further,

> The ideas and techniques of polar oceanography have now progressed to the point where sharply focused new oceanographic investigations are both necessary and possible. These deal with problems of major consequences in both a Canadian and a global context; for example, with the variability of ice cover and climate. Indeed, oceanographers attach extraordinary importance to the ice island as the only likely means by which a number of these investigations may be carried out over the next 1–2 decades.

LOREX and CESAR had been one-time exercises—the ice occupied upon return of sunlight, then quit before the onset of serious melt (late June), with its puddling, ponds, open water, and breakup. (Today, still, most fieldwork takes place during the twenty-four-hour light of spring and summer.) Accordingly, activity had been crammed into eight or so weeks, with high initial cost and the threat of fracturing. In contrast, an ice island might proffer several orbits, as T-3 had—and cancel the worry of breakup. Exploiting a semipermanent base, moreover, the price of supporting one researcher for one day would be slashed—being amortized over a longer span and more projects. And men could deploy off-island as well as *on* it. And so, granted a fair wind and good drift,

> The ice island will offer an excellent opportunity to add significantly to our knowledge of oceanic conditions off Canada's northern coast. The major advantage of the ice island over ice camps on sea ice is that studies can be done year round. The proposed research programs will allow us to answer basic questions regarding the oceanography of the Arctic Ocean in support of wide ranging national and international climate studies; understanding the seasonal variability of oceanographic phenomena in our most northern ocean; and assessing the structure and function of the Arctic marine system as a basis for predicting the potential impact of hydrocarbon exploration, exploitation or transport in the Arctic.[16]

Sovereignty, Hobson reminds, is a persistent Canadian theme. As well as probing his nation's "geological roots," an ice-island station offered a potential database for securing Canada's sovereign offshore limits.

Install a remote field camp, and scientists will come; outposts attract them. In that sense the proposed Canadian facility was no gamble. Along with firm support from the research community, "Industry has also indicated a considerable interest in the study of this ice island since it may have ramifications in the Beaufort Sea and the potential for hydrocarbon production in that area." Hostage to the gyre, the berg (soon labeled Hobson's Choice) was expected to log an initial drift path over the ice-rich continental shelf, along the wastes of the Canadian polar margin. "The ice island will be revisited in early August," Hobson records, "... at which time a decision will be made as to where an airstrip and a camp can be established." The first meeting of the Steering Committee for the Ice Island Project took place on 4 October 1984, at the Wellington Street offices of PCSP, Ottawa. Principal topics: the priorities of the scientific program and navigation requirements.

END OF T-3

Receding beyond the pole, dropping latitude, T-3 was exit-bound. Unmanned since 1974, the slab had been rediscovered by a searching C-117 within two years. A buoy was deployed. NARL then had tracked the beast as it shuffled past Point

Barrow. October 1979 found T-3 about two hundred miles out, to the northeast, after which the island began to sprint westward. Its trajectory—along the 74° line—suggested capture by the Transpolar Drift Stream. Near the longitude of Wrangel Island, NARL elected to fly a drop-in.

> The former camp was found to be in good condition. Much of the equipment left there from the pre-1974 occupation is available and in working condition. A four-member NARL team flew to T-3 November 13 [1979]. On board the NARL Twin Otter were pilot Lloyd Zimmerman, Alain "Frenchy" Le Cloirec, John Bitters, and Frank Akpik. [Bitters and Akpik] remained on T-3 to make an inventory of all usable supplies and equipment and to mark their location to facilitate retrieval in the snow.[17]

Rather than simply track it, the intent was to reoccupy the platform. With time and transfers, apparently, no one was left responsible for this well-known ice. The project died, the errant berg forgotten. "That was never the intention," Captain Shoemaker (former NARL director) insists; T-3 was yet useful. "It is a shame because if we had re-manned T-3 . . . we could have legitimately rode [sic] it all the way to Greenland where it finally broke up." On 20 August 1983, the last confirmed position for Fletcher's Ice Island was recorded: 83.1° N, 004.2° W—on approach to Fram Strait. T-3 is estimated to have exited late that September or in October.[18] In high North Atlantic waters, the peripatetic iceberg melted away to nothing.

—·—·—·—·—·—

Along the opposing coast Soviet aircraft again pressed to seaward, opening yet another construction cycle (1984). That spring, SP-26 floated about 1200 kilometers (745 miles) northeast of Ostrov Zhokhova. A fuel paradrop helped replenish the station. That same sector saw a fifteen-man party parachute onto ice, to prepare an airstrip. On 2 June, *Severnyy Polyus-27* (1984–87) went operational 370 kilometers northeast of Ostrov Zhokhova, just off the 80° meridian. No apprentice, camp leader Yu. P. Tikhonov was embarking on his *eighth* wintering at a drifting station. Mobilized later that year, air-expedition *Sever-36* would resupply "27" near the so-called Pole of Relative Inaccessibility. And that autumn, a relief plane found itself unable to land at SP-26, "so once again stores were parachuted in."[19]

HOBSON'S CHOICE AND SPS

Elsewhere, the enticing target now engaging Polar Shelf had shifted southwest to a position off the mouth of Nansen Sound—about where T-3 had grounded a decade before. In September 1984, a decision was taken:

IN ORDER TO ESTABLISH A CANADIAN PRESENCE ON THE ICE ISLAND AND PROVIDE A SMALL BASE FOR SCIENTIFIC STUDY OF THE CANADIAN ARCTIC POLAR SHELF, EMR IS ESTABLISHING A SMALL BASE CAMP IN SEP 84 USING TWIN OTTER/DC3 AIRCRAFT . . .

On the Canadian continental shelf, camp-erection work yielded seven buildings the first month. (Shipping costs are horrific. The annual Sealift resupply is commercial; from staging-point Resolute, *everything* had to be flown in by charter. The deployment cost of *each* hut that year was $80,000.) With time, various upgrades to the main scientific camp would be incorporated to meet the demands of the environment and user communities. The 1992 field season (February–September) would find Ice Island equipped to handle twenty-eight guests "comfortably," mess and lodging for a total of forty. To hand were power, radio communications, satellite and radio navigation aids, and meteorological sensors, as well as laboratory and auxiliary facilities. The 1985 field season saw a dozen huts as well as Parcolls (framed canvas and nylon structures) ready to serve as workshops and for storage.[20] About thirty-five personnel were aboard initially in 1985, most for the erection work. A handful, though, were preparing for the kickoff scientific assay—two to three weeks preparing a continuous reflection seismic along the polar rim.

The Twin Otter is an exceptionally versatile platform. But its payload is small, especially for drummed fuel. The ski-equipped LC-130, the largest ice-capable bird, requires relatively thick ice *and* a well-groomed strip 1600 meters (5568 feet) long and 50 meters (175 feet) wide—a major project. Unlike icebergs but like floes, ice islands are markedly horizontal and free of pressure ridges. Low parallel rises are characteristic, however, so the new runway was installed along this ridge system. Seawater was pumped up through holes bored through the island with hot-water drills, and one of the troughs flooded. The first drill hole was 44.5 meters (146 feet) in depth. Thus, in semidark, laboring round-the-clock, field crews consumed about a month fashioning a suitable surface.

A base manager, a mechanic, and a cook comprised the minimum permanent staff. The presiding cook was Tom Krochak. A professional chef, he had long maintained a flow of hearty and refined fare onto high-latitude tables. He could do nothing for tired bones but, oh, those growling stomachs. Krochak had been sought out that August (1985). The government wanted a permanent floating ice camp; could he join the project? When convinced that the call was serious, he accepted. "The ice island to me," he recalls, "was one of man's last frontiers." He arrived to find the Parcolls up, the cabins under construction. "And DC-3s were flying in like a yo-yo." Lumber and supplies off-loaded, each bird promptly set course for Eureka to ferry the next load out. Krochak's floating kitchen was a four-burner stove in the larger Parcoll.

Conveyed southwest and close inshore, along the margin of the Queen Elizabeth Islands, Ice Island lay off Phillips Inlet late in December at approximately 81° N, 97° W. Canada's at-sea outpost had migrated about one hundred kilometers since May. "It is expected to move," Hobson wrote, "down the coast in a manner similar to the movement of T-3."

In March 1985, in Moscow, Mikhail Gorbachev was made general secretary by the Politburo. Under the liberal policy of *glasnost*, the tremors of a genuine change—at first imperceptible—were to lead with undreamed quickness to a political earthquake.

That March and April saw upgrades at the Canadian facility. An evolving "Main Street" was to boast six sleeping cabins, an office, navigation quarters, a generator hut, a workshop, and (somewhat later that season) a combined kitchen and dining area, this last a godsend to chef Krochak, who had been persuaded to return with a promise of modern workspaces. Arriving in early March, staff was moving snow for a new kitchen. Already the outpost housed fourteen, with another dozen to fourteen investigators inbound along with ancillary personnel—pilots, crewmen, others. Krochak would have to make do—again—with four-burners and no helper. "It was long, hard work, and tiring, but at the same time rewarding," he says. Duration of his 1985 tour: sixteen grinding weeks.

Ottawa's mobile polar outpost was open for business.

Moderately well-used during its working life, the facility at times grew crowded. The 1985 field season was a pacesetter, with investigations by several federal and private organizations. The research syllabus included structural geology, oceanography, and seismic investigations. A refraction survey was run, and 420 kilometers of "good data" recorded. As well, gravity and bathymetric data were taken beneath the seismic profiles. Samples of the water column and bottom samples were collected, the latter in the form of piston cores, dredges, and bottom grabs. Geophysical investigation persisted, including mapping bottom topography along with sub-bottom stratigraphy. Continuous depth profiles (CDP) of reflection seismic data were dutifully recorded.

—·—·—·—·—·—

In the presatellite era, countless surface-, air-, and submarine-based missions had probed the Northern Ocean, its peripheral seas and fringing shores. Most had been funded by the military—the services needing to understand the environments in which they now operated. (An example: submarine-mounted acoustic mapping of pack-ice thickness.) The practical fruits of these myriad of programs—whatever the mission—had begun to intersect with research applied to questions attending *climate-related* concerns.

Sea ice is an amalgam: solid crystals, liquid brine, and pockets of gas. Ice emits microwave radiation, as do all materials, the exact spectrum dependent on whether the ice is newly frozen (first-year) or has deformed and thickened.

Ice and liquid water have very distinct emission signatures, primarily because of differences in the way molecules are arranged in each. Moreover, different ice samples can have distinct signatures, partly because of temperature differences and partly because of differences in texture and impurity content. Snow density, grain size, surface roughness, brine content [as new ice forms, brine is rejected] and the degree of wetness all influence the radiated energy, and they influence it differently at different wavelengths.[21]

Visible wavelengths hold limited utility for monitoring the Arctic Ocean–a cracked canopy screened by summer cloud. Satellites are capable of imaging through cloud and darkness, facilitating all-weather, high-resolution repeat imagery. Back in 1961, photographs from TIROS II had helped route a navy transport through difficult ice into a Newfoundland harbor. Satellite imagery has been exploited operationally for ice analysis ever since. Satellite-visible imaging (from *Nimbus 5,* 1972) was the first source of global information on sea-ice extent. Increasingly sophisticated instrumentation has been a boon to polar research; the 1990s, indeed, marked a new age of observations, satellites center stage.[22] Nonetheless, ground-based field observations ("ground truth") are indispensable to interpreting and understanding space-derived imagery.

National data banks (one IGY legacy) archive and digest a virtual Niagara of data and images vital to regional and global climate supercomputer models. Their output predictions are strongly dependent on how mobile sea ice is incorporated within them. Still, sea-ice processes persist as ciphers. "Our increased awareness of the role of polar ice in modulating and responding to world climate, in controlling sea level, and in modifying ocean properties has exposed our lack of understanding of ice behavior."

We do not know [1986] whether the ice sheets of Greenland and Antarctica are growing or shrinking; we do not know how much snow falls on these ice sheets, how much surface melting there is, or how much ice flows into the ocean to form icebergs; we cannot calculate accurately the forces that drive ice motion; and we cannot assess how strongly sea ice affects heat transfer between the ocean and atmosphere and hence influences both ocean properties and climate. We do know, however, that the polar regions are not dormant and unchanging. . . . We also know that predictions from climate models are strongly dependent on how sea ice is incorporated within the models.[23]

The National Aeronautics and Space Administration (NASA) uses space-borne sensors to determine the characteristics of the ice canopy and to decipher "how these are influenced by and in turn influence the atmosphere and ocean." A 1987 report tenders this assessment:

> Our immediate concern is to improve our capability of measuring from space the extent, type, movement, and surface characteristics of the polar ice cover. ... Both global weather and long-term climate change are strongly linked to ocean behavior. In the past 2 decades, observations from meteorological satellites have significantly improved our ability to forecast weather. However, improving our ability to predict climate requires a better understanding of the role of the oceans in the global climate system than we possess at present. New global information available from satellites, coupled with data from the ocean interior and our greatly enhanced computer capability can meet this need.[24]

Two decades later, intense research notwithstanding, a professional journal can still lament the "gross lack of understanding of the [Arctic Ocean] area."

—•—•—•—•—•—

For the 1986 field season, Canada floated bedding-down accommodations for up to thirty.[25] Power, radio communications, satellite and radio navigation aids, meteorological sensors, and limited lab space lay to hand for client-scientists and visiting technicians. For Ottawa, Ice Island represented an immense expenditure: the 1986 Polar Shelf budget was $5.1 million, of which approximately $2 million was devoted to Ice Island—nearly 40 percent. (See Table 5.) Economizing essential, Hobson in March cautioned his station managers thus:

> Unused buildings should be closed down to save fuel. I remind you that fuel costs at least $14.00 per gallon and is not that easily transmitted to the island. Eight people should not occupy 4 buildings. I would strongly suggest that 8 people occupy 2 buildings, thereby shutting down and saving fuel on the other 2.[26]

For ice-based science and navigation, position is critical: researchers must know where they are, so as to plot their data precisely—and to ensure rescue. Navigation for the geophysical profiles and for locating the moving encampment was maintained through Siledus, GPS (ground positioning system), and SATNAV (satellite navigation), along with astronomical "shots"—the conventional, time-tested method. Camp managers served as ice-island "navigators." These men were heavily burdened, responsible for all aspects of logistical infrastructure, accommodation, messing, safety, coordination of transportation, support staff, and for logistical support in collaboration with a fellow manager ashore, at base Resolute. (See Appendix 6.)

Ottawa's 1986 at-sea field season resembled that of the year before. In mounting daylight that March, a runway extending 1000 meters (3300 feet) was constructed. Refraction seismic surveys continued. And three new oceanographic projects got under way. Explosive shots were detonated at regular intervals for a reflection seismic project. The geology program was expanded to include piston cores, grab samples, bottom images, and "Sparker" and echo-sounding records. (The latter penetrate more than fifty meters into the sea bottom.) Before the season's close, a geothermal project and several heat-flow stations were added to the suite of activities. "Recording gravimeter and regional seismograph stations continuously recorded data," Hobson noted with satisfaction, "while the ice island was operational." Hung over an island edge, at least one Defence Department hydrophone was listening. Ally Canada's possible military role in the northlands was (again) becoming a matter of widespread discussion.

TABLE 5.
CHARTER AIRCRAFT RATES FOR LOGISTICAL SUPPORT, POLAR CONTINENTAL SHELF PROJECT, ARCTIC CANADA (MARCH 1986)[1]

PLATFORM	RATE (CANADIAN DOLLARS)[2]	RATE (U.S. DOLLARS)[2]
205-A[3]	$2,320 per hour with fuel	$1,630/hour
206-L-1[3]	$907 per hour with fuel	$637/hour
Twin Otter	$2,381 per hour with fuel	$1,673/hour
	$834 per hour [out of Resolute][4]	$586/hour
HS-748	$2,675 per hour with fuel [out of Resolute]	$1,880/hour
DC-3	$1,260 per hour with fuel [out of Resolute]	$885/hour
C-130 *Hercules*	$22.50 per mile with fuel	$15.81/mile

1. Courtesy Polar Continental Shelf Project.
2. These rates were used by Ice Island camp managers for the calculation of field-related expenses. One Canadian dollar (1986) = 0.70274 U.S. dollars.
3. Rotary-wing aircraft.
4. The base at Resolute Bay is located on Cornwallis Island just shy of 75° N, central Canadian Archipelago.

The increasing military use of the Arctic region by both US and Soviet forces has raised the possibility of it becoming a site of superpower confrontation. Canada's geographic position between the two superpowers, and our collective security agreements, of course make Canadian involvement in any superpower conflict unavoidable, and recognition of this reality has prompted renewed interest in the strategic importance of the Canadian Arctic and of the adjoining oceans and airspace.[27]

SP-28

A world away, at the airfield in Tiksi, a Lena River terminal on the Russian Arctic coast, intense preparation gave way to waiting; machinery and equipment and men stood ready. On 15 May 1986, an AN-12 aircraft pressed to sea. At the chosen site, 700 kilometers (435 miles) northeast of Ostrov Zhokhova—a rocky dot on the shelf of the East Siberian Sea—two parachutists settled onto ice. In their turn, a pair of IL-76 aircraft gained the place, whereupon four pallets loaded with freight, 120 barrels of fuel, and a tractor plumped heavily down. (The huts, scientific equipment, and food supplies had come via helicopter.) But the site was soon beset. Abrupt shifts of floes, breakups, and hummocking attacked the camp-floe, conflicting preparations and forcing several moves. Laying down a runway proved no less punishing, as cracks repeatedly canceled progress. On 21 May, the call sign for SP-28 (1986–89) was broadcast: the station was operational.[28]

Six weeks earlier, two degrees nearer the pole, SP-26 had decamped when its third shift of personnel (thirty men) was airlifted clear to be conveyed back to "big land."

The Leningrad institute was again fielding superb logistical support. Still, this sprawling enterprise stood disadvantaged by a scarcity of good instrumentation and gear—not least of all, computers. In part, a lack of funds was responsible—one problem with a command-administrative system reliant upon ensnarling, security-obsessed (even hostile) bureaucratic machinery. In this context, then, SP-28 boasted very special equipment.

> For the first time a floating ice station will have a portable computer unit with an automated data collection and processing system and with transmission via a satellite communications channel. This will make it possible for the station to supply weather forecasts efficiently to the on-shore weather services and to the marine operations headquarters of the Northern Sea Route.[29]

The season's last aircraft took its departure on 22 June. The camp was then at 81°12′ N, 166°59′ E. "The station was sponsored by the Komsomol [youth] organisation and the research programme was chiefly meteorological... 60 tonnes [66 tons] of stores were dropped in December 1986–the first time this had been done in the polar night. The leader for this period was A. Chernyshev."[30]

Along the ice-rich shoreline of Arctic Canada, 15 June found Ottawa's ice-island base at 80°52′ N, 97°49′ W, a position northwest of Axel Heiberg Island and still on the continental shelf, but southwest of its position when first discovered. The berg was drifting as hoped. "The ice island," Hobson wrote to his files, "continues to move in a southwesterly direction down the coastline of the Queen Elizabeth Islands. There has been minimal movement in 1986 because of cooler temperatures throughout the entire Arctic."

On-island, a typical field season embraced the months of February into September; during the dark season, the outpost lay unoccupied. (At 80°, early April brings continuous daylight, October total darkness.) Had the island-raft shifted into Alaskan-Siberian waters, it might have been exploited the full boreal year. When the 1986 occupation closed out, Polar Shelf boasted two fertile research seasons (1985–86) of varied customer usage. That November, base manager Michael Schmidt offered a prospectus for Ice Island:

> The Ice Island has to date provided an excellent opportunity to develop the technology and means to operate a remote scientific research station in the Arctic Ocean. If, as envisioned, the island drifts around this vast ocean, the time, effort and money devoted at this time will be repaid by an efficient and safe environment in which to conduct scientific experiments. This last season has provided a solid foundation upon which to build.[31]

As it happened, the project was curtailed during 1987 to limit most of the science to the span March–June. The four-man crew that opened camp (on 3 March) found its buildings surrounded by drifts up to two meters high. Eight days later the first parties began to arrive. From 12 March through 6 April, a 1.13-kilometer (3,700-foot) runway was constructed on a close-by frozen lake surface. Thaw-season ablation and ponding thoroughly vexed station staff–the worst melting of three seasons atop the island ("All six cabins have settled...").

> The runway... becomes unusable around the middle of June due to excessive melting and the consequent accumulation of water. Therefore from mid-June until the latter part of August it is continuously necessary to relocate the Twin Otter strip to other locations, based on the prevailing surface conditions on the island. The process of site selection and preparation is lengthy.

Note the flexibility afforded by the STOL (short takeoff and landing) performance of the DHC-6 Twin Otter. Available in various configurations (wheels, wheel-skis, tundra tires, floats), the Twin has a minimum takeoff and landing requirement of fourteen hundred feet. There's room for nineteen passengers or mixed passenger and freight loads. This thaw season, a skeleton staff would remain encamped till closure (mid-August) attending to facilities and minimizing losses from encroaching meltwater. Despite an abbreviated field season, a total of 1393 person-days or 157 individuals were supported. (See Table 6.)

The camp received most of the visiting scientists during the mid-March to mid-May time frame with additional parties working to the end of June. Projects included heat flow measurements, marine geology, chemical oceanography and historical research. The Ice Island's track was monitored for the duration of the season.[32]

SP-29 AND SP-30

In spring 1987, *Sibir* logged a spectacular first. Ordered into the central Arctic, the icebreaker would conduct multidisciplinary studies in oceanography, sea ice, meteorology, geophysics, marine geology, and geodesy. As well, her operational orders included the geographic pole and some unusual elements: to evacuate the SP-27 ice station and to deploy its successor camp. SP-27 floated at 86° W, 360 kilometers (224 miles) off the geographic pole. Its drift-track vector was south via the transpolar drift. When one of its stations neared the Fram Strait outlet, Leningrad elected to pack it up, and preparations were undertaken to evacuate. Elsewhere, far inside the pack-ice edge, SP-28 recorded its own useful work. An airdrop that April–courtesy of *Sever-39*–debouched twenty-seven staff, along with two dozen fuel platforms, onto the site.[33]

Sibir had stood out from Murmansk on 8 May. The research expedition (for such it was) then crossed the ice edge of the Barents Sea. Until 17 June, *Sibir* navigated entirely through ice (much of it multiyear); on the 18th, she approached the floe rafting SP-27. Broadside on, *Sibir* recovered the camp's equipment, instruments, and stores after which she took aboard its personnel. SP-27 was officially closed on the 19th.

The icebreaker now steamed eastward. Maintaining a track north of 85°, guided by satellite imagery, scientific studies were conducted along ship's course. When the forecasted ice divergence (leads and ridging) near the pole (87° to 90° N) was confirmed by helicopter reconnoiter, the course was set for Ninety North. On the afternoon of the 25th, *Sibir* reached the place. "In the immediate area of the pole," Ivan Frolov, scientific leader, writes, "an entire series of studies was carried out, the flag of the USSR hoisted and a meeting of the expedition partici-

TABLE 6.
OCCUPANCY AT CANADA'S ICE ISLAND DRIFTING STATION, 1987 FIELD SEASON (POLAR CONTINENTAL SHELF PROJECT)

MONTH	PERSON DAYS	NO. OF INDIVIDUALS[1]
March	317	35
April	470	61
May	183	22
June	198	21
July	71	3
August	133	8
September	21	7
Totals	1393	157

1. Another twenty-four individuals were housed in a separate camp established by the Canadian Hydrographic Service during March and April. April taxed ice-island facilities: camp accommodations were stretched to the limit, with extra bunks placed where possible. Effective the end of June, only a skeleton staff was on-island, to maintain the facilities. One final project—at the end of August—was undertaken prior to camp closure.

pants held." Following a seventeen-hour loiter, *Sibir* cleared the geographic pole.

Station SP-29 had been slated for the area northeast of Severnaya Zemlya (North Land), at about 80° N, 113′ E. The planned position was in a zone of intense ice dynamics. *Severnyy Polyus-29* would in fact prove short-lived (1987–88). Still, the history of its deployment and occupation is illustrative of Soviet prowess and determination—as well as of the exigencies of far-northern logistics.

Typically, the process began a month in advance. Occasionally, one aerial sortie was needed to locate a suitable floe; sometimes, weeks of camp search. Success is much easier in springtime; indeed, if executed by air rather than surface ship, searches are restricted to that season. (Icebreakers, though, enhance flexibility in confronting conditions prevailing in the sea area of interest—in this instance, very difficult ice.) Sea search is more difficult in the autumn, given its weather and the likelihood of fog. As the pack unscrolls, the heads of *Sever* and the station-to-be study the canopy in concert with ice specialists. Multiyear ice has a specific color; also, such ice invariably is veneered in old snow. So the search is fundamentally visual. Although the direction of ridges is an indirect clue to the age of ice, the most important indicator is ridge *smoothing*. Upon touchdown, the landing crew decides where to set camp and prepare a suitable runway(s). (Coring reveals no specific ice qualities save thickness.) An experienced *polyarnik* knows the signs; he need not core in order to decide—but he does. There must be additional good floes in the vicinity—for insurance, in the event of fracturing. Plus young ice, for airstrips.[34]

After the foundation is chosen, an aerial ferry ensues. Overall logistics is largely the responsibility of the attending *Sever* expedition. The lives of the entire *Severnyy Polyus* party rely upon the station leader. Well before deployment, he has picked his team from nominees, taking into account proposals made by the various institute departments. As might be expected, the character and skills of each man are absolutely vital; any leader must know his men. As for the science program, usually this is assigned. Still, given the vagaries of weather and chance, a station head might exercise his initiative during the yearlong sequestration.

In the end, a good outcome is dependent on timing, instinct, and luck.

Prior to its SP-29 drifting station, the institute had exploited multiyear ice. In this instance, a one-month air search had realized *ten* candidate floes, none deemed suitable. (Additional ice reconnaissance had found seven.) A further month turned up two pans appropriate for a station, one of them an iceberg. As were the others, the berg was rejected: the ice fields enclosing a berg would compromise pack-related investigations.

The season well advanced, "29" was surface-deployed. Having taken on equipment and instruments for the nascent station, *Sibir* stood out from Dikson on 4 June. She crossed the Kara Sea then, in the Laptev, entered close pack ice. On the 6th, following a helicopter reconnaissance, a floe measuring about 700 by 885 meters was selected by Valery Lukin, head of the *Severnyy Polyus-29* drifting station expedition. After its thickness had been confirmed (2.5 to 3 meters), a second MI-8 sortie deployed a radio beacon, but onto another floe because of fog. A three-day passage through hummocked ice delivered *Sibir* to the beacon. A search for the chosen pan found it broken. Thought usable anyway, it was found upon inspection to have numerous small cracks and fractures. And so a "special meeting" was convened aboard ship to analyze the situation.

Ice action in this sector poses great risk. Further, the available pan was small, precluding a runway. (Helicopters could have reached the station, if necessary.) Still, a small pan might prove most stable. On Lukin's decision, the expedition settled for the floe onto which a beacon had been set and commenced off-loading. Officially, SP-29 became operational on 10 June near 80° N, 113° E—a blank in the northern Laptev Sea. Never before had so small a floe conveyed a long-haul Soviet station. The decision proved wise: SP-29 was to drift—unchanged, productive—for fourteen months of self-imposed exile.[35]

Sibir stood pierside in Murmansk on 19 June. "The experience gained during the May–June 1987 voyage of *Sibir* to the North Pole," Frolov concludes, "convincingly proved the possibility of carrying out research expeditions aboard nuclear icebreakers in high latitudes during any time of year."[36]

That autumn, northeast of Wrangel Island and due north of Bering Strait (171°), *Severnyy Polyus-30* (1987–91) was deposited onto floe ice by *Vitus Bering*, *Yermak* assisting. Functional on 9 October, this station hosted three shifts (1987–91) of *polyarniks*. Its predecessor, thoroughly beset, was closed out within a year of "30" occupying its own ice. On 19 August 1988, SP-29 ceased operating. Shifted less than five degrees poleward within heavy pack, "29" *had* sprinted across nearly fifty-seven degrees of longitude (112°59′ to 56°34′ E) far to seaward–north of 80°–of central Siberia.

APLIS 88/OPAL 88

In fading dark, the sun rising to ever-higher elevation-angles, aircraft out of Prudhoe Bay scouted the drear sea-ice cover off North Alaska. They sought a suitable floe north of 72° for the seasonal APLIS 88 drift. On day One of camp-search, reconnaissance located a raft that satisfied criteria. Set-down took place on 1 March 1988, on a nearly four-by-four-mile (six-by-six-kilometer) multiyear floe ranging from five to eight feet in thickness—more than ample to support structures (sleeping, mess, power, science, logistics) but not too thick for the drilling of hydroholes. A further advantage was a large refrozen lead embedded in its middle. The two-month bivouac would be established and then maintained by personnel from the Applied Physics Laboratory (APL).

> The camp was established at the edge of a refrozen lead, which was essential for building a runway for aircraft, the only means of transportation to and from APLIS 88. The finished runway was used almost daily by a Twin Otter and a CASA 212-200. It was made long enough to handle the landing and takeoff of a Lockheed Hercules (stretch version) that was used twice for delivery of fuel in drums. Multiyear pressure ridges and rubble fields surrounded the lead and the camp. All parts of the floe were easily accessible by snowmobile until, in the latter part of the exercise, one end of the runway was lost when high winds opened up a new lead.[37]

The average population was about fifty–sixty-two at the height of activities. Weather information was recorded several times daily. In addition to current and ice-related research, a noise-measurement program was conducted emphasizing underwater acoustic propagation. Sound-speed profiles are essential to understanding acoustic phenomena, so conductivity, temperature, and depth

(CTD) casts –"the most important basic measurement for underwater acoustic studies"–(forty-two in all) were a pressing chore. Adrift in the Fram Strait sector this spring, OPAL 88 floated its own acoustics-related investigations.

—•—•—•—•—•—

That October of 1988, Ottawa's northernmost bed-and-breakfast floated (still) close inshore, off the Arctic Islands. The ice island had not moved as hoped. Having shifted some 805 kilometers (500 miles) to the southwest off the Queen Elizabeth Group, wind and current now conspired to push the berg *south*. This unlikely, unwanted course would shuffle the slab into the archipelago where, denied the gyre, its worth as research raft would cease. "The Ice Island evidently moved some 60 kilometers [37 miles] in a northerly direction during this last month," Hobson recorded in November, "which may get it back out into the mainstream of Arctic life from the backwater in which it has resided for the past year or so. . . . I sure hope that it gets out there and starts down the coast and out into the Arctic Ocean where it belongs."[38]

CEAREX

That autumn, in the Norwegian Arctic, the drift phase of the Coordinated Eastern Arctic Experiment put to sea. An Office of Navy Research–sponsored investigation, CEAREX was designed to understand how air, sea, and ice work as a system during the onset of fall freeze-up, and in the dark winter period. A huge venture, data collection would exploit in situ and airborne measurements along with satellite sensors. From September 1988 through May 1989, a series of field studies were staged from a Norwegian ice-strengthened ship and two ice-based camps. In addition to sled-mounted instruments, six buoys were helicopter-deployed about the vessel to obtain ice-motion data. (These recorded unusual stresses.) A flight program augmented the experiment's drift phase: a NOAA (National Oceanic and Atmospheric Administration) P-3 aircraft based out of Norway. Its mission was a series of instrumented flights to the vicinity of Fram Strait collecting data on temperature, humidity, and wind; fluxes of heat, moisture, and turbulence; information on solar radiation, gas, and aerosol; and the properties of surface ice (important to heat exchange).

The primary objective: to understand the structure and function of large- and small-scale mechanisms controlling the transport and distribution of oceanic heat from warm Atlantic Water advected into the basin through Fram Strait. (Other heat source: absorption of shortwave radiation through leads and polynyas.) Escorted by the U.S. Coast Guard's *Northwind*, *Polarbjorn* was allowed to freeze in on 16 September. Position: 82°41′ N, 32°26′ E, north of Svalbard. The ship's complement was twenty scientists supported by eight crewmen. The vessel lay at the edge of two adjacent pans designated Alpha and Beta. Focusing on the transition to winter dark, CEAREX investigators observed air, ice, and ocean together, as a *system*. In the words of one drift-group report:

> Specific observations were carried out under independent research programs by investigators from a number of organizations. The concurrent information provided an opportunity to measure and understand the relationship between stresses observed in individual ice floes and the geophysical driving forces and overall ice conditions, and to identify and describe noise generated by different processes. The coordinated effort has resulted in a [unique] complete set of data for helping to understand the dynamic interactions among ice, ocean, and atmosphere.[39]

A series of Twin Otter ferry flights (Norway base) rotated personnel for each leg of this unusual seagoing experiment.

A complex, coordinated program, CEAREX proved fruitful despite fall–winter operations and (in November) contrary wind forces, which pushed the pack against a small island of the Svalbard archipelago. (See Table 7.)

> The ice deformation that occurred when strong northerly winds pushed the pack against Kvitoya Island created extremely rough surface conditions.... As the *Polarbjorn* was dragged through the shear zone surrounding [the island], extremely heavy ridging was observed. The amount of ice crushing that occurred during the formation of ridges was larger than expected and may suggest new constitutive laws for describing ice stress.[40]

Deformations obliged early retrieval of equipment off surrounding ice. One parameter, ambient noise, had been monitored nearly continuously from about 10 October through 18 November, after which ridging and rubbling prevented use of the hydrophone cables. Breakup canceled further observations, so most of the scientific party was retrieved on 12 December. But the meteorological and bathymetric observations persisted as conditions allowed. All drift operations stood complete by mid-January 1990.

Despite its abbreviation, CEAREX's expert logistical support and chief scientists—experienced professionals all—had realized success. At the University of Washington, Andy Heiberg summarized the Polar Science Center's contract support for fiscal 1989 to the chief of naval research with marked satisfaction.

> This nine-month project was staged from northern Norway, and with its ambitious objectives and wide participation, it offered many new and demanding challenges for everybody. Nevertheless the operation was carried out close to plan, within budget, and came to its scheduled conclusion in

TABLE 7.
Operations Overview for the CEAREX Drift Experiment, 1988, Eastern Arctic Ocean—Eurasian Basin[1]

The measurements made at the ice camp included:
a. A horizontal and vertical array of discrete temperature, conductivity, and velocity measurements sampled at high frequency.
b. High vertical resolution profiles of temperature and conductivity and profiles of temperature, salinity, and velocity microstructure.
c. Acoustic Doppler current profiler measurements of water-velocity profiles.
d. Turbulence measurements in the mixed water layer and pycnocline.[2]
e. An acoustic scintillation array for measuring velocity patterns in the mixed water layer.
f. An array of ice tilt and strain measurements.
g. Mapping of the ice underside around the measurement arrays.
h. Standard meteorological measurements, including radiosondes (balloon-lofted instruments) and a micrometeorology study.
h. A suite of bio-optical measurements in and under the sea-ice canopy.
i. An airgun operated at the main ("O") camp to serve as a source for underwater acoustics measurements at a companion ice-borne acoustics ("A") camp.

Three programs made the following measurements deployed away from the ice camp:
j. A helicopter-borne CTD (conductivity, temperature, depth) survey conducted in an area within 75 kilometers (46.6 miles) from the ice camp.
k. A line of thermistor chains and current meters established between two camps. The line was also sampled as part of the helicopter-deployed CTD cast program.
l. SOFAR (sound fixing and ranging) equipment and drifter array installed between the two camps.[3]

1. The Coordinated Eastern Arctic Experiment was designed to observe atmosphere, ice, and ocean behavior concurrently, during the transition (freeze-up) and dark winter periods. Main platform (frozen-in): the Norwegian ship *Polarjorn*. Along with supporting ice camps and research aircraft, autonomous buoy arrays were deployed to record ice-motion data and underwater sound.
2. Pycnocline refers to the steep density gradient between the upper, less dense layer and the deep denser water. A function of both temperature and salinity, the pycnocline in the Arctic Ocean is found at depths of from fifty to two hundred meters. In other oceans, it is customary to refer to the thermocline—a steep density gradient in temperature.
3. Most field programs ended when wind forcing caused ridging of surrounding ice, obliging removal of all ice-rafted equipment and termination of the experiment.

Source: CEAREX "operations overview," 21 March 1988.

May. . . . During this time, one research ship and two drifting stations provided support to more than 6,000 person-days of science. Aside from the wealth of new information gathered, a major accomplishment, therefore, is that everybody returned safely and that no one suffered any lasting physical harm. For this, credit goes to everybody that participated. Our fair allotment of good fortune is also acknowledged.[41]

A subsequent workshop reviewed results and helped stimulate interdisciplinary analyses and plan publications. CEAREX had realized the most comprehensive air-ice exchange data set to that date.

—•—•—•—•—•—

The Intergovernmental Panel on Climate Change (IPCC) dates from 1988. Its reports, each painstakingly peer-reviewed, have shaped the global agenda on climate change. Enjoying a reputation for credibility and consensus, the Geneva-based IPCC has been called the closest thing to a global scientific consensus in climate forecasting—an arcane, contentious, and intensely politicized subject.

Global warming was now [post-1988] firmly in place as an international issue. In many countries it was hotly debated in national politics. The scientific community was taking up the topic with far greater enthusiasm than ever. Conferences proliferated, demanding time from researchers, government officials, and environmental and industry lobbyists. As one conference delegate put it, the "traveling circus" of the greenhouse effect debate had begun.[42]

Understanding of the climate system is tested by how accurately climate models reproduce not only present but also past conditions and natural variabilities. Long-term observations are foundational to accepted predictive outputs. Yet for direct data, forecasters can access a mere one hundred-year-long record of near-term climate—too short a span to use in distinguishing long-term trends. For knowledge of the total system, indirect proxy indicators are therefore exploited for reconstruction of the deep record: lake and sea-level changes, fossil corals, long-lived trees and pollen records, ice cores, deep-sea microfossils.

Climate involves interactions among the atmosphere, the oceans, the land surface and its vegetation and hydrology, and the cryosphere. It naturally varies on time scales ranging from interannual (El Niño) to millennia or longer. The instrumental record of a hundred years or so is clearly inadequate to help us understand these processes. . . . Paleoclimate reconstructions fill this void.[43]

By showing how the system once operated, by going back thousands even millions of years, researchers can deduce probable future changes. Within the inherent chaotic variability of climate (nonlinear dynamics), what is the stability of pack ice to perturbations as Earth's temperature rises; what percentage of the water now covered by ice will give way to open water? In a 1989 editorial "The Arctic: A Key to World Climate," the journal *Science* framed the unresolved issues pertaining to the sea ice-climate system:

> The Arctic is part of a great global heat engine. Changes in the arctic atmosphere, ocean, sea ice, and permafrost are early precursors to climate change elsewhere. In the past, those changes have been drastic. Only 18,000 years ago, virtually all of Canada and some of the United States were covered by a thick layer of ice.
>
> One of the obstacles to confidence in predicting the future of the arctic climate is an imperfect knowledge of the past. We know that 70 million years ago, the climate was mild and the Arctic Ocean was ice-free. Sediments formed about 5 million years ago contained glacially related materials. We know little about what happened in the long interval, and knowledge concerning more recent events is sketchy. . . . The importance of understanding the past, present, and future of the arctic climate requires that support for [sediment studies] have a top priority. Desirable efforts include more studies of fossils, an international program of deep drilling in the Arctic, more weather monitoring, and additional satellite surveillance of the polar region.[44]

Increases in computer power, ever more sophisticated models, and new measurement technologies have helped mightily. Despite the uncertainties, climate scientists have advanced their understanding of what they don't understand. Complex linkages and feedback processes between the atmosphere, hydrosphere, cryosphere, biosphere, and lithosphere (Earth's outer shell) are now well recognized. Yet a physical understanding of the planet's fluid systems stands incomplete. Monitoring continues unabated, courtesy of both repeated field programs and remote sensing—a global observing system. Gathering observational evidence, ships steam and aircraft sortie, men deploy, buoys drift, satellites orbit and image. Mountains of archived data are "crunched," the climate models (there are fifteen or so) are constrained, calibrated, and verified—and simulations run. Still, the uncertainties that undermine the reliability of the results persist.[45] But a connection between human activity and climate change *has* been teased clear; the greenhouse signal, in other words, has emerged unequivocally from the "noise" of natural climate fluctuations.

The Polar Research Laboratory was again on ice in spring 1988, this time encamped close to the pole. On 26 April, CRYSTAL floated at 88°12′ N, 72°67′ E longitude—130 miles removed, in the sector claimed by Moscow. A din heralded visitors. An MU-8 Aeroflot twin-turbo helicopter pounded in and, after circling the main encampment, landed. Twenty-eight Russians disembarked—nearly the entire third shift of SP-28. Their camp floated at 89°26′—just off the pole and mere miles from CRYSTAL. Although brief, the surprise drop-in was warm, with "smiles all around." Permission from Washington granted, PRL logged a reciprocal visit. The American contingent found itself well received: "They were very gracious and extremely friendly," Beau Buck recalls.

> We had brought along an American flag that all of us had signed in ballpoint and Capt Dick Porter [chief naval officer's O-in-C of the East Arctic operation, based at Nord] presented this to their camp leader. He was so touched by this gesture that he solemnly kissed the hem of the flag and immediately ordered one of his men in Russian, who ran out of the mess hall post haste, lowered their hammer and cycle [sic] (a really large one that always flew over their camp) and brought it to the camp leader, who presented it to Porter. Dick kissed it and we all had another round of vodka. Touching moment.[46]

At mid-year, U.S. president Ronald Reagan put his arm around President Mikhail Gorbachev and declared the evil empire an oddment of the past. Communist ideology had run its course. Autumn 1989 witnessed a political earthquake: the collapse of the Soviet empire in Eastern Europe—the defining event of the decade, when Gorbachev truly ended the Cold War. Meantime, Leningrad launched the Soviets' thirty-first drifting station. On 22 October 1988, *Severnyy Polyus-31* (1988–91) was established almost due north of Barrow, less than three hundred miles out. Led by V. C. Sidorov, head of the station, this first shift—thirty-nine personnel—would not see its at-sea reliefs until 23 December 1989—a thirteen-month sequestration.

Off the Canadian littoral, Ice Island's Krochak was logging a protracted tour of his own—twenty-two weeks, from February to mid-July. "It was a long time, a long, long time," the chef recalls. "Busy. And long. I'll never do it again." Thereafter, nine- to twelve-week tours were on order before a three-week respite. Ice Island fed well. "Sometimes there are five salads," one tenant records, admiringly, "laid out beside the main course, which could be prime rib, fillet of sole, steak and shrimp, or veal cutlets. There are often three or four pies, a cake, some cookies, and a basket of muffins beside the coffee machine. The fridge is full of yogurt, milk and juice."[47]

Profound isolation plays upon the psyche. Physical stamina is not enough. During analogue missions in harsh, isolated environments like Antarctica, NASA has found that more than 10 percent of subjects develop serious psychological adaptation problems, and up to 3 percent experience symptoms of psychiatric disorders such as major depression.[48] The National Science Foundation is understandably cautious as to candidates performing scientific research on the frozen continent.

> Before traveling to Antarctica with the U.S. Antarctic Program, all persons must be certified to be in good health; after the [research] proposal is approved, a physical examination is administered for this purpose. Candidates for work at antarctic stations during the long winter isolation also must pass a psychological screening.[49]

The Canadian camp floated remote from any habitation, farther still from family, friends, trees, grassy scents, free-flowing water. Long-duration isolation—living and working in confined spaces—can affect individual performance as well as the whole mission. A pressing sense of aloneness, of isolation, is intensified by

THE ICE ISLAND OCCUPIED BY CANADA, 13 AUGUST 1989. ITS CAMP IS BARELY DISCERNABLE (CIRCLE). AS EXPECTED, "HOBSON'S CHOICE"—NAMED FOR THE THEN-DIRECTOR—DRIFTED ALONG THE CANADIAN CONTINENTAL SHELF, SOUTHWEST. NOTE HOW THE SLAB CONTRASTS WITH THE ENCLOSING PACK ICE. MELT PONDS ARE CONFINED BY THE SLAB'S RIDGE-AND-TROUGH SYSTEM. (MICHAEL SCHMIDT, CANADIAN GEOLOGICAL SURVEY)

outside forces. Radio blackouts are common and may persist five or six days. ("When it's bad," Dave Maloney advised, "you can't raise anyone.") High winds or blizzards and the scarcity of agreeable weather may ground all airlift resupply. In 1990, a five-day blow with winds stronger than forty knots kept everyone confined to their ice-island cabins. In 1985, no plane arrived for ten consecutive days. Waiting on the weather, with the science slowed down (or stopped), can grind on scientist and staff alike. "It's hard on your mind," Krochak says. "As you start counting the days you're asking for a lot of trouble 'cause your mind can play horrendous tricks on you—it can really do you in."[50]

Taken-for-granted conveniences are absent. The only water: melted snow. Toilets: stainless steel pails lined with plastic bags. Living quarters consisted of plywood huts equipped with bunks—bring your own sleeping bag. The amenity most missed, perhaps: a shower. A geophysicist termed this the "hardest part" of her stay. (The year 1986 saw a shower installed.) Many made do with sponge baths, heating a miserly volume in an impromptu stall. Others did without. "Some guys would wait a month—two months," Krochak recalls, laughing. "You could always tell 'em cause they were the ones who ate alone."

Life-on-ice is not all tedium and discomfort, hard work and inconvenience. Investigators are engaged in their work—they would not be on ice otherwise. Contributing to one's own profession is rewarding, even heady. And the splendor of the Arctic enfolds one. It is a haunting world of violent beauty light-years removed from the hustle and noise of workaday experience. The author knows this place. The horizontal scale is immense, so large that camp—any outpost—shrinks to absurd insignificance. Snowmobiling to an edge, one confronts a soundless, formless circuit of white beyond the range of vision. "You go to the end of the island and you sit there," Krochak mused. "You see the ocean—summertime—you see open water as far as the eye can see—and it's black, mean, sinister." Cold hits like something solid. "Here, in this arid cold," a visitor wrote, "ice seems to have lost all relation to water, to have forgotten it was ever wet. It bears no resemblance to the sea than coal to a tropical swamp. It is a whole landscape of some new gemstone, too abundant to be precious but far too magnificent to be common."[51]

"I like my camps to be a joyful place," Chef Krochak insists. "You have to become almost like a small family." People make it a memorable, communal experience. "These are total strangers you meet at the beginning of the season. But when you spend twelve, fourteen, sixteen weeks together, they get to know a lot of private things of your life that very few people know—that maybe you or your wife share.... It becomes very close." And "When a program is ending and the people are going home I try to make a special dinner. Something that they'll remember. I dress up the salads. . . . nice prime rib dinner with Yorkshire pudding . . .

something like that you tend to remember, especially *where* you ate it. [Laugh.] And the next morning [I] make them eggs Benedict before they hit the plane."[52]

THE END OF SP-30 AND SP-31

Exploiting its system of logistics and support bases with profound expertise, detachments of Soviet personnel have prospected vast sectors of the white-ocean Arctic–a research crossroads of numerous paths. Such was the case in 1990. Far to seaward of North Alaska, on 24 April, the third shift for SP-30 relieved their comrades–by air. Closer in, SP-31 had welcomed *its* second shift three days before. Teamwork, organization, and know-how had realized a record of performance that "in its scale, persistence and consequences dwarf those of any other state.... Whatever our judgment on the details of Soviet activity in the Arctic, in the ensemble they can only be called outstanding."[53]

That November–December, a cargo paradrop via IL-76 was made onto the platform for SP-30 and that conveying SP-31.

On 8–9 January 1991, high-latitude air expedition *Sever-42* transported "scientific appliances" to this floating pair of camps, along with food products and scientific and operating personnel. In all, 110 tons of cargo and nearly 100 individuals met the ice. Seventeen personnel–the next shift–had reached "30" camp, twenty-three to SP-31. But ice conditions left the strip supporting SP-30 too narrow for direct deliveries by large planes, so the sister camp was exploited–receiving the deliveries of necessary supplies and staff. The airlift platforms were two AN-2s along with an IL-14 based temporarily at "31." Deploying from Pevek and Cape Shmidt in the Siberian Far East, the AN-12s conveyed fuel. The IL-12, for its part, air-dropped drums onto SP-30 for retrieval, then delivery by the shuttling smaller pair.

Elsewhere on the canopy, "jumping" detachments assigned to *Sever-42* (four ski-equipped aircraft) conducted oceanographic survey work over the continental shelf, weaving 150 hydrological stations into the fabric of the year's operations plus another eleven in Long Strait (the corridor between Wrangel Island and the mainland used by coastal Sea Route traffic).[54]

The spring of 1991–the final gasp of Soviet power–saw SP-30 cease operations. On 4 April, its flag was lowered at coordinates 82°31′ N, 126°26′ W–a blank well inside the pack, far to the northwest of the Canadian littoral. The occupation had known drama. And much labor. Fracturing had twice necessitated a shift of the main encampment, once across a wide lead. The move consumed a month. Subsidiary camps floated free, cut off from base. Most vexing, perhaps, the runway length had been repeatedly sliced. One May-to-March span had no landing whatsoever; seventeen airdrops had substituted. In all, airlift support exacted six runway constructions, a few of which received only a single plane before fracturing to uselessness. (A too-short runway is no runway.) The last strip–impossibly broken–had necessitated a light-plane evacuation: an AN-2 aircraft exploiting SP-31 as an intermediate airstrip.[55]

In the Soviet "North Pole" series of drifting stations, a single airstrip has seldom proved ample. Still, the half dozen needed to sustain "30" camp was unusual. The run of *Severnyy Polyus-31* would prove no less dramatic.

Hostage to the large-scale circulation of the gyre, the floe conveying SP-31 had shifted more or less east, then vectored south. This spring, the camp was tracking westward, more or less parallel to North Alaska perhaps ninety miles off. On 20 April, the next–third–shift encamped to take on its roles and responsibilities under station leader Nikolai Vinogradov, a veteran *polyarnik* of the long-serving SP-22. The station log for *Severnyy Polyus-31* now held twenty-four months of at-sea entries.

Within three months, however, Vinogradov, his men, and the station itself were at peril. The enclosing ice field had shifted into the region influenced by a relatively warm layer–the Alaska branch of the in-flowing Pacific current. One outcome was intense ice thawing. So began the crushing and disintegration that would scuttle the expedition. By July, sea-surface coverage had declined to eight-tenths concentration. Its foundation compromised, SP-31 was riding into desperation. Emergency evacuation was ordered on 23 July. The station floated at 73°33′ N, 161°04′ W–about 338 kilometers (210 miles) to the north-northwest of Point Barrow.

> As a result of these processes, the basic ice field was broken down into a few dozens of swimming fragments measuring 50 x 100 m. Further work of the station became impracticable, so on 23 July the decision was taken to urgently evacuate the station. . . . From 25 to 29 July the personnel of the station and the most part of the equipment were removed from the crushed ice fields with the help of the "Murmansk" icebreaker and MI-8 helicopter. For the first time since 1954, on 25 July 1991, the last scientific research drifting station came out of being in the Arctic Ocean. After closing of the drifting stations, the AARI [Arctic and Antarctic Research Institute] lost the unique ground for conducting its investigations into the physical processes in the atmosphere-ice-ocean system.[56]

The collapse of Soviet Communism came swiftly–the biggest event in Russian history since 1917. That July, Russian president Boris Yeltsin was inaugurated. And in August, hard-liners launched their abortive coup attempt in Moscow.

In Leningrad, the institute found itself in trauma. Yoked to problematic funding, starved for support, its internal agenda "notably declined" that portentous year. AARI yearned to return to the ice, to establish the successor in a lineage dating back to Shmidt and Papanin. Before 1991 was out, preliminaries were under consideration for the next station in the "North Pole" series of expeditions. In *Polarnie Novosti (Polar News)*, Valery Ippolitov records the mood:

> It is intended (if financing is available) to land in autumn 1992 a new drifting station "North Pole-32." The AARI Department of research expeditions has started the preparatory work to arrange all matters concerning this station. The setting up . . . will require a lot of financial expenses but its existence is badly needed for scientific and practical purposes.[57]

It didn't happen: *Severnyy Polyus-32* did not deploy. Evidently, Leningrad (now St. Petersburg) had exhausted all available funding for offshore initiatives. Had the last ice miles been drifted?

Inside the former Soviet Union, instability sapped morale, creating a sense of crisis that pervades all aspects of Russian life. Uncertainty replaced the suffocating—yet predictable—Soviet times. "Russia is now [1992] experiencing a spiraling economic collapse, with the great majority of its citizens reduced to the most abject poverty, the diet of most citizens limited to sufficient carbohydrate calories to sustain life itself."[58] Among the institutional casualties: research programs and the civilian scientific establishment.

> Science as a whole appears to have become a low priority item, and the public respect for science and its recognition as a fundamental component of modern society, so traditional in the [former Soviet Union], are fading rapidly. The psychological impact of having been pushed from the pedestal of a privileged class into an abyss of uncertainty is particularly demoralizing for the [FSU] scientists. When I asked: "Now that winter is over and spring is in the air, what is your main worry?" they replied in remarkable unison: "next winter!"[59]

"At this time," the observer added, "the cost increases in fuel, electricity and domestic transportation alone present a nightmare to research institute managers." One devastating outcome: economic migration. At the Arctic Institute that year, a deputy director remarked ruefully, "We want to keep our brains, our experience." No ice camps were operational, so staff were concentrating instead on writing up data, publishing papers, presenting at symposia. Still, the Northern Sea Route is a "very important road"; so ice reconnaissance and polar stations remain vital.[60]

Shorn of empire, Russia is a proud, humiliated country. For Western researchers, the demise of the Soviet Union has granted unprecedented opportunities for bilateral research, for tapping into Russian know-how. Its scientists and science officials stream to the world's lone superpower seeking aid and offering plans and proposals for collaborative work.

Proposals for specific projects with Federal agencies have resulted. Other agencies have taken the initiative to develop their own contacts and programs in Russia. Revelations about environmental contamination in the Russian Arctic and efforts to "rescue" scientific data from the former Soviet Union have been the principal motivations behind much of this activity.[61]

THE END OF HOBSON'S CHOICE

Inside Arctic Canada, Ice Island had blundered deeper into the Queen Elizabeth Islands. November brought a halt. But the following summer's breakup saw an unwanted detour resumed—another 30 kilometers (18–19 miles) south. Then yet another pause, this time off the east coast of Ellef Ringnes Island. Interisland movement stalled through the winter of 1989–90. But from May to late August 1990 (months when camp again was operational), the island shifted but a short distance from its more or less fixed winter coordinates.[62] Further position data were not recorded until August 1991, when a rapid drift southward began—motion aided by a pack breakup more extensive than normal.[63] "The ice island has not gone into the gyre as we had hoped," Hobson laments, "but has evidently moved down the east coast of Ellef Ringnes Island. I have not seen a track plot lately so I cannot confirm this. . . . I am sorry to say that it has not gone according to my plans. But then I was not really in control."[64] "They're doing some good science there," Ken Hunkins opined, referring to the wayward Canadian facility, "but it's been of limited use compared to what I think they hoped it would be."

The platform split during October and November. Officials delayed a dismantling, anticipating a reversal of ice motion. Hope was denied. During March 1992, in bitter temperatures, Polar Shelf began shuttling Twin Otters onto the berg (five men deployed), to retrieve about $3 million (Canadian) of scientific and other equipment. Hobson's Choice floated mere kilometers off Cornwallis Island. "We have to get everything off," James Godden told newsmen. Manager-in-charge at Resolute and responsible for the teardown, he was logging twelve-hour days, every day. "If it had gone where we thought it was going to go," the tall and lanky Godden added, "it would have been an ideal research station. Now we'll have to wait for another one." "It's cheaper to get it all off than to go out and buy it all new again." The final remnant to be retrieved was the Canadian flag.

All in all, the entire experience had proven costly. Yet, "It was a great effort for a few years," Hobson summarizes, "and actually some good science was done from that platform. So be it. It is now in the past."[65]

ICE STATION WEDDELL

Polar water masses profoundly affect the deep world ocean. The present-day pattern of circulation is driven by the sinking of dense, cold water in the high North Atlantic. Saline water cooled at the surface becomes relatively dense, sinking and starting an equator ward flow that forms deep water to fill the major oceanic basins. In turn, this sinking flow causes warm surface waters to be pulled northward.

Full understanding of the process can be gained only by studying the Arctic Ocean and the Southern Oceans themselves. On 1 January 1991, a startling Russo-American venture was announced: a joint manned drift off Antarctica—a first for the Southern Ocean. Wreathing the frozen continent, the Southern Ocean serves as a conduit—via the fast-flowing Antarctic Circumpolar Current—for climatic signals between the Pacific, Atlantic, and Indian Oceans.[66] Perennially ice-covered, the Weddell Sea has resisted both exploration and science. Planned for a large floe near where Sir Ernest Shackleton's ship *Endurance* had been beset in 1915, Ice Station Weddell was expected to move north, parallel to the Antarctic Peninsula—as Shackleton and his men had. Adrift for four to six months,

> Using icebreaking vessels, satellites, helicopters, airplanes and experiments on and below the ice floe, 10 American and 10 Russian scientists and 12 support personnel will study the relatively unexplored western Weddell Sea. The delicate balance there between the air, ice and ocean has critical effects on the world's climate and ocean currents; understanding their complex interactions is crucial to determining how changes such as the greenhouse effect might disturb the Earth's climate.[67]

American and Russian scientists had met repeatedly to hammer out plans for a joint expedition conceived in 1988. It would collect the first extensive data on the sea-air-ice interactions in the perennial sea ice of the Southern Ocean. The chosen floe would raft roughly eighty tons of equipment along with two research parties plus support personnel. For shelter, the ice bivouac used Soviet-model preassembled huts, as well as tents. Equipment and provisions were American: heaters, sleds, tracked snowover vehicles, food, even a microwave oven. "One worry," mused Peter Wilkniss, Ph.D., the National Science Foundation's director of U.S. projects in Antarctica, "is that the project could be affected by changes in the Soviet Union. The Soviet program will be under the Arctic and Antarctic Research Institute in Leningrad."[68]

Embarked in Montevideo, Uruguay, the expedition loaded aboard *Akademik*

Fedorov, an icebreaking ship. Having then steamed thirteen hundred miles south of Cape Horn, the chosen floe was in sight on 6 February 1992, at position 71°35′ S, 50°01′ W.

> More than 80 tons of equipment and gear were transported onto the floe [6.5 feet thick] and ferried by bulldozer and helicopter from the edge to the interior, taking precautions to distribute weight evenly and avoid triggering breaks in the ice. Food, fuel, survival gear and living quarters were also widely deployed, in case a break-up occurred. Some small cracks did open, but the floe did not break apart.[69]

Ice Station Weddell logged a 117-day, nearly four-hundred-mile track roughly astride that of Shackleton. Making common cause, sixty investigators logged research time atop the floe, some rotating via aircraft or ship, others remaining for the full mission. Antarctic bottom water–the frigid, nutrient- and oxygen-rich flow vital to fisheries and the global climate–got special scrutiny. Arnold Gordon, Ph.D., a physical oceanographer at the Lamont-Doherty Geological Observatory, was coordinator of the two-nation science program. The formation and dispersal of this water is, in his phraseology, "the refrigerator that makes the world's oceans cold."

U.S. and Russian icebreakers rendezvoused with the drift camp on 4 June, to complete the mission. Next day, the thirty-two scientists and support personnel began five days of decamping, loading gear, instruments, and personal effects aboard *Nathan B. Palmer* and *Akademik Fedorov.* Nothing was left behind.

Definitive climate-change conclusions cannot be made based on any one experiment or expedition. Nonetheless, the Weddell venture, Gordon enthused, was a "fantastic success." "Already an early assessment of the data suggests that we will have to alter our views of how this region fits into the global climate system," he reported. There were problems; there always are. Among these were malfunctioning heaters and less-than-what-was-promised toilets.[70] Contaminated fuel clogged the heaters and temporarily grounded helicopters, causing a major headache and necessitating a fuel drop to the floe. Still, good science resulted. Investigations of sea-ice dynamics by Russian investigators, for example, impressed U.S. researchers. "Scientifically," Gordon summarizes, "we [the United States] will do an ice station again–no doubt about that. Its the only way to go in that area." A frozen-in vessel, he muses, would escalate comforts, enhance scientific returns, and prevent premature closure due to breakup.[71]

Are joint drifts in the offing in high northern latitudes? The Russians, Gordon remarked, are "definitely keen on joint work in the Arctic." The "first steps" have been taken, the author was assured. So stated Sergey Priamikov, former head of department of international science cooperation at Leningrad's AARI. The realization of such plans lies in working cooperatively, he insisted, building upon a superb legacy of achievement.[72]

SCICEX

Gorbachev resigned as general secretary of the Communist Party and dissolved its Central Committee on 24 August 1992, effectively ending the Bolshevik era. A long-deteriorating Soviet Union and the bipolar world of the Cold War vanished. Competition now blindsided circumpolar science, like most every field of endeavor in which the former Soviet Union played a role.

Until the demise of the Soviet Union, the U.S. Navy's undersea fleet had been secure in its mission. Then the perceived threat vanished. Denouement involved changing national, hence naval, priorities: reduction, consolidation, controlling costs, new missions. Civilian oceanographers (and some forward-thinking officers) had long pushed for nuclear submarines as platforms for basic research. Given the national study program on global change, submarine-based research projects—civilian, unclassified—had become opportune. Fortuitously, the navy had begun to retire its *Sturgeon* (SSN-637) class of attack boats—the only class that can operate in shallow ice-covered seas. The Science Ice Exercise (SCICEX), a program variant on SUBICEX, was conceived.

Oceanographers were thus offered an opportunity to work with the service to save and also refit one *Sturgeon*-class SSN—a platform with year-round Arctic surveillance capability. A transit requiring months by icebreaker could be logged in mere days, collecting—quickly, cleanly—data and observation useful to oceanography, bathymetry, geology, ice cover, atmospheric science. "Nothing compares, when you want to explore the ocean," notes Dennis Conlon, head of ONR's High Latitude Dynamics Program. Ray Sambrotto, an oceanographer at the Lamont-Doherty Earth Observatory, was chief scientist aboard the 1996 SCICEX sub-ice cruise. "It suddenly gives you an access to an area that previously was inaccessible," he remarked. Max Britton could only marvel. "It was unheard of in my years, which I consider pretty recent, you know. We couldn't dream of getting anything on a submarine in the way of research while I was still around. But here you have a five-year program—a submarine *dedicated* to research."[73]

SCICEX became a seagoing program in mid-1994. By a memorandum of agreement between the chief of naval operations, the commanders of the Atlantic and Pacific submarine fleets, the chief of naval research, and the directors of NSF, NOAA, and the U.S. Geological Survey, a singular, eclectic ferry service into the Arctic Ocean was established and made operational. The prospect had NSF intrigued.

The recent availability of a U.S. Navy nuclear submarine to support a research cruise in the central Arctic Ocean has provided a significant boost to access in this remote area. . . . Continuation of such a project would prove extremely valuable to all elements of Arctic science.[74]

The kickoff mission stood out in 1993–a trial cruise, to demonstrate the concept. That August, a five-man team of civilian scientists was welcomed aboard USS *Pargo* (SSN 650)–part of the first-ever nonmilitary, unclassified scientific mission. George Newton, a former sub driver, was a member of the U.S. Arctic Research Commission. "This is sort of like a blind date," he said that July, preparations wrapping up. "Neither the navy nor the scientific community has done this in the recent past and therefore they're both anxious for it to go smoothly." In the central Arctic, *Pargo* operated under summer ice for three weeks. Participating scientists (subject to physical and security qualifications) systematically surveyed over 5600 miles along a track from the pole to the North Alaska shelf. Fifteen surfacings were logged, along with air measurements and water samples. As well, a half-dozen meteorological and oceanographic buoys were deployed.

In Washington, this single mission had NSF gushing:

Never before has so much scientific data been collected on one operation in the Arctic Ocean. The cruise has supported the data requests of 45 individual U.S. scientists, five of whom were on board. This result is considered extraordinary for an operation that was planned and executed on seven months' notice.[75]

Still, under-ice missions are risky. "The Arctic is so dangerous our subs rarely go there now that the Cold War with Russia has faded," a retired captain of the Center for Defense Information remarked. "Maneuver wrong and you could get icebound, killing everyone on board."[76]

Late in 1993, a committee of the American Geophysical Union summarized its concerns relative to basic geophysical research on understanding a threatened planet.

As we close in on the 21st century, we face significant global challenges. These include a burgeoning world population that requires more energy, minerals and water; increasing risks of losses from earthquakes, volcanoes, floods, droughts, and soil erosion; pollution of our air, water, and land; and the consequences of global climate change. . . . Humans may become agents of their own destruction unless they learn to manage planet Earth as a nonrenewable resource. But to do so requires scientific understanding of the

Earth's interconnected natural processes.... Policymakers at all levels—from global to local—face momentous decisions on a range of problems in the next few decades.[77]

Theory must be consistent with observation. Instrument and proxy records over a range of time scales are key to model input and to model validation—a reality check for climate models. Certain critical data requirements, such as ice motion fields, are met by the continuation of the Arctic Ocean Buoy Program.[78]

In 1995, the two thousand scientists of the Intergovernmental Panel on Climate Change issued their landmark report. The signal of greenhouse warming, the panel was convinced, had been teased from the inherent "noise" of global-climate data. The risks attending warming were spelled out, including possible shifts in vegetation and storm patterns—and the havoc such might wreak. Shifts in the water cycle will be perhaps the single most disruptive aspect, on both human populations and ecosystems.

Climate models grow ever more refined. Still, "Experts on all sides of the debate acknowledge that the models are imperfect, and even proponents of their use say their results should be interpreted cautiously." Perhaps the biggest and most recalcitrant uncertainties confronting projections is the role of clouds and pollutant hazes (aerosols, including soot and dust) that can either mask or enhance greenhouse warming.[79] Clouds interfere with the distribution of short-wave and long-wave radiation over sea ice, and thereby strongly influence the surface energy balance in the atmosphere/ice-ocean system.

SHEBA

During the late nineties, a project titled Surface Heat Budget of the Arctic Ocean (SHEBA) was funded by the NSF, with additional support from ONR's High Latitude Physics Program plus the Canadian and Japanese Governments. The ice-albedo feedback mechanism (heat exchange) is a key component in the heat and mass balance of the ice cover and the upper ocean. The 1997–98 SHEBA experiment would provide "ground truth" (calibration), in order to correctly assess the surface-heat budget using earth-orbiting satellites. Donald K. Perovich, Ph.D., was chief scientist. "We didn't have good numbers for computer-model simulations of global climate," he explains. "A major objective of the field experiment was to obtain a complete time series of parameters defining the state of the SHEBA column—the imaginary cylinder—over an annual cycle in perennial pack, from summer-melt through winter ice-cover conditions. Observations in the column included radiative and turbulent fluxes; cloud height, thickness, and properties; energy exchange in the atmosphere and ocean boundary layers; snow depth and ice thickness; and upper ocean salinity, temperature, and currents."[80]

SHEBA dusted off a lingering notion. Rather than a pack-ice camp and pared-down facilities, the heart of SHEBA was a frozen-in ship—the Canadian icebreaker *Des Groseillers*. "*Des Gros*" would log a yearlong winter in the Beaufort, covering the October-to-October seasonal cycle. "At the very beginning," Perovich recalls, "while trying to find an ice floe to be the experimental site, we discovered that the ice was approximately 1 m thinner than expected. A concurrent submarine survey [SSN 678 *Archerfish*] confirmed this."[81] An unsettling find: what it meant for Arctic ice and world climate was—and continues to be—unclear.

On 2 October 1997, freeze-in was logged 320 miles to seaward of Deadhorse, a coastal village southeast of Barrow. *Des Groseillers* was the first vessel thus immobilized since the Norwegian *Fram*. A base of operations, she off-loaded an adjunct camp and instruments, onto the ice. The largest Arctic project initiated by NSF, Ice Station SHEBA was supported by research aircraft, helicopter surveys, two nuclear submarines, and a Canadian satellite (Radarsat). As many as forty-five scientists would participate, representing Canada, Japan, the Netherlands, and the United States. Workaday pursuits resembled those at any science outpost. Still, for researchers at workstations nearby or servicing remote sites, the red-hulled drifter tendered a reassurance unthinkable at conventional drift camps.

> The *Des Groseillers* is action central. Its steep gangplank gets lots of traffic. At any given moment some of the 35 or so scientists and technicians or members of the logistics squad and ship's crew are likely to thump aboard or emerge on deck in the breathtaking cold wearing superheavy parkas and overalls, carrying radios, rifles, scientific equipment. They go through the polar-bear-proof cage at the end of the gangplank and take one of the snowmobiles parked in the "lot" or head out on foot for research sites. Most . . . are within a half-mile of the ship, and the farthest . . . is only three miles away. . . . Seen from the icebreaker's bridge, the people working at the research sites seem to plod along in slow motion. "No one runs in the Arctic," says Perovich.[82]

The upper, shallow portion of the water column was found to be warmer and less saline than had been recorded twenty-two years earlier—a troubling sign. The implication was that considerable sea ice had melted the previous summer. The melt at SHEBA proved long—nearly eighty days (beginning with rain on 29 May). Melt seasons observed at several SP stations had averaged fifty-five days, with ranges from twenty to eighty-three days. During its drift, SHEBA recorded air temperatures ranging from −42° C (in the floodlit dark of late December) to a relatively balmy 1° C during July, when the horizon-circling white lay pocked with ponds plus gray, softening snow. "During the summer melt season," Perovich notes, "the sea ice cover undergoes profound changes in its physical state and optical properties. As incident solar radiation increases and air temperature warms, the ice cover evolves from a highly scattering, snow-covered medium to a darker combination of bare ice, melt ponds and leads."[83]

In Eurasia, August 1998 saw collapse of the ruble, thus ending Russia's free-market frenzy. The post-Soviet economy continues to track a wayward course. Foreign investors (for instance) have been reluctant to join in developing Arctic Russia's energy resources because of the country's unstable history (hence risk), shifting tax laws, and a reputation for violating contracts and reneging on debt.[84]

Between 1990 and 1999, Russian industrial production fell by more than half. Though the economy has recovered somewhat in the past couple of years, the Russian gross domestic product is still well below where it was when the Berlin Wall came down. Poverty rates are much higher, life expectancy has fallen (almost unprecedented in a developed country), and much of Russia's industry is in the hands of former Communists and gangsters.[85]

Off North Alaska, the ice field conveying SHEBA programs would move

ICE STATION SHEBA (SURFACE HEAT BUDGET OF THE ARCTIC), APRIL 1998. MOTIVATED BY CONCERN FOR CLIMATE CHANGE, THE SHEBA FIELD EXPERIMENT (A FROZEN-IN ICEBREAKER) ASSESSED FEEDBACK PROCESSES THAT GOVERN THE THERMODYNAMICS OF THE ICE PACK, TO DERIVE BETTER NUMBERS FOR LONG-TERM COMPUTER SIMULATIONS (CLIMATE MODELS). THE EXPEDITION HAD STOOD OUT FROM WESTERN CANADA IN FALL 1997. (D. K. PEROVICH)

over 1740 miles (2800 kilometers). *Des Gros* passed the 80th parallel (from 75° N) on 18 September 1998–fall freeze-up. Within weeks, the huts and other temporary buildings were being dismantled and stowed on board *Des Groseillers* or its support ship, *Louis S. St.-Laurent*. Engines tested, the icebreaker got under way for Prudhoe Bay on 9 October.

Field data in hand, the project shifted to analysis and computer modeling. SHEBA had realized a superb return: 1.5 teribits of high-quality data, "the best field experiment I ever saw," ONR's Conlon marveled.

> Overall [Perovich writes] we obtained an extraordinary data set describing in detail the properties of the atmosphere, ice, and ocean over an entire annual cycle. We now know a great deal about a particular place for a particular year. The true legacy of the project will lie in how these data are used to understand the feedback mechanisms and to improve models so that we can accurately simulate any site in the Arctic Ocean for any year.[86]

The biggest feedback uncertainties for key model parameters involve clouds; water vapor is a powerful greenhouse gas. Among the SHEBA conclusions: the frequent cloud ceiling makes the ice melt *faster,* not slower.

As for ship-borne drift science in Arctic waters, not everyone concurs as to the cost-benefits and field efficacy:

> My thoughts on SHEBA [one expert opines] are that science and the Navy would have been better served by a series of small manned ice camps at the North Pole supported with automated buoys over a large area and replaced as they drifted south over a period of 2 years. Freezing in a ship in this day and age is an expensive luxury with unnecessary small-scale complications such as water pollution, air-pollution and the sail-effect of the ship.[87]

Chief scientist Perovich cannot agree.

> Arctic research has many wonderful observational tools: satellites, ice camps, autonomous buoys, and ships. The particular science questions and requirements of a program dictate the best combination of these tools to be used. SHEBA had the complex goals of understanding the ice-albedo and cloud-radiation feedbacks. This resulted in an extensive set of requirements, including a large interdisciplinary field team, an array of sophisticated state-of-the-art instruments, and an uninterrupted yearlong dataset. A ship-based experiment provided the best way to achieve the goals and objectives of SHEBA.[88]

All ships are tight for space. Still, the SHEBA program was well served. Instead of tents or huts, *Des Groseillers'* comforts were abiding, not least her French-style mess—"a fabulous restaurant." More to the point, SHEBA realized good science. As at Amundsen-Scott South Pole Station with its range of fields spanning the physical sciences, researchers need working space, electrical power, equipment, computer systems, and paraphernalia to do what attracts them to the top (or the bottom) of the world. For the SHEBA drift, investigators had the luxury to "focus your concern on your experiments, not on safety."[89]

The political winds off SHEBA blew strong through Ottawa. A long-festering concern is the issue of scientific sovereignty, specifically, American domination of research, especially in Canadian territory. The biggest single project in the Canadian Arctic in 1997–98, SHEBA returned the issue to boil. Funds for Arctic science, researchers lamented, had shrunk to $65.7 million for all government departments. Canadian research, a university biologist observed, could be characterized as "almost shut down." "Arctic science is in trouble in Canada. We just don't have the resources," a federal official lamented. Nor, many agreed, the political vision and commitment.

> The lobby volume rose last summer [1998] when U.S. scientists dominated the biggest research operation in the Arctic, even though the project was based on a Canadian icebreaker frozen into the polar icepack for a year. . . . This climate-change project meant that U.S. institutions last year probably financed more scientific research related to the Canadian Arctic than did Canadian agencies.[90]

In April 1999, Ottawa announced a near-doubling of support funds, the monies shifted to Polar Shelf from other programs within its parent federal department. "I hope this will encourage more young people to get into Arctic science," Bruce Rigby opined. Executive director of the Nunavut Research Institute, Rigby, a botanist, headed the panel that reviews research proposals applying for support from the logistics agency.[91] (In 2003, the budget for Polar Shelf was increased by $2 million, up from $3.2 million in 2002.)

SCICEX END

In August 1998, in support of SHEBA, USS *Hawkbill* (SSN 666) had surveyed ocean properties as well as ice thickness and bottomside topography. Spring 1999 had *Hawkbill* berthed at Pearl Harbor–Scientific Ice Expedition 99 completed. Diving under pack ice on 25 March, the boat had logged almost two months of research activities, surfacing several times. On board (since 1998) were the Seafloor Characterization and Mapping Pods (SCAMP)–the largest and most expensive civilian research system ever installed aboard a submarine. Among the deepwater areas surveyed (swath bathymetry) was the Gakkel Ridge, considered the world's slowest spreading sea-floor ridge. It is the deepest and most remote portion of the mid-Arctic Ocean system. Immense gashes in Earth's crust, these hydrothermal rifts extend tens of thousands of miles, encircling the planet. Here crustal plates are pulled apart and new seafloor formed. Water samples were collected, further geophysical bottom-mapping surveys compiled. As well, the ashes of Dr. Waldo Lyon–U.S. Navy Chief Scientist for Arctic Submarine Technology for a half century–were scattered at the North Pole.

After five active years (211 days in data collection), the plug for SCICEX was pulled. Determined to shed its aging *Sturgeon* boats, the fast-attack fleet was given over largely to the *Los Angeles*-class. Despite faded prospects for dedicated cruises, the notion of a follow-on program persists. Boats from other nations have been discussed, "accommodation" cruises tendered–using portions of classified navy sorties to collect high-priority scientific data. Meantime, an invaluable access to the Arctic Ocean has been forfeited.

> Despite excellent scientific results, the U.S. science community was not able to find the funding to support a conversion. And the Navy was not able to provide full budgetary support for conversion or annual operating costs. The *Mendel Rivers* [SSN-683], the last *Sturgeon*-class boat capable of under-ice operations [was] retired in January 2001. . . . A unique opportunity has been lost.[92]

Arctic Ocean ice has lost 40 percent of its volume in less than three decades. Is this a temporal change or an essential change? Is natural polar-climate variability the cause or, rather, the greenhouse-gas burden? Perhaps both? Whichever it is, retreat and thinning imply–within decades–an ice-free Arctic during the summer.[93] In the far high latitudes, cold water below the cover is a key element of ocean circulation patterns. Snow-covered polar ice is one of the planet's most reflective surfaces, bouncing back into space up to 80 percent of the solar radiation reaching its surface. In contrast, the open ocean reflects 10 percent of the solar energy and absorbs the rest. This trapped thermal energy may alter heat and mass exchanges with effects on ocean and atmospheric circulation (and weather) extending into the mid-latitudes.[94]

> Thus, an increase in the sea ice cover cools Earth's surface, thereby promoting further advances in sea ice; retreating ice warms Earth's surface, promoting further retreats in sea ice. This positive [destabilizing] feedback is believed to

be the main reason for the polar amplification of the warming observed in climate model scenarios for the 21st century.[95]

Covered by dark, heat-absorbing ocean, the Earth would lose its ability to regulate itself. Further, storm patterns and ocean circulation could drastically alter, particularly in the North Atlantic—a vital sector. The implications for Western Europe are profound. Further, sea ice (and the ice edge) serves as a critical habitat for many marine species. One oceanographer summarizes, somewhat colloquially, "If you stop forming ice in the Arctic, it screws up everything."

The IPCC released its third assessment report in 2001. In it, new projections were given for global-mean warming in the absence of policies to limit climate change, based on a new set of emissions scenarios. The full warming range over 1990 to the year 2100 is 1.4° to 5.8° C—a range significantly higher than that presented in the IPCC second assessment report.[96] Humanity's exact contribution to the warming remains problematic.

> Future emissions of greenhouse gases, their climactic effects, and the resulting environmental and economic consequences are subject to large uncertainties. The task facing the public and their policymakers is to devise strategies of risk reduction, and they need a clear representation of these uncertainties to inform their choices. Absent this information, policy discussion threatens to deteriorate into a shouting match.[97]

VIEW FROM SHIP'S BRIDGE AT SHEBA, APRIL AND AUGUST 1998. DURING THE ABLATION SEASON, AS MELTING INTENSIFIES AND LIQUID MOISTURE INCREASES, THE SNOW COVER AND THEN THE ICE SURFACE METAMORPHOSE. UNDERSTANDING OF THE PROCESSES AND FORCES INVOLVED IN THE GROWTH, DEFORMATION, AND DECAY OF SEA ICE ELUDES RESEARCHERS STILL. (D. K. PEROVICH)

EPILOGUE

> Arctic research provides a critical component of virtually every science element in the U.S. Global Change Research Program.
> —ARCTIC RESEARCH OF THE UNITED STATES (2001)

A CENTURY AFTER FRIDTJOF NANSEN and his *Fram,* much is yet to be learned. Remote and inhospitable, expensive and daunting, the deep-ocean Arctic constitutes one of the world's least understood precincts. The geology and geophysics have proven hugely complex. And fascinating. Significant gaps exist in knowledge of every applicable discipline of science: oceanography, geology and geophysics, sea-ice studies, atmospheric science, glaciology, ecology, hydrology, hydrochemistry, climate science.

It is still possible to be an explorer.

The Arctic Ocean is a natural laboratory of Earth's own rhythms, a repository of environmental records. Evidence of the history of climatic change awaits discovery in proxy indicators: sea ice and marine sediments, snow, glaciers, ice sheets. Knowing how systems once operated, by tracing natural variability back tens of millions of years, paleoceanographers will deduce probable near-future fluctuations.[1] Thus does present-day science grapple for answers. "By no means is the scientific research completed up there," Lamont's Dr. Hunkins says. This is true as well for the scientific programs that operate on Antarctica year-round— forty-five stations manned by eighteen nations.

Access is costly. (The average reimbursable charges for a Coast Guard cutter approximate $13,000 per day.) Closed to conventional surface-passage, radio, the icebreaker, aviation, and the microchip have transformed polar logistics. Aircraft opened the central Arctic Ocean to basic exploration, systematic inquiry, and quasi-permanent occupation.

The geographic poles have inspired men to penetrate unstudied realms. In the scientific process, the first step is to see what's there. Modern technologies deployed for new missions and programs have led to quantum leaps in knowledge and understanding. As we have seen, national security has tended to supersede pure science. Now, environmental data-gathering is the rationale for U.S. interest in the Arctic. (See Appendix 7.)

Since the late 1980s, U.S. policy in the Arctic has focused primarily on environmental protection. Human-induced climate change is one of the most important issues facing society worldwide—an engine for high-latitude science and research. Interdisciplinary collaboration transcends national boundaries. In the hypercomplex climate system, with its interconnectedness and interaction, and its myriad factors all fluctuating simultaneously, controlled experiments are impossible. Thus, insights into mechanisms derive largely from theories and from simulations performed with computer models. But what is the best model? Further, good simulations must be grounded in actual data. Regional and global forecasting models, as well as various climate models, devour basic atmosphere-ice-ocean observations; the demand is insatiable. Accordingly, systematic campaigns of oceanographic, cryospheric, and atmospheric measurements persist.[2] Happily for investigators of circumpolar patterns and processes, air-ice-ocean sampling has been revolutionized by new and emerging technologies—space-based, aircraft-borne, ground-based. The work cuts across interdisciplinary fields. "For the first time," a hydrologist observed not long ago, "we are setting out to understand the [whole] earth system."

> Just a few hundred years ago, that is, during the last instant of geologic time, we began a phase of exploration that provided, for the first time, the opportunity to comprehend the entire surface of the globe on which we live. We now know, as countless earlier generations did not, what the entire Earth's surface looks like. During this past century, we have revealed the nature of the seafloor in unprecedented detail, and we have developed geological maps of the surface of the continents almost everywhere. Earth exploration is not nearly complete, however, and we can anticipate continuing exploration, extensive mapping, and related activities throughout the 21st century.[3]

Global studies of temperature trends show that natural climate variability cannot explain the atmosphere's warming over the past century. Emissions of greenhouse gases and aerosols must be included in climate models to reproduce the observed warming, especially for the second half of the twentieth century.[4]

Epilogue

The consensus: *most* of the warming is caused by buildup in greenhouse gases. Ice-covered areas are receding; snow and ice lines are retreating; freeze seasons are shorter; species and ecosystems are shifting. Model scenarios predict a marked warming at high latitudes—and, indeed, instrument records show that warming is under way in much of the boreal wilderness.[5] Average ice thickness in the central Arctic Ocean has decreased from 10.2 feet in the 1957–76 era to 5.9 feet in the nineties. Not only is sea ice thinning, the canopy appears to be melting back. Recent computer models point to changing atmospheric circulation as the culprit for the abrupt thinning in the 1990s. Still, the decrease is real—in the 10 percent to 15 percent range. The world, it seems, is melting from the top. Ashore, negative mass balances from 1976–98 have led to substantial glacier thinning. The retreat of small alpine glaciers is nearly universal.[6]

Change is ubiquitous in Earth's history. And as Earth's climate changes—its harbingers are most vivid in the polar regions. The causes are (largely) identifiable, the consequences of inaction ranging from serious to catastrophic.[7] "Of course, whether to act is not a scientific judgment, but a value-laden political choice that cannot be resolved by scientific methods."[8]

> One of the great scientific challenges of the 21st century is to forecast the future of planet Earth. As human activities push atmospheric carbon dioxide (CO_2) and methane concentrations far beyond anything seen for nearly half a million years (prompting the strongest statement yet from the Intergovernmental Panel on Climate Change that human activities are warming the world), we find ourselves, literally, in uncharted territory, performing an uncontrolled experiment with planet Earth that is terrifying in its scale and complexity.[9]

Policymakers tend to be insensitive to ambiguity; they crave certainty. Greenhouse warming was officially declared to be fact in 2001, in the IPCC's third report. Unprecedented global warming, it reported, is "very likely" under way. Further, it is "likely" human-induced. In the confusing, sometimes contradictory study of climate, uncertainties persist.

> About all that climate researchers can say with any confidence concerning global warming is that the world has warmed during the past century and that much of that warming is probably due to humans pouring greenhouse gases into the atmosphere. How bad could things get as the world continues to warm? Scientists' bottom-up approach—trying to understand the role of every part in the dizzyingly complex climate machine—has left that question unanswered.[10]

By mid-year, the issue seemed clearer:

> The science itself is not in doubt. Of course there are continuing uncertainties about the proportion of natural to human-driven change, but the existence of human-driven change is clear. The conclusions of the Intergovernmental Panel on Climate Change and the main national academies of science (including that of the United Stares) represent a broad international consensus with little serious dissent.[11]

And, as 2003 commenced,

> The scientific evidence on global warming is now beyond doubt. Readers of [*Science*] during the past couple of years have seen one careful study after another documenting the role of anthropogenic sources of carbon dioxide and other greenhouse gases in global warming; describing the impact of past and present climate change on marine and terrestrial ecosystems; and measuring rates of glacial melting in the Arctic, the Antarctic, and on the tops of low-altitude mountains.[12]

The 41,000-member American Geophysical Union concurs. "Human activities are increasingly altering the Earth's climate," reads a 2003 position statement. "Scientific evidence strongly indicates" that humans have played a role in the rapid warming of the past half century. And it is "virtually certain" that increasing greenhouse gases will warm the planet. "The unprecedented increases in greenhouse gas concentrations, together with other human influences on climate over the past century and those anticipated for the future, constitute a real basis for concern."[13]

Currently, various international organizations and individual nations are planning for the International Polar Year 2007–2008. Like its International Geophysical Year predecessor, this IPY will address research in both polar regions and will be truly international in scope. The "year" will kick off during the Arctic spring of early 2007 and extend through the Antarctic fall of early 2008.

> The IGY will be a hard act to follow. But the half-century of polar science it ushered in has only deepened scientists' appreciation of the complexity and importance of polar processes. What happens at the poles is inextricably tied to patterns of cold and warmth, rainfall and drought. To have any hope of understanding what might happen in the future, scientists need a better picture of conditions at the poles and how they interact with and influence ocean and air currents.[14]

The oceans are a driving force of climatic variations, exerting both controlling and mitigating influences. The history of the North Atlantic deep-water formation is fundamental. Interacting with the atmosphere, the Arctic Ocean and its marginal seas—the marine cryosphere—play a key role in controlling weather patterns and longer-term, global climate. Much of the deep water of the world ocean originates in the Arctic-North Atlantic region. These cold, oxygen-rich flows "ventilate" the seas and, as well, provide an important mechanism for heat transfer and the cycling of carbon and nutrients. Regional changes in climate such as increased precipitation might slow or collapse this crucial ocean conveyor-belt circulation—with grim implications for the hemisphere.

In an era of space-based sensors,[15] drift-ice stations retain relevance and application. "Drifting stations," the Encyclopedia of Oceanography concluded nearly four decades ago,

> together with submarines and aircraft, are the only vehicles for carrying scientists into the Arctic Ocean. Drifting stations are particularly valuable for long and detailed observations which are not easily carried out from submarines or aircraft. Many of the special Arctic phenomena, such as auroras, magnetic effects, and the role of the polar region as a heat sink for the atmospheric and oceanic heat engines, will probably continue to be studied from drifting stations in the future.[16]

Nansen's groundbreaking work has been built upon. In the wake of that giant, the Soviet Union nurtured a maritime trail. Its Arctic and Antarctic Research Institute proved especially assiduous, mobilizing and deploying near-annual expeditions. Skill, experience, and stubborn determination realized thirty-seven years (1954–91) of continuous at-sea, on-ice investigation. One need not admire a discredited Soviet system to recognize the originality and commitment to the enterprise, and the intensity of its pursuit.

In a time of astounding technology, where solutions to logistic problems include airborne observations, submersibles, ice camps, remote imagery, and satellite navigation systems, drifting stations may seem an anachronism. Hostage to wind and current forcing, slow and transient, its track only broadly predictable, pack ice is a platform of opportunity. Movement is utterly uncontrollable—Ottawa's Ice Island weighed in at a billion tons. Such camps are largely useless for relatively precise investigations—to assess a particular feature, say, or to define a desired profile. But because the basic, obvious offshore features have been defined, focused studies are the order of the day. Abbreviated bivouacs like Canada's LOREX expedition are one answer.

Fraught with impediments and perils, ice camps confer sustained access, thereby facilitating continuous, long-term, real-time observations. The sea-ice canopy effectively limits inquiring "views" into the ocean from remote sensors deployed from above and, as well, obscures ice-atmosphere interactions from platforms operating below the pack. Drift ice literally floats between two worlds of inquiry. In Dr. Hunkins' view, the utility of drifting stations has diminished but not vanished. "So I'd say, the ice-station era has come and gone–though not entirely gone." The National Science Foundation is terse as to ice-borne bivouacs: "Drift stations and other ice platforms including Russian and Canadian opportunities will be utilized as research needs dictate."[17] And despite the politics of funding and the confusion of national scientific priorities, the United States boasts a small but highly professional cadre of Arctic experts and specialists (as does Russia).

So, are more U.S. ice camps in planning? "Nobody's talking manned drifting stations these days," Untersteiner remarked as the century ended. The emphasis is on high technology: airplane drones, unmanned submersibles, autonomous buoys.[18]

In all this, Russia–its territory is twice as large as that of all the other republics combined, including the whole Asian Arctic–is a natural ally in collaboration. "When the Arctic was a strategically significant arena for superpower military competition," the NSF had written, "all other aspects of Arctic international relations were constrained by that reality." No more. In the new spirit of East-West accord, harmony trumps rivalry. "The strategic gridlock of Cold War Arctic confrontation has ended."[19] Russia is home to colossal resources: trained personnel; priceless archives and collections; environmental and historical data sets–plus vast geographic areas of geophysical, biological, and ethnographic interest, including about one-half of the world's Arctic regions.[20] It is in the West's best interest to provide advice and assistance, to help secure the survival of the Russian scientific enterprise. Opportunities abound for collaborative research, for scientists to join their international colleagues in collecting and sharing data.

As well, trading-partner Canada remains a reliable ally and collaborator. In April 2003, the country's first dedicated research icebreaker was announced, the converted Canadian Coast Guard ship *Sir John Franklin*. North of the border, still, wariness of investigators from beyond its far-flung, three-ocean perimeter continues.

> To the foreign scientific community, Canada, particularly the north and remote regions, is a vast attractive laboratory. The infrastructure is modern and reasonably economical, and scientific activities are supported generally. Moreover, specimens are available for study and qualified "coolies" are ready to help with technical and menial tasks. Well financed foreign scientists see

this attractive environment as an excellent opportunity to further private and institutional interest with few obligations to the host.[21]

For its part, Russia–the USSR's true successor state–is straining. Most Russians are barely subsisting, anxious or desperate for extra rubles. Science and higher education also fare poorly. The breakup of Soviet-era scientific establishment in a state shouldering the loss of empire is among the most sweeping structural changes in the science of our time.[22] "Science there," an observer notes, "is in danger of becoming vestigial: laboratories are poorly equipped, research projects are regularly scuttled or scaled back and workers are underpaid or unemployed." In the unraveling, post-socialist economy in which admirals barter scrap from their fleets, shortages of every description abound.

> Russia has paid an enormous price, not just in terms of loss of its superpower status and its essential economic tie with the other republics, but also on the sense of identity and purpose. Since the time of Peter the Great, Russia has had no experience of living as anything other than an empire. The need to maintain and strengthen the empire, the pride of being a great power, the multiethnic composition of the ruling elite united by devotion to the vast and mighty state became key components of Russian political culture.[23]

Slashed budgets have led to the departure or dismissal of more than half of Russian scientists and engineers compared with the number active in 1990. Of those who remain, many have been obliged to work outside their institutions, in cooperative and private activity, to supplement salaries. Less than one-third of Russians with a science or engineering education were actively working in their specialties.[24] In Antarctica, station staffs have been reduced and support operations cut to the bone.

> The economic situation in Russia has unfortunately made us interrupt a more than half a century cycle of observations from the drifting ice of the Arctic.... To organize a research station on drifting ice is not only a very expensive undertaking. It also requires enormous experience, which our country gained over the years of the "NP" station activity more than anyone else in the world.[25]

Cooperative programs notwithstanding, Russians stayed mostly ashore. "They," a U.S. researcher observed, "have more important things to do–like eat." "Since 1991 we have not been on ice for NP purposes," an AARI scientist observed a decade later, further drifts unrealized.[26]

... the basic institutions of public administration and welfare are in catastrophic condition, wide-scale corruption is the economic norm, and the guarantors of liberty are, at best, in jeopardy.... The country is now so demoralized that it has readily accepted [President Vladimir] Putin's implicit grand bargain: to suspend further democratic transition in favor of essential order.[27]

Ending a twelve-year hiatus, the "cycle" resumed: SP-32 was established in March 2003, a dozen explorers encamped under station head Vladimir Koshelev, an expedition veteran. Meteorological, oceanographic, and ice studies were conducted, along with investigations of the ocean-atmosphere processes and pollution, and hydrobiological observations. Sea ice is ever fickle: "32" was emergency evacuated by helicopter in March 2004 because of breakup. Still, funding for "33" seemed secure. And SP-33 was deployed to sea.

The institute's newest drifting station, SP-34, was established on 19 September 2005. Five months adrift found "34" floating at 88°01′ N, 96°12′ E—coordinates placing it deep within permanent pack, not far off the pole. Its ice rafts a dozen men with two dogs. Telegramming his position on 9 February 2006 (at 2100 hours), T. V. Petrovskiy, station head, dutifully noted temperature (–39°C); barometric pressure; wind speed and direction; general drift direction (northwest); and the day's drift—eight kilometers (five miles). "Ice situation is without changes," he advised St. Petersburg. "All staff is well, Petrovskiy."[28]

The programs and missions of the United States in the North may appear desultory and episodic by comparison with Russian programs and persistence—intermittent, opportunistic bursts of furious activity interspersed by spells of relative quiescence. The Soviet Union (to exaggerate slightly) has *operated* atop the canopy whereas the United States has *experimented*.

—•—•—•—•—•—

Northern Siberia has no all-weather land link; road and rail networks are limited. The region must therefore depend on waterborne transportation for movement of bulky goods. Operations along the Northern Sea Route command center stage. More than one hundred ice-strengthened merchantmen ply the white seas of northern Siberia, supported by a fleet of about forty icebreakers, some nuclear-powered ("priceless in a region of nonexistent refueling facilities"). Above the obstinate pack as well as ashore, aircraft serve as trucks and buses, especially on short-haul flights. "In the Russian Arctic, flight safety regulations were to be taken into consideration and disobeyed with caution, but not followed blindly: a natural consequence in a country where rules were too numerous and usually impracticable."[29]

The former state airline, Aeroflot, has mutated into hundreds of smaller lines. Service stands less than exemplary:

Passengers frequently have to carry their own baggage aboard and dump it at the rear before scrumming for seats, and those who lose out end up standing for the duration of the flight. . . . Timetables are often fantasy-based and many flights are delayed, often for hours, without explanation. Rerouting of aircraft *in mid-flight* is not uncommon, as many airports are now demanding hard-currency payment for landing fees.[30]

At isolated outposts, the bonds are at once communal and lonely, persistent, powerful, enduring. "I am always restless on the continent," a *polyarnik* records. "I visit here and there and then go back to work, to the ocean, onto the ice. . . . Friendship and dedication to duty are not measured [in the Arctic] in rubles"–the many-sided stories behind the science.

Today, St. Petersburg[31] presents a decayed (if restored) European charm–monolithic architecture in endless procession, wheeling traffic, bustling sidewalks, lovely parks, shabby facades, pushing trolleys, intricate canal bridges. Housed in a late-Soviet-era concrete pile, the Arctic and Antarctic Institute conducts ice forecasting, long-range weather prediction, and oceanographic work; currently, climate, ozone, air-sea interaction, meteorology at the poles, and polar medicine (human adaptation to cold) are emphasized, among a host of disciplines. A large department is devoted to ice and ocean physics. A number of research vessels are operational. However, uncertainties haunt every layer of post-Soviet society: priorities, funding, survival. Polar science has been blindsided by events and must compete–like most everything else in the new Russia.

Interests have converged. Confrontation has shifted to cooperative programs, burden sharing, alliances, resource transfers. Russian professionals–a well-educated, motivated, talented force–are eager to share tradecraft. The West, for its part, has computers and other high-tech instrumentation much needed inside the former Soviet Union.

Papanin's on-ice canvas home is on permanent exhibit. Its lettering bold and proud, this icon of cultural patrimony stands in the Arktika Musee equipped as it was in 1937–laid out as Papanin last arranged it. Nearby is the original radio log along with Fedorov's book, *On the Drifting Ice* (Russian and English editions). A transportable wooden hut used by later stations also is on view, fully equipped. And over the street entrance hangs a *Shavrov* SH-2, one of the aircraft deployed to support the exalted *Chelyuskin* airlift.

As for scientists and logistics-support staff, field time remains an outsized, intense experience. John Burroughs remarked that without a sense of the unknown and unknowable, life is flat and barren. "I have a feeling," a *polyarnik* writes, "that you never experience more vividly and acutely than in the North. I am talking about my responsibility to the people, when everything can come to a grinding halt without my work and when my success is the success of our entire fraternity." A yearning to return, to experience again the beauty, color, and forms is (I'm told) "like some kind of illness." The experience satisfies, somehow, the sensual part of a *polyarnik*'s nature. They know raw hardship, yet also the subtle seductions of the circumpolar.

Mainland pleasures notwithstanding, many do return, as if heeding the Russian song: "Once, once again, many, many more times. . . ."

APPENDIX 1

ARCTIC OCEAN DRIFTING STATIONS SINCE THE ICE SHIP *FRAM* EXPEDITION OF 1883–86, 1937–2004

[SEE APPENDIX 3 FOR U.S. ICE CAMPS DEDICATED PRIMARILY TO ACOUSTICS]

STATION NAME	DEPLOYED	EVACUATED	COMMENT
Soviet Union			
"North Pole"	21 May 1937	19 Feb. 1938	air-deployed; rescue via surface ships
Sedov	23 Nov. 1937	13 Jan. 1940	involuntary freeze-in (ship)
SP-2	2 April 1950	11 April 1951	floe changed (fracturing)
SP-3	15 April 1954	19 April 1955	floe camp approached N. Greenland
SP-4	3 April 1954	19 April 1957	floe camp transected the basin
SP-5	21 April 1955	8 Oct. 1956	floe camp drifted onto Siberian shelf
SP-6	15 April 1956	14 Sept. 1959	ice-island raft (IGY); transect of basin
SP-7	23 April 1957	11 April 1959	floe approached N. Greenland (IGY)
SP-8	19 April 1959	19 March 1962	floe entirely broken
SP-9	21 April 1960	28 March 1961	floe considerably broken
SP-10 [I]	17 Oct. 1961	29 April 1964	deployed by *Lenin*; floe neared pole
SP-11	12 April 1962	20 April 1963	floe considerably broken
SP-12	30 April 1963	25 April 1965	floe approached Arctic Canada
SP-13	22 April 1964	17 April 1967	floe entirely broken; near-transect

265

SP-14	1 April 1965	11 Feb. 1966	floe collided with DeLong Islands
SP-15	29 March 1966	22 March 1968	camp shifted to another floe; transect
SP-16	9 April 1968	22 March 1972	four personnel "shifts"; near-transect
SP-17	29 April 1968	16 Oct. 1969	floe approached Fram Strait, near-transect
SP-18	9 Oct. 1968	24 Oct. 1971	ice island occupied by SP-19; neared pole
SP-19	7 Nov. 1969	14 April 1973	approached Fram Strait; full transect
SP-20	11 April 1970	10 May 1972	field program completed
SP-21	1 May 1972	25 May 1974	field program completed
SP-22 [I]	13 Sept. 1973	8 April 1982	ice island; nine "shifts"; near-transect
SP-23	5 Dec. 1975	1 Nov. 1978	ice island; approached Greenland
SP-24 [I]	23 June 1978	19 Nov. 1980	ice island; transect of basin
SP-25	16 May 1981	20 April 1984	floe approached Beaufort Sea
SP-26	21 May 1983	9 April 1986	three "shifts" of camp personnel
SP-27	2 June 1984	20 May 1987	floe evacuated by *Sibir;* near-transect
SP-28	21 May 1986	23 Jan. 1989	floe camp approached Fram Strait
SP-29 [I]	10 June 1987	19 August 1988	deployed by *Sibir;* program completed
SP-30 [I]	9 Oct. 1987	4 April 1991	floe-camp approached Beaufort Sea
SP-31 [I]	22 Oct. 1988	25 July 1991	third "shift" emergency evacuated
SP-32	April 2003	March 2004	deployment ended a twelve-year hiatus
United States			
Jeanette	6 Sept. 1879	13 June 1881	involuntary freeze-in and drift (ship)
USAF ice camp	20 Feb. 1950	10 March 1950	first U.S. experimental floe-ice station

Target-3 (T-3)	19 March 1952	April 1974	ice island; near-continuous occupation
ALPHA	4 April 1957	7 Nov. 1958	first U.S. floe-based IGY camp; cracked
CHARLIE	13 April 1959	7 Jan. 1960	ALPHA replacement; floe cracked
ARLIS II	12 Sept. 1961	11 May 1965	ice island; continuous occupation
AIDJEX	March 1975	March 1976	pilot studies 1970–74; main experiment
CAMBRIX	1978	1978	
FRAM I	11 March 1979	15 May 1979	Eastern Arctic; studies of Nansen Ridge
FRAM II	14 March 1980	1980	basin flanking Nansen/Lomonosov Ridges
FRAM III	13 March 1981	13 May 1981	focus on physical oceanography
FRAM IV	1982	1982	
MIZEX-83	1983	1983	
MIZEX-87	1983	1984	Greenland Sea
AIWEX	1985	1985	
CEAREX	16 Sept. 1988	5 Jan. 1989	research ship, two winter-ice stations
SHEBA	2 Oct. 1997	9 Oct. 1998	research ship; heat-budget focus

Canada

S. Storkerson	14 April 1914	11 Oct. 1914	probable ice island, Beaufort Sea
"North Pole"	6 May 1967	14 May 1967	geophysical (gravity) expedition
"North Pole"	2 April 1969	3 May 1969	geophysical expedition
LOREX	21 March 1979	1979	studies of Lomonosov Ridge
CESAR	30 March 1983	23 May 1983	studies of Alpha Ridge
Ice Island	Sept. 1984	March 1992	wind-shoved into Arctic Archipelago

Norway

Ice Ship *Fram*	22 Sept. 1893	13 August 1896	epic drift, the first for modern science

| *Maud* | 1922 | 1925 | ice drift confined to Siberian shelf |

a. Soviet drifting stations are termed *Severnyy Polyus* ("North Pole"), numbered consecutively. The Shimdt-Papanin drift of 1937–38 was simply "North Pole." If the ice raft did not deform, the next twelve-month shift of personnel was deployed to continue the work of a station.
b. The annual high-latitude "jumping" expeditions (*Sever*) were air-deployed, resupplied, and (often) based at the semi-permanent "North Pole" stations.
c. [I] = Icebreaker deployed.
d. Annual sea-ice export from the Arctic Ocean is primarily through Fram Strait, between Svalbard and eastern Greenland, conveyed by the Transpolar Drift Stream into the far North Atlantic.
e. ARLIS = Arctic Research Laboratory Ice Station
f. As a program, AIDJEX–Arctic Ice Dynamics Joint Experiment–endured from 1969 to 1977. The autonomous U.S. data-buoy program was seeded by AIDJEX; it continues today as the international buoy program.
g. MIZEX = Marginal Ice Zone Experiment
h. CEAREX = Coordinated Eastern Arctic Experiment
i. SHEBA = Surface Heat Budget of the Arctic Ocean
j. LOREX = Lomonosov Ridge Experiment
k. CESAR = Canadian Expedition to Study Alpha Ridge

APPENDIX 2

CIVILIAN SCIENTIFIC INVESTIGATORS, U.S. IGY DRIFTING STATION ALPHA JUNE 1957–NOVEMBER 1958

Station APLHA in the Arctic Ocean was the first floe camp established and maintained by the United States for the International Geophysical Year. (BRAVO camp exploited an ice island.) The scientific program embraced meteorology, ice physics, oceanography, geophysics, geology, underwater acoustics, and marine biology. The number of on-ice residents ranged from twenty-five to thirty, about half of which were U.S. Air Force personnel.

Scientific Leaders[1]

F. Badgley	University of Washington
G. Cvijanvoich	Lamont Geological Observatory (Columbia Univ.)
M. Davidson	Lamont Geological Observatory (Columbia Univ.)
N. Untersteiner	University of Washington
F. van der Hoeven	Lamont Geological Observatory (Columbia Univ.)

Scientific Investigators

A. Assur	Cold Regions Research and Engineering Laboratory
R. Edwards	Woods Hole Oceanographic Institution
T. English	Arctic Institute of North America
J. Farlow	Woods Hole Oceanographic Institution
G. Frankenstein	Cold Regions Research and Engineering Laboratory
A. Hanson	University of Washington
K. Hunkins	Lamont Geological Observatory

B. Isaacs	Lamont Geological Observatory
E. Kelly	U.S. Navy Underwater Sound Laboratory
E. Langdon	U.S. Weather Bureau
G. Latham	Lamont Geological Observatory
R. Neufer	Arctic Institute of North America
J. Scholten	Argentine Antarctic Institute
W. Schwarzacher	University of Washington
W. Senior	U.S. Navy Oceanographic (Hydrographic) Office
K. Staack	U.S. Weather Bureau
T. Stetson	Woods Hole Oceanographic Institution
O. Wattenbarger	U.S. Weather Bureau

Short-Term Investigators

J. Buckley	University of Alaska
R. Carpenter	U.S. Weather Bureau
H. Fleming	Columbia University
E. Herrin	Southern Methodist University, Seismological Observatory
R. Iverson	University of Wisconsin
T. Laudon	University of Wisconsin
T. MacDonald	U.S. Weather Bureau
L. Rowinski	University of Alaska

Source: Gerry H. Cabaniss, Kenneth L. Hunkins, and Norbert Untersteiner, compilers, "US-IGY Drifting Station Alpha Arctic Ocean 1957–1958," Air Force Cambridge Research Laboratories special report no. 38, Terrestrial Sciences Laboratory, Office of Aerospace Research, U.S. Air Force, Bedford, MA, November 1965.

1. Length of tour project-dependent, varying from a few days to more than twelve months. A scientific leader was appointed by the U.S. National Committee for the International Geophysical Year. His role: to represent the investigators to the air force in matters of common interest such as research and personnel requirements. Leaders reported on activities at the ice station to the U.S. National Committee.

APPENDIX 3

PRIMARILY ACOUSTICS U.S. DRIFTING STATIONS, ARCTIC OCEAN, (1962–96)

LOW FREQUENCY

NAME	CAMP LOCATION	SUPPORT AIRCRAFT	ON ICE
GMIS I	Beaufort Sea	ARL Cessna 180s	April 1962
GMIS II	Beaufort	"	April 1963
ARLIS I [I]	Beaufort	ARL Cessna 180s	
ARLIS III	Beaufort	ARL R4Ds, Cessna 180s	March–April 1963
ARLIS IV	Beaufort	"	March–April 1965
GM acoustics camp on Fletcher's Ice Island, T-3			1966–1970
ARLIS V	Beaufort	ARL R4Ds, Cessna 180s	March–May 1970
ARLIS VI	Beaufort	"	March–May 1970
MIZPAC-71 [I]	Chukchi Sea (two camps: APL and PRL)		August 1971
MIZPAC-72 [I]	Chukchi (unmanned camps emphasized)	helo	1972
MIZPAC-73	North Bering Sea (short-term camps)	helo	1973
AIDJEX	Beaufort (pilot experiments 1972–74)	R4D, Twin Otter	March–April 1975
PITCHER	Chukchi		March 1976
MUMMERS	Beaufort		April 1976

Name	Camp Location	Support Aircraft	On Ice
Eastern Arctic (Eurasian Basin) Program, Fram Strait sector:			
RUBY 77	Pioneer PRL camp for East Arctic	Twin Otter	April–May 1977
PEARL 77	Pioneer PRL camp (north of strait)	Twin Otter	April–May 1977
BOXER 78 [I]	First U.S. autumn pack-ice station (two camps: APL and PRL)		October 1978
OPAL 79		TriTurbo-3	April 1979
OPAL 80		"	April 1980
SAPPHIRE 81		"	April–May 1981
EMERALD 82		TT-3 (with parajumpers)	April–May 1982
CRYSTAL 83	Deployed to north pole	TT-3, Cessna 185s (on station)	March–April 1983
OPAL 84			March–April 1984
AREA 85	Three ice camps	TT-3 (with parajumpers), Twin	March–April 1985
AREA 86	Four camps, 60 men on ice	Twin, Caribou, C-130, C-141	March–May 1986
AREA 87	Two camps	TT-3, Twin, helo	March–April 1987
AREA 88		"	March–April 1988
AREA 89	Two camps		March–April 1989
AREA 90	Four camps, 60 men on ice	Twin, C-130, TT-3, helos (3)	March–May 1990
AREA 91	Three camps, 45 men on ice	TT-3, Twin, C-130, C-141, Bell 212	March–April 1991
AREA 92	Four camps, 80 men on ice	Twin, C-130, C-141, Bell 212	March–April 1992
AREA 93	Three camps, 48 men on ice	Twin, C-130, C-141, Bell 212	March–April 1993
AREA 94	Three camps, 30 men on ice	Twin, C-130	March–April 1994
AREA 95	One camp	C-130	March–April 1995
AREA 96	Two camps	Twin, C-130, Bell 206	March–April 1996

Appendix 3   273

HIGH FREQUENCY

APLIS Ice Camps

1971	Chukchi, Beaufort [I]	HH-52	July–August
1972	Chukchi [I]	HH-52	August
1972	(UARS testing, off T-3)	L-100-385 (*Hercules*)	March–May
1973	Bering Sea [I]	HH-52	March
1974	Beaufort	Huey, Cessna 182	April
1975	Beaufort	Twin Otter	March–April
1976	Chukchi	Twin Otter, FH-1100	March–April
1977	Chukchi, Beaufort	Cessna 182, Bell 205	May–July
1978	Chukchi [I]	HH-52	October
1979	Kane Basin	C-141, C-130	March–April
1980	Bering Sea [I]		March–April
1980	Beaufort [I]	HH-52	Sept.–October
1982	Beaufort [I]	HH-52	Oct.–November
1984	Beaufort [I]	C-130, Chinook	Sept.–November
1986	Beaufort	L-100-385, C-130, CASA 100, 200, Twin	March–May
1987	Beaufort	CASA 200, Twin Otter	March–April
1988	Beaufort	L-100-385, C-130, CASA 100, 200, Twin	March–April
1989	Kane Basin	C-141, Twin Otter	March–April
1990	Beaufort	L-100-385, CASA 200	March–April
1991	Beaufort	Twin Otter, CASA 200	March–April
1992	Beaufort	Twin Otter, Bell 212	March–April
1993	Beaufort	Twin Otter	March–April

[I]	Icebreaker deployed
AIDJEX	Arctic Ice Dynamics Joint Experiment. The AIDJEX 1972 pilot experiment exploited SHRAMS (Short-Range Automatic Measurement Station). That May, the then-abandoned main camp was exploited by Polar Research Laboratory as an acoustically quiet camp.
APL	Applied Physics Laboratory, University of Washington, Seattle
APLIS	Arctic Polar Research Lab Ice Station. Camp sizes grew from 1982 on—one response to the under-ice Soviet "boomer" threat.

	Forward support bases, usually Point Barrow or Prudhoe Bay.
AREA	Arctic Research in Environmental *Acoustics* (later, *Activities*)
ARL	Arctic Research Laboratory (later NARL), Point Barrow, Alaskan Arctic
ARLIS	Arctic Research Lab Ice Station. III–IV primarily research quiet-camps, two-month occupancy.

GM acoustics camp, Ice Island T-3, was manned for navy experiments with submarines on all-season basis.

GMIS	General Motors Ice Station. Two-man/two-week, GMIS I was the first such pack-ice camp.
MIZPAC	Marginal Ice Zone–Pacific. MIZPAC 71 was the first U.S. summer pack-ice station.
MUMMERS	Manned-Unmanned Environ. Research Station. Helo-deployed from main AIDJEX camp.
PRL	Polar Research Laboratory, Santa Barbara, California
TriTurbo-3	modified, long-range R4D for deployment/support of ice camps distant from forward bases.
UARS	Underwater Arctic Research Submersible System (autonomous underwater vehicle)
Low-frequency acoustics research	Very long-range detection and tracking of under-ice submarines.
High-frequency acoustics research	Under-ice weapons-related research, that is, mines and torpedoes.

NOTE: Selected non-APLIS and all APLIS ice camps were elements of submarine ice exercises (SUBICEX). The sponsoring organizations were many, from Naval Research Laboratory, Office of Naval Research, and Naval Underwater Systems Center to DARPA (Defense Advanced Systems Research Projects Agency), the National Science Foundation, and the U.S. Coast Guard. COMSUBLANT and COMSUBPAC provided ship assets—U.S. nuclear submarines.

ICESHELF (not listed) was an acoustic experiment conducted in 1991–96. A joint U.S.-Canadian program, this involved an underwater autonomous vehicle (UAV) bringing a fiber-optic cable to an array deployed at an ice camp. Logistics for its operations: Polar Associates, an experienced contractor.

Appendix 4

RESEARCH PROJECTS CONDUCTED ON ARLIS II, MAY 1961–MAY 1965

Project/Principal Investigator	On Ice	Sponsoring Institution
Acoustics and VLF Measurements		
Mr. Guy S. Harris	5 months	USN/Underwater Sound Laboratory
Dr. Robert H. Mellen	10 months	USN/Underwater Sound Laboratory
Arctic Basin Acoustic and Seismic Studies		
Dr. Kenneth L. Hunkins	12 months	Lamont Geological Observatory/ Columbia Univ.
Arctic Basin Airborne Remote Sensing		
Mr. Walter Wittmann	2 months	U.S. Navy Hydrographic Office
Arctic Basin Heat Flow		
Dr. Arthur H. Lachenbruch	2 months	U.S. Geological Survey
Arctic Basin Marine Biology		
Dr. John L. Mohr	48 months	University of Southern California
Arctic Oceanography		
Dr. Phil E. Church	16 months	University of Washington
Dr. Kou Kusunoki	18 months	Hokkaido University

Project/Principal Investigator	On Ice	Sponsoring Institution
Arctic Underwater Acoustics		
Mr. Beaumont M. Buck	5 months	General Motors Defense Research Laboratories
ARLIS II Photography	2 months	U.S. Navy
Geology and Glaciology of ARLIS II		
Dr. David Smith	4 months	Louisiana State University
Gravity and Magnetics		
Dr. George P. Wollard	21 months	University of Wisconsin
Dr. Ned A. Ostenso	27 months	University of Wisconsin
Gravity-Microbarograph		
Mr. Walter Wittmann	3 months	U.S. Navy Oceanographic Office
Lt. Leonard A. LeShack		U.S. Navy Oceanographic Office
Sea-Ice Micrometeorology		
Dr. Phil E. Church	48 months	University of Washington
Strain Measurements on Pack Ice		
Mr. Walter Wittmann	32 months	U.S. Navy Oceanographic Office
VLF Radio Navigation		
Dr. J. Stanborough	7 months	Columbia University

Source: John L. Schindler, "The Impact of Ice Islands–The Story of ARLIS II and Fletcher's Ice Island, T-3, Since 1962," in *Arctic Drifting Stations: Report on Activities Supported by Office of Naval Research*, proceedings of symposium, John E. Sater, Coordinator, Warrenton, VA, 12–15 April 1966 (published November 1968).

APPENDIX 5

POLYARNIK PERSONNEL AT *SEVERNYY POLYUS-22*, FOURTH SHIFT OF THE DRIFTING STATION, 15 APRIL 1976 TO 16 APRIL 1977[1]

PERSONNEL	FUNCTION/ SPECIALIZATION[2]
Vinogradov, N. D.	head of the station
Maximov, G. A.	meteorologist (atmosphere, weather, climate)
Evdoxushin, G. I.	meteorologist (atmosphere, weather, climate)
Napoikin, U. I.	aerologist (atmosphere)
Borzenko, B. M.	aerologist (atmosphere)
Alexandrov, V. V.	aerologist (atmosphere)
Transhel, U. G.	aerologist (atmosphere)
Ivanov, U. I.	ion concentration measurement specialist
Vogtev, E. N.	ion concentration measurement specialist
Zayzev, A. F.	oceanologist (oceanographer)
Kochelev, B. A.	hydrochemist
Pyatibrat, O. M.	magnetic field measurement specialist (magnetologist)
Sveshnikov, A. M.	ozone measurement specialist (upper atmosphere)
Borgmon, L. S.	physician
Kyxikov, S. I.	radio specialist
Bokanov, B. B.	radio specialist
Tchvetkov, P. A.	mechanic
Bykov, A. F.	mechanic
Dobryakov, I. S.	cook

1. Installed on an ice island not sea ice, "North Pole-22" was operational for nine years (1973–82), rafting nine shifts of *polyarnik* personnel as well as basing two *Sever* "jumping" air

277

expeditions, in 1975 and 1976. A diplomatic dustup was incited when, in 1977, it penetrated the Canadian Arctic sector—waters claimed sovereign by Ottawa.

2. Each Soviet drifting station operated all day, every day, yet airlift resupply limited the participants on ice. Versatility was therefore expected from each man: duties during any drift expedition did not necessarily coincide with training and specialization. The magnetologist usually served also as astronomer, for example; the physician doubled as housekeeper. And "all lend a hand in hacking a runway out of the ice."

APPENDIX 6

ICE-ISLAND REGULATIONS, POLAR CONTINENTAL SHELF PROJECT, GOVERNMENT OF CANADA, 1990

See Next Page

Polar Continental Shelf Project

ICE ISLAND REGULATIONS

Welcome to the Canadian Ice Island Research Station. In order to ensure the efficient and safe use of the facilities it is necessary that you respect the following rules:

Camp Radios Channel 1: There is a PT-300 located in each building in the camp. This serves as a communications and alarm system. Please ensure your radio is turned on and operating. Radio communications between 2200 and 0600 should be kept to an absolute minimum as others are sleeping.

Accidents: Report all accidents, injuries, or illness to the Base Manager immediately. The infirmary is located in cabin 6.

Firearms: Guns are kept in selected locations within the camp. Familiarize yourself with these. The weapons are to be used in life-threatening situations only. Notify staff immediately. If you require practice or instructions, schedule it with the Base Manager. Unauthorized use of firearms is prohibited.

Fire: Detailed fire orders are posted in each building, please familiarize yourself with them.

Runway: Keep clear of the runway and helicopter pad when aircraft are taking off or landing. Stay at least 100 metres to the side of the runway and approach. The approach extends 1000 metres past either end of the runway. Note that engines are often kept running during loading and unloading. Aircraft emergency procedures are posted in each building; please familiarize yourself with them.

Fuel Usage: If you require fuel you are to request it from the base staff. Unauthorized use of fuel is prohibited.

Skidoos: If you require a skidoo you are to request it from the Base Manager. Skidoos are essential to the operation of the base and are to be used for work purposes only. Do not overload the skidoo; one person per machine is the rule. During cold season, skidoos must be stored inside. Give your machine time to warm up prior to driving it; do not leave it idling for more than 10 minutes. Whenever leaving the base you are to advise the Base Manager as to where you are going, how long you will be and ensure you have a radio, gun, tool kit, spare plugs and belt, emergency kit and that you are properly clothed in cold weather gear. If you require a briefing or instructions on operating a skidoo, please contact the Base Manager.

Energy, Mines and Resources Canada

Énergie, Mines et Ressources Canada

Canada

- 2 -

<u>Meal Hours</u>: Breakfast: 0700-0800)
                 Lunch: 1200-1300) Please be prompt
                 Supper: 1700-1800)

Note:

1. The staff eats 15 minutes prior to these hours as this is often the only time we can get together during the day. We would appreciate privacy during this period.

2. If you will be gone during a meal hour, please advise the kitchen staff well in advance of your requirements.

<u>Use of Equipment, Tools and Staff</u>: Any such requirements will be authorized by the Base Manager <u>only</u>. If you have any requirements, please discuss them with the Base Manager.

<u>Smoking</u>: Prohibited in the kitchen and in bed.

<u>Showers</u>: Water is a rare commodity - shower with prudence, and remember there is a lineup behind you.

<u>Washer & Dryer</u>: For staff use only; any other use must be approved by the Base Manager.

<u>Litter</u>: Place all litter in the garbage Komatiks. <u>No cans, no plastic</u>; separate these and give to staff.

<u>Drinking Water</u>: Your drinking water comes from the surface of the island; therefore, any contaminants, i.e., yellow snow, fuel, litter, will end up in the drinking water.

<u>Duties</u>: You may be called upon to perform some tasks to assist with the smooth operation of the base camp; please pitch in.

<u>Departure</u>: Prior to departure, please clean and vacuum your cabin and empty your honey bucket. Project leaders are to check out with the Base Manager prior to departure to ensure all equipment has been returned and all documents have been signed.

<u>Freight</u>: All items shipped from the Ice Island are to be weighed, labelled with weight, strapped and packaged for shipment. Provide the Base Manager with an itemized list well in advance of shipment.

<u>Alarm System</u>: The camp uses a radio system to notify everyone of an emergency; we have also installed a horn/siren. When you hear the horn, return to the base assembly area outside the entrance to the kitchen and await instructions from the staff. No one is to leave the base area for any reason unaccompanied or without advising the Base Manager.

- 3 -

Please appreciate that this camp is in a very remote location. In the event of an accident, every effort is made to evacuate the injured party(s). However, we cannot control bad weather or radio blackouts which can last up to 10 days. Exercise extreme caution and good judgement in your daily work and activities during your stay on the Ice Island.

Thank you for your cooperation and have a pleasant stay.

APPENDIX 7

ARCTIC RESEARCH BUDGETS, INDIVIDUAL U.S. FEDERAL AGENCIES, 2000–2004 (IN MILLIONS OF DOLLARS)

AGENCY	FY 2000 ACTUAL	FY 2001 ACTUAL	FY 2002 ACTUAL	FY 2003 ESTIMATE	FY 2004 ESTIMATE
Department of Defense (DOD)	23.3	20.4	19.4	15.2	17.1
Department of Interior	43.9	43.9	47.2	55.7	53.9
National Science Foundation (NSF)	67.5	74.2	86.0	91.0	93.5
National Aeronautics and Space Admin.	46.6	34.2	38.5	36.1	37.5
National Oceanic and Atmospheric Admin.	29.7	30.7	47.4	35.2	35.4
Department of Energy (DOE)	4.7	4.2	15.8	16.3	4.5
Department of Health and Human Services	13.8	15.9	21.2	21.5	21.5
Smithsonian Institution	0.5	0.5	0.5	0.5	0.5
Department of Homeland Security[1]	-	-	10.4	5.9	8.2
Department of Transportation (DOT)	6.3	10.9	-	-	-

AGENCY	FY 2000 ACTUAL	FY 2001 ACTUAL	FY 2002 ACTUAL	FY 2003 ESTIMATE	FY 2004 ESTIMATE
Environmental Protection Agency (EPA)	0.7	0.7	0.4	0.6	0.5
Department of Agriculture	4.8	4.9	3.3	3.0	2.9
Department of State	0.0	0.0			
TOTAL	241.9	240.4	290.0	280.8	275.4

1. These data represent the U.S. Coast Guard. As part of the government's reorganization, the service was transferred in 2003 from Transportation into the Department of Homeland Security.

Sources: Charles E. Myers, Ph.D., personal communication, and *Arctic Research of the United States* Interagency Arctic Research Policy Committee, Office of Polar Programs, National Science Foundation (Fall/Winter 2001).

APPENDIX 8

PLOTTED TRACKS OF SELECTED DRIFT STATIONS

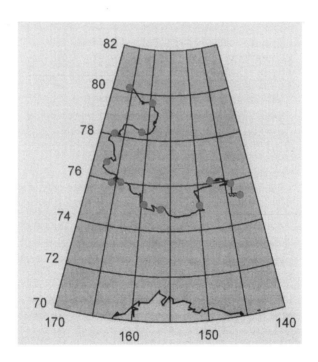

A. The drift track of ice station SHEBA. The gray area is the coast of Alaska. Dots denote months from October 1997 to October 1998.

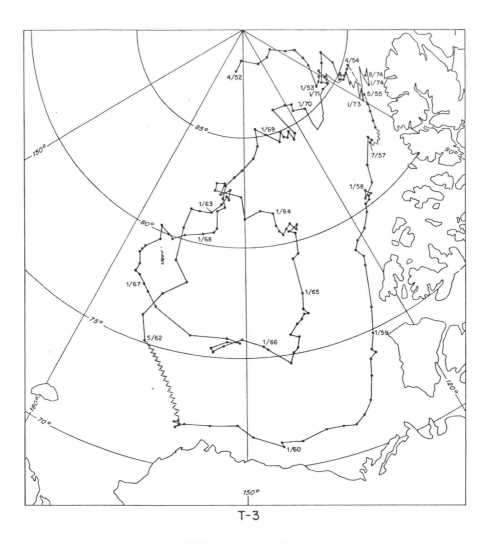

B. Drift track of the T-3 occupation, 1952–74. Evacuated spring 1974, the ice island was tracked intermittently but not remanned. The last confirmed position (August 1983) had T-3 nearing Fram Strait—en route to the high North Atlantic.

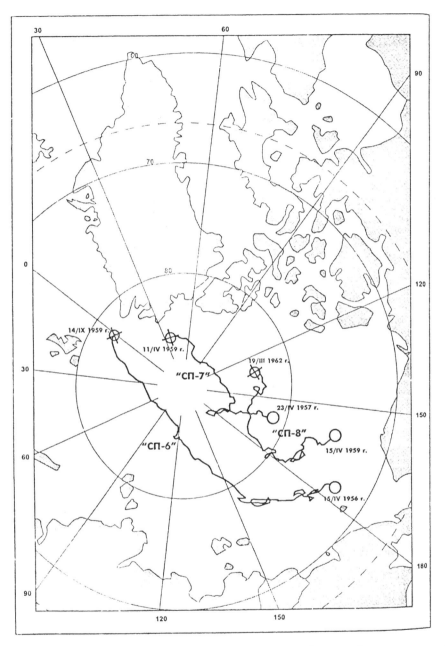

C. Drift tracks for *Severnyy Polyus-6,* SP-7, and SP-8 (1956–62). For its "North Pole" series, Leningrad's Arctic Institute often deployed into waters off eastern Siberia. From that sector, long-haul stations might transect the basin toward Fram Strait–the deepwater, high-energy connection to the far north Atlantic. SP-6 exploited an ice island, "7" and "8" floe ice.

NOTES

CHAPTER 1

1. In this respect, the Arctic differs radically from the continent of Antarctica, an uninhabited land and ice mass surrounded by oceans.
2. The ice pack of other oceans is a seasonal phenomenon; for example, the extensive winter cover of the seas surrounding Antarctica shrinks to a narrow fringe in the austral summer. In the north, the land masses surrounding the Arctic Ocean and the Norwegian-Greenland Seas acted as loci for northern hemisphere ice sheets—with profound impact offshore.
3. *Arctic Research of the United States,* Interagency Arctic Research Policy Committee, Office of Polar Programs, National Science Foundation, 14 (Spring/Summer 2000), 5.
4. By ventilating it is meant that water descends to the abyss after contact with the atmosphere and its gases, most notably oxygen. Geologically dating the onset of this large-scale circulation at depth between the Arctic and North Atlantic basins is critical for modeling climate change in the Northern Hemisphere. As part of the global heat engine, the Antarctic has a major role in the transfer of energy. Its ocean-atmosphere system is now known to be both an indicator and a component of climate change.
5. Water is the only liquid whose specific gravity decreases as it approaches its freezing point, which allows ice to float. For seawater, the addition of salts suppresses both the freezing temperature and the temperature of maximum density.
6. At its winter maximum, sea ice accounts for approximately two-thirds of Earth's entire ice cover.
7. Vilhjalmur Stefansson, *Arctic Manual* (New York: MacMillan, 1944), 362–363. On land, ice deforms by creep under gravitational stresses.
8. Though notable additions to the maps were made and valuable data collected, the prime objective was to plant a northernmost flag. Though doubt continues to cloud Peary's claim to the geographic pole, "he at least put an end to the terrible waste of money and energy expended in trying to reach it, and for that alone he deserves our thanks." (Moira Dunbar and Wing Commander Keith R. Greenaway, Royal

Canadian Air Force, *Arctic Canada from the Air* (Ottawa: Defence Research Board, 1956), 435.)

9. Fridtjof Nansen, Ph.D., *Farthest North: Being the Record of a Voyage of Exploration of the Ship "Fram" 1893-96 and of a Fifteen Months' Sleigh Journey by Dr. Nansen and Lieut. Johansen* (2 vols.) (New York: Harper and Bros. Publishers, 1897), 1:7-8.

10. Christened the DeLong Archipelago, these lifeless rocks were to bedevil later penetrations. In 1893, Nansen's *Fram* (intentionally) was beset not far from the group, and, in our century, the icebreaker *Sedov*. A pair of Soviet "North Pole" camps would be evacuated in those same waters.

11. Lofted into ridges, the churned, broken surface offers hideous obstacles to sled parties, to say nothing of men yoked to boats. "Some can be avoided," one ice traveler wrote, "but more often it is necessary to struggle through. Alluring alleys end in cul-de-sacs, great balanced fragments must not be disturbed and there are snowed-over fissures in which one might break a leg. Dogs struggle along, traces become tangled and komatiks [Eskimo sleds] jam and upset. Eskimos meet these obstacles patiently and with strength and skill work their way through." Guy Blanchet, *Search in the North* (New York: Macmillan Co. of Canada Ltd., 1960), 108.

12. Nansen, *Farthest North*, 1:14.

13. A water-sampling bottle designed by Nansen was in use for oceanographic work until recently.

14. Nansen, *Farthest North*, 1:24, 37, 29.

15. Both quotes from *Farthest North*, 1:41, 48.

16. Standard canon held that seafloors were flat, featureless. Bathymetric maps from the 1800s portray fanciful bowls—smooth, elongate, shallow—hosting uncharted lands and islands. Instead, bottom and sub-bottom structure are enormously varied. Contemporary charts depict a noble terrain: interposing between the deep-sea Canada and (smaller) Eurasian Basins is the Lomonosov Ridge, a towering partition that snakes from the Siberian shelf across the pole to Canada. Flanking it are plains of abyssal ooze, escarpments notched by immense canyons, plateau rises, jutting seamounts, and zones of deep-crustal fracture. Such complexities appear nowhere on the bathymetric map of the Arctic Ocean compiled by Nansen himself and published in 1928.

17. Nansen, *Farthest North*, 1:270–271, 293.

18. Sverdrup, Otto, "Report of Captain Otto Sverdrup, The Drifting of the 'Fram' From March 14, 1895," in *Farthest North*, 1:667.

19. Nansen, *Farthest North*, 1:246–249, 388.

20. Nansen, *Farthest North*, 1:497.

21. Sverdrup, "Report," 644, 647.

22. Sverdrup, "Report," 663.

23. Sverdrup, "Report," 685.

24. Col. Joseph P. Fletcher, U.S. Air Force, another pioneer (see chapter three), referred to the nineteenth-century penetration as "probably the most efficient and productive Arctic expedition ever conducted."

25. The term "sweepstakes" was used by Lt. Karl Weyprecht, a physicist and an officer in the Austro-Hungarian navy. In his view, the poleward race was an activity wherein "immense sums were being spent and much hardship endured for the privilege of placing names in different languages on ice-covered promontories, but where the increase in human knowledge played a very secondary role." In 1872–73, as co-leader of the German North Polar Expedition and scientist-captain of the *Admiral Tegetthoff*, he sailed into the basin. Beset off Novaya Zemlya, the vessel meandered

northward–the Arctic's earliest drift for science. Franz Josef Land was thus discovered and named. Internationalist in his views ("science is not a territory for national possession"), Weyprecht argued for a synchronized yearlong program of observations concerning meteorology, magnetism, tides, and aurora and electromagnetic phenomena at locations in the north- and south-pole regions and at established observatories in lower latitudes. His ideas would help inspire the International Polar Year of 1882–83.

26. In area, the Siberian continental shelves are the world's widest and shallowest. The central Arctic is deep, oceanic, perennially veneered.

27. "For the first time, Soviet scientists began to carry out investigations in a planned, multidisciplinary way, using icebreakers, aircraft, radio, meteorological rockets, and other technical means." "Arktika," *Soviet Military Encyclopedia*, vol. 1A, Moscow, Voyennoye Izdstelœtovo, 1976, trans. CIS Multilingual Section, National Defence Headquarters, Ottawa, 169.

28. Capt. O. P. Araldsen, Royal Norwegian Navy, "The Soviet Union and the Arctic," Naval Institute *Proceedings*, no. 93 (June 1967), 51.

29. T. A. Taracouzio, *Soviets in the Arctic* (New York: Macmillan Co., 1938), viii.

30. Russia's polar stations remain vital to the safety of Northern Sea Route navigation by keeping its sailors updated on their location and supplying information (including satellite data) on ice-related and other conditions. This thread of stations functions as an organic whole extending from Murmansk in European Russia (the largest city north of the Arctic Circle) 30,590 kilometers (19,000 miles) eastward across northern Siberia to the Kamchatika peninsula.

31. Ernst Krenkel, *RAEM Is My Call-Sign*, trans. R. Hammond (Moscow: Progress Publishers, 1978), 57, 61.

32. The eastern islands of the Svalbard Archipelago, which lie within this triangle, were accepted by other nations as belonging to Norway. In 1928 and again in 1950, the Soviet press was to advance a claim to the open polar seas, including its drifting ice. The entire triangle was shown on Soviet maps as defining Soviet territory.

33. Although it was advocated in certain quarters, the United States has never endorsed the Soviet claim or, indeed, the sector principal upon which it is based. Why? Acceptance would tend to validate such claims in Antarctica as well, undermining the international controls favored by Washington for the White Continent.

34. Krenkel, *RAEM*, 143.

35. Decree of the Council of Peoples' Commissars of the USSR of December 17, 1932, on the Organization in the Council of Peoples' Commissars of the USSR of the Central Administration of the Northern Sea Route. Quoted in Taracouzio, *Soviets*, 383. The Northern Sea Route Administration has sometimes been compared with the East India Company and the Hudson Bay Company. "[The two] were intended to extract a speedy profit by ruthlessly exploiting the rich natural resources of these colonies and their population's labour. This was systematic and barbarous plunder, not construction and development." Yevgeny Fedorov, *Polar Diaries* (Moscow, Progress Publishers, 1983), 7.

36. Anonymous, "History of Arctic Exploration," *Sourcebook on Submarine Arctic Operations* (San Diego: U.S. Navy Electronics Laboratory (NEL), 27 April 1966), 2–66.

37. In 1958, the task of organizing research in Antarctica was assigned, effecting a change to the current name, Arctic and Antarctic Research Institute.

38. The value of regular communications to wintering parties may be imagined. Thanks to radio, "news, lectures, instruction on various topics, as well as concerts, not to mention private communications between winterers and their families, are now

[1937] regularly transmitted from the centers to those serving the Soviets at the farflung outposts in the North." Taracouzio, *Soviets*, 116.
39. Russia today has a six-strong fleet of nuclear icebreakers, operated by the Murmansk Shipping Company. "The power of modern ships gives us the impression of great ability. Remember that the wind and current can move more ice in a few hours than the most powerful icebreaker can penetrate in several days." Cdr. R. S. Crenshaw Jr., U.S. Navy, *Naval Shiphandling* (Annapolis: U.S. Naval Institute, 1955), 307.
40. H. P. Smolka, *40,000 Against the Arctic* (New York: William Morrow and Co., 1937), 248.
41. Ernest Frederick Roots, "International and Regional Cooperation in Arctic Science: A Changing Situation," Proceedings of the Nordic Conference on Scientific Research, 2–8 August 1984, in *Rapport fra Nordisk Vilenskapelig Konferanse om Arktisk Forskning* (Trondheim, Denmark: Universitet i Trondheim Press), 133. The IGY of 1957–58 is a direct descendant of the IPY. (See chapters three and four).
42. Krenkel, *RAEM*, 145.
43. On modern icebreakers, the propellers assist by chewing blocks once the vessel's forepart has broken into the pack. Without this aid, the ice would re-close the wake far more quickly, posing a threat to following vessels. Built in Glasgow 1909 for the Newfoundland sealing trade, *Sibiryakov* was in fact an icebreaking ship, not an icebreaker.
44. Ilya P. Romanov, Ph.D., interview by the author, audiocassette, Arktika Musee, St. Petersburg, Russia, 19 May 1992.
45. To Soviet planners, Wrangel represented a future base for air-highway traffic across Asiatic Russia. "Tokyo–New York passengers," one airman predicted, "will breakfast there, together with people flying from Moscow to San Francisco." Lying above the Bering Strait, Wrangel's strategic importance was plain enough. Determined to assert sovereignty, Moscow was resupplying plus resettling the island.
46. Ivan Alexandrovitch, quoted in Smolka, *40,000*, 83.
47. This is not as improbable as it might first appear. Charles Lindbergh (among others) had assessed locations for Arctic airfields, which, he foresaw, would help nurture intercontinental air transport. In 1934, for example, in a long brief prepared for Pan American Airways regarding the Greenland-Iceland transatlantic air route, he writes, "It seems rather certain that the Arctic will eventually be crossed by air routes wherever time and distance can be saved by northern flying. With every year that passes more aircraft are used in the Arctic for local services; and the very fact that ground and water transport is so difficult makes their use relatively more advantageous."
48. Quoted in William Barr, "Imperial Russia's Pioneers in Arctic Aviation," *Arctic* 38 (September 1985), 226-227.
49. The first aircraft ever to fly over any part of Arctic Canada was a seaplane belonging to the U.S. government and attached to a summer expedition in 1925, based in North Greenland. Its pilot: Richard E. Byrd. (Richard Finnie, "Flying Beyond Sixty," *Canadian Aviation* 12 (February 1939), 9.)
50. M. V. Vodopyanov, *Wings Over the Arctic* (Moscow: Foreign Languages Publishing House, 1957), 8.
51. With the end of the Big Bull Market, a general disillusionment with capitalism had taken hold in America, enhancing the allure of the Russian experiment. "In the summer of 1929 Russia had seemed as remote as China; in 1931, with bread-lines on the streets, the Russian Five-Year Plan become a topic of anxious American interest. The longer the paralysis of industry lasted—and how long it lasted!—the

more urgent became the demand for some measure of American economic planning which might prevent such disasters from recurring, without handling over undue power to an incompetent or venal bureaucracy." (Frederick Lewis Allen, *Only Yesterday: An Informal History of the Nineteen-Twenties* (New York and London: Harper and Bros., 1931), 355.)

52. Even though they undoubtedly exaggerated their accomplishments to meet Stalin's formula, the Soviets accomplished a very great deal. An astounding progress was plain. "The achievements have been considerable and striking," a British editorial noted. Yet an all-pervading militarism was mortgaging the private sector. "The hardships of the people, chief of which is the shortage and poor quality of the food, have been heavy." *Flight* 24 (1 December 1932), B.

53. "By 1933, while flying was still generally regarded as a hazardous enterprise in a more southerly clime, the aeroplane had already become a commonplace factor in the life of the [Canadian] Arctic, along with the boat and dog-team. It has been said that the aeroplane robbed the North of its romance. Certainly the aeroplane robbed the North of much of its terror and its chill inaccessibility." (Richard Finnie, "Flying Beyond Sixty," pt. 2, *Canadian Aviation* 12 (March 1939), 19.)

54. John Grierson, "A Flight in Russia," *Flight* 24 (10 November 1932), 1044. A Fodor's *Soviet Union 1986* offers this observation: "The individual traveler isn't really welcome, only tolerated, and pays more."

55. Teresa D. Safon, "Stalinist Myths and the Saga of the Arctic Explorers," paper presented at the Southern Slavic Conference, Jacksonville, FL, 1991, 22, courtesy Carl J. Bobrow.

56. *Izvestia*, 3 November 1936, quoted in Taracouzio, *Soviets*, 244.

57. Smolka, *40,000*, 85–86.

58. The ANT-25–all Soviet, including engine–had been designed and constructed by the Central Aero-Hydrodynamical Institute under the supervision of A. N. Toupolev, whose initials mark its designation.

59. Brig. Gen. George C. Marshall, commanding a brigade at Vancouver Barracks, and his wife had spent that evening listening to radio reports of the approaching aircraft. In landing at Pearson, the airmen became–abruptly–Marshall's responsibility. Their machine under guard nearby, the Russians staggered into the Marshall home at 8:20 a.m. "They wore huge parkas of fur," Mrs. Marshall remembered, "only their faces showing and these were so streaked with oil and dirt, so haggard and covered with beards, that the men hardly looked human." Given two rooms, "They immediately got into their baths, but none of them knew how to work the modern bath-room fixtures and none of them spoke English. However, with the help of a Russian-born doctor on the Post and a CCC [Civilian Conservation Corps] boy of Russian parentage we discovered that they were calling for cognac. I had sent up orange juice, bacon, eggs and coffee, which they ate while still in the bathtub." A circus of attention engulfed the airfield. "All this excitement seems strange now," she adds, "but at that time this flight across the pole was considered little short of a miracle." (Katherine Tupper Marshall, *Together: Annals of an Army Wife* (New York: Tupper and Love, Inc., 1946), 25–28.)

60. Smolka, *40,000*, 103.

61. See Everett A. Long, "Has the Levanevsky Mystery Been Solved?" *Air and Space Smithsonian* 2 (December 1987/January 1988), 52–53. On 6 October, assisting the search, Vodopyanov flew an ANT-6 from Rudolf Island over the geographic pole to 88°33' N, 122° W–the first time 90° north had been crossed during the twilight period. Operating out of the Western Arctic that August and September, Sir Hubert

Wilkins and his party flew almost to the pole in a twin-engine consolidated flying boat during a "brilliant but fruitless search" for the vanished transpolar party. Richard Finnie, "Flying Beyond Sixty," pt. 3, *Canadian Aviation* 12 (April 1939), 24.

CHAPTER 2

1. "From the drifting ice all movements of the water—the horizontal currents as well as ... vertical oscillations of the layers—may be continually and carefully studied at all depths in an ideal manner which is not possible in the open ocean; and many of the greatest problems of oceanography may thus be solved." (Fridtjof Nansen, "The Oceanographic Problems of the Still Unknown Arctic Regions," *Problems of Polar Research*, American Geographical Society Special Publication No. 7 (Worcester, MA: Commonwealth Press, 1928), 12.)
2. I. P. Tolmachev, "The Geology of Arctic Eurasia and Its Unsolved Problems," in *Problems of Polar Research*, 75.
3. Dense, cold surface waters are now known to sink and expel southward, flooding into the deep basins that cup the Atlantic and Pacific as relatively strong, narrow currents. Counterflows of warm waters are conveyed poleward. Via Bering Strait, nutrient-rich water circulates in. On the Chukchi shelf this inflow then spreads throughout the Canada Basin, where it forms a distinct layer between the floating ice cover and warmer, deeper water. Thus do the Pacific waters insulate the sea-ice cover of the western Arctic from melting as they inoculate the upper ocean.
4. Fedorov, *Polar Diaries*, 201. The Soviets were keepers of the Nansen flame, and their audacity would soon elevate Fedorov to international celebrity.
5. Nansen had died in 1930, his notion unrealized. Revived during planning for the second polar year, a North Pole expedition had faded to disappointment, foreclosed by worldwide financial chaos.
6. "Planes Kept Busy All Over Russia," *New York Times*, 11 July 1937.
7. Otto Shmidt, "We Landed at the North Pole," in Lt. Col. C. V. Glines, U.S. Air Force, ed., *Polar Aviation* (New York: Franklin Watts, Inc., 1964), 89.
8. A. F. Treshnikov, "The Soviet Drifting Station SP-3, 1954-55," *Polar Record* 8 (September 1956), 222. The author Treshnikov was leader of the third manned drifting camp of Soviets.
9. George H. Wilkins, "Polar Exploration by Airplane," in W. L. G. Joerg, ed., *Problems of Polar Research*, 404.
10. Vodopyanov, *Wings*, 152-153. "And I began to develop my ideas about establishing a drifting station on the North Pole," the airman continues. "It was pure fantasy but founded on the accomplishments of Soviet aviation. In my view, aircraft alone could land a scientific expedition well provided with food, scientific instruments and equipment in the heart of the Arctic. I thought several machines should participate for then the pilots could help one another in an emergency." (79)
11. The regime's sensitivities were hyperacute. As well as pure scientific exploration, what the enterprise could contribute to international prestige influenced all of Soviet science. In our time, political and fiscal factors continue to afflict high-profile "big science." The space programs of the superpowers are obvious examples.
12. Vodopyanov, *Wings*, 96.
13. Krenkel, *RAEM*, 295.
14. "What is done by a telephone call in England requires hundreds of papers here," physicist Piotr Leonidovich Kapitza complained in 1935. "You are trusted in nothing.... People do not trust each other at all. They only trust paper—that is why

paper is so scarce! Bureaucracy is strangling everyone." Quoted in Aleksei Kozhevnikov, "Piotr Kapitza and Stalin's Government: A Study in Moral Choice," Office for History of Science and Technology, Univ. of California, Berkley, *Historical Studies in the Physical and Biological Sciences*, 22, pt. 1, 141.
15. Vodopyanov, *Wings*, 137.
16. R. E. G. Davies, *Aeroflot: An Airline and its Aircraft* (Rockville, MD: Paladwr Press, 1992), 30–31.
17. Fedorov, *Polar Diaries*, 44.
18. Nikolai Yaganitsin, interview by author, audiocassette, Arktika Musee, St. Petersburg, Russia, 19 May 1992.
19. Fedorov, *Polar Diaries*, 219. Golovin had gained 90° N at 4:23 p.m. In response to reported deteriorating conditions back at base, the flight crew radioed, "Flying over the North Pole. Proud to have reached the top of the world in our orange bird. But unfortunately the Pole is obscured. Impossible to break through. Returning. Not worried about the weather on Rudolf Island. Petrol will more than suffice."
20. Krenkel, *RAEM*, 306; Vodopyanov, *Wings*, 229.
21. Vodopyanov, *Wings*, 229–233.
22. The actual point of touchdown was 89°48' N, 78°40' W–19 kilometers (12 miles) off 90° N latitude. Because of atmospheric lows that sweep in from the northern Atlantic, the most suitable time for a sea-ice landing in that region is not May but April. Further, Vodopyanov had set down through heavy overcast. As everyone knew, the melt and its more hostile ice conditions were imminent–the expedition had either to proceed or abort. Touchdown was under marginal, even dangerous, conditions. Credit must be granted the pilots for their skill and boldness.
23. Ivan Papanin, *Life on an Ice Floe* (New York: Julian Messner, Inc., 1939), 7.
24. The drift-station technique, it should be emphasized, was made practicable by aircraft and reasonably safe by radio. Once on their own, the polar party lay beyond any landfall; the quartet would therefore have to rely on regular transmission of their position and well-being to potential rescuers.
25. Vodopyanov, *Wings*, 238.
26. Papanin, *Ice Floe*, 7–8.
27. To Krenkel, this was a masterpiece. "The word 'tent' was not really appropriate to this marvel: it should have been called a small house. Its basis was a light, duralumin framework clad with three 'skins.' The first was made of light, rubberized material. The second was uniquely beautiful and was essentially of a sky-blue, silk quilt. No less than 17 kilograms of eiderdown were used in making this quilt and our pajamas–also silk–which, incidentally, we did not wear very much. . . . The third and final "skin" was a black tarpaulin impregnated with a water-resistant substance . . ." (Krenkel, *RAEM*, 295.) Today, in St. Petersburg's Arktika Musee, the tent–an icon of Russian national memory–stands fully outfitted, a holy relic.
28. R. G. Barry et al., "Arctic Sea Ice-Climate System: Observations and Modeling," *Reviews of Geophysics* 31 (November 1993), 399–400; Spencer Weart, "Ocean Currents and Climate," in Weart, The Discovery of Global Warming, http://www.aip.org/history/climate, August 2001, p. 12 of record copy at American Institute of Physics, College Park, MD.
29. Undated *(Pravda?)* article, courtesy of Arktika Musee, St. Petersburg.
30. "Moscow, May 21.–Daring Soviet aviators today carried the first aerial landing party to the North Pole. An expedition of eleven men, having flown over the Pole, was landed safely on an ice floe about thirteen miles from the top of the world." "Fliers Alight at North Pole to Establish Soviet Air Base," *New York Times*, 22 May 1937.

31. Krenkel, *RAEM*, 323. When Gromov neared in July, "Greetings to conquers of the Arctic," crackled into Krenkel's earphones, "Greetings to Soviet eagles" was transmitted back to the plane. Any hunger for newspapers and letters notwithstanding, no drop was made to the ice.
32. The aim was to define regional as well as local anomalies and to study variations and disturbances in Earth's magnetic field.
33. The Arctic Ocean is dynamically coupled to the Atlantic. By the mid-fifties, Leningrad was studying the spreading of this Atlantic water layer into the central Arctic, based on observations by drifting stations and high-latitude air expeditions. First reported in 1993, the observed warming and shallowing of this inflow has been accompanied by an alarming shift of the front separating Atlantic and Pacific water masses—a possible response of the system to global climate change.
34. In some respects, little has changed. "Science [on sea ice] is mainly a grueling process of packing, loading, getting there, unloading, unpacking, waiting, rushing, breaking things, repairing things and always improvising.... There is the cold, which for this expedition rarely rose above 10 below zero and can cause liquid crystal displays on instruments and laptops to solidify.... Then there is the distance..." ("Doing Science at the Top of the World," *New York Times*, 13 May 2003.)
35. Papanin, *Ice Floe*, 80.
36. A yellow, rock-like slab was pea soup; another—red—was jelly. Coffee, tea, and cocoa came in briquettes; eggs and milk were powdered. Toast was fortified with 30 percent meat powder. "Wherever one looked," Krenkel observed, "there were calories and vitamins: five thousand hens had perished in order to fly to the Pole in dried, powdered form." (Krenkel, *RAEM*, 296.)
37. The solution was simplicity itself. "Instead of washing the tableware—that is, aluminum saucers—we scratched the name of the user on each one with a knife, in order not to confuse the utensils. The struggle against dirt went no further. Each man ate from his own bowl, into which soup was poured, followed by *kasha* and finally by stewed fruit. After the meal the bowls were placed out on the ice and before the next meal each of us would take his own bowl and knock the remains of the food out of it." (Krenkel, *RAEM*, 320.)
38. Research has since established lack of water for washing as one of the most irritating aspects of prolonged stint in small, closed environment. Others are a sense of isolation, the inability to move freely, and boredom. See Richard Harding, *Survival in Space* (London: Routledge, Chapman and Hall, Inc., 1990). One is reminded of a remark by Sir Edmund Hillary, first to scale Mount Everest: "Danger is one thing, but danger plus extreme discomfort for long periods is quite another."
39. Papanin, *Ice Floe*, 56–57.
40. Papanin, *Ice Floe*, 72.
41. Papanin, *Ice Floe*, 137. Medical support had been agonized over. An eighteen-month drift was planned—a year and a half with no doctor. A fifth man would have demanded a larger tent as well as more food, clothing, and so on, thus pushing the enterprise over its weight limit. The solution had been to train one of the four. At the State Institute of Advanced Medical Training, the hydrobiologist was taught the intricacies of diagnosis and the cure of diseases most common in the north, the care of wounds, making injections, even performing minor surgery. An acute medical crisis would still pose a risk, especially if surgery involved procedures of any intricacy.
42. By the thirties, Krenkel writes, the Arctic had "changed completely" from a few years before. "The romantic character of polar exploration had made it extremely popular [in the USSR this was particularly the case] and it was not surprising that journalists

hurried to cross the Arctic Circle with such determination that the Promised Land might have lain there." Two months into the drift, a *Pravda* correspondent radiogrammed the Papanin floe. "The whole country has its eyes on you," he reported, "people experience with you the puddles, the compression and the drift. Write more and get Shirshov and Fedorov to write about their scientific findings, because so-called 'experts' have already appeared here." (Krenkel, *RAEM*, 147, 329.)
43. Papanin, *Ice Floe*, 22; Krenkel, *RAEM*, 339.
44. The expedition's supply of alcohol had been left behind. For the preservation of specimens, Shirshov distilled alcohol from cognac.
45. Papanin, *Ice Floe*, 129.
46. A *second* (unplanned) drift for science had begun. The freeze-up of 1937 had come early, trapping twenty-six vessels along the Sea Route, among them seven of eight serviceable icebreakers—a fiasco that much diminished Shmidt's stock. *Sedov*, an icebreaking ship, shut down her engines on 23 October but kept her boilers stoked until 7 November. Within ten weeks, *Sedov* was conveyed into a region never before visited by steamers. With the return of light (1938), air operations were to pluck away most of the stranded human cargo, after which this beset fleet—manned by skeleton crews and reprovisioned—was left to wind and current. By the simple expedient of declaring her one, *Sedov* became a scientific drifting station. Roughly paralleling *Fram*'s track, she would ride the Transpolar Drift fully across, recording observations until freed near Greenland by a relief ship—3800 miles over 812 days. See William Barr, "The Drift of Lenin's Convoy in the Laptev Sea, 1937–1938," *Arctic* 33 (March 1980), 3–20.
47. "In brief," Krenkel summarized, "working conditions were perfect. It was no wonder that we could listen to the entire world with our little receiver: and listen we did, to every continent." (Krenkel, *RAEM*, 341.)
48. Krenkel, *RAEM*, 339; Papanin, *Ice Floe*, 161.
49. Papanin, *Ice Floe*, 217.
50. Papanin, *Ice Floe*, 232.
51. Papanin, *Ice Floe*, 243, 256.
52. In the Soviet Arctic that winter, passengers and superfluous personnel were airlifted to the mainland from various vessels logging an enforced wintering at sea. The repercussions would prove severe. See Barr, "The Drift of Lenin's Convoy in the Laptev Sea, 1937–1938."
53. Vodopyanov, *Wings*, 295.
54. Krenkel, *RAEM*, 360.
55. Douglas Botting, *One Chilly Siberian Morning* (London: Travel Book Club, 1965), 149. For an account of the life and work of Andrei N. Tupolev, especially the purge years, see L. L. Kerber, *Stalin's Aviation Gulag: A Memoir of Andrei Tupolev and the Purge Era* (Washington, DC: Smithsonian Institution Press, 1996). "It was lethal, for example, to be head of any kind of scientific institute that failed to solve immediately some pressing problem. During this period, the secret police arrested *thirteen* successive directors of the Academy of Sciences in Kiev." (viii–ix) Stalin's foreign minister made no apologies. With enemies everywhere, Vyacheslav Molotov deemed the purges a mere extension of the revolution in a time of "complicated international circumstances." "[The purges] were necessary," he insisted. David Remnick, *New York Review of Books* 40 (25 March 1993), 33–34.
56. Safon, "Stalinist Myths," 26.
57. Safon, "Stalinist Myths," 35.
58. Although a hero of the Soviet Union, Papanin did not have a career entirely with-

out strain. The vicissitudes of Stalin's favor were notorious; anyone might sink abruptly into disfavor—and worse. As an acquaintance of the author put it, Papanin was subject "to the rising and falling as it used to be during the Stalin regime." (Boris I. Silkin, Ph.D., Academy of Sciences of Russia, Geophysical Committee, Moscow, letter to author, 18 June 1991.)

59. At the Eighteenth Party Congress, in 1939, Papanin, speaking of the military significance of the Northern Sea Route, said, "In an emergency, if the enemy dares to attack us from the west or from the east, we shall be able, undisturbed and in a short time, to transfer warships from one sea border of our great Soviet Union to the other." As if to underscore the remark, in September 1940, the German commerce raider *Komet* made passage from west to east in three weeks. Papanin had been thinking of Japan; but his words were put into action against a new mortal enemy.
60. Fedorov, *Polar Diaries*, 200.
61. Boris I. Silikin, letter to author, 18 June 1991.
62. P. A. Gordienko, "The Arctic Ocean," *Scientific American* 204 (May 1961) 92.
63. Productivity is a function of the available nutrients and light. The latter is sensitive to ice thickness, snow cover, and (in summer) meltwater pools. The rate of primary production is comparatively low. Sea ice itself hosts a thriving microbiology. "Beneath the snow lies a unique habitat for a group of bacteria and microscopic plants and animals that are encased in an ice matrix at low temperatures and light levels, with the only liquid being pockets of concentrated brines . . . their prolific growth ensures they play a fundamental role in polar ecosystems." D. N. Thomas and G. S. Dieckmann, "Antarctic Sea Ice—a Habitat for Extremophiles," *Science* 295 (25 January 2002), 641. "There is a somewhat different species compositions between [boreal and austral] systems, but all that we discussed in the paper could be said about Arctic ice, with the exception that krill is not dominant. Rather than krill, amphipods and the polar cod are the main grazers at the underside of ice floes." D. N. Thomas, e-mail to author, 15 August 2002.
64. Data from the Soviet on-ice bivouac had proved of great value to the U.S. Weather Bureau. Almost from the day camp was pitched (21 May 1937), its report-summaries were relayed four times daily. These ceased on 29 January 1938.
65. Vodopyanov, *Wings*, 299.

CHAPTER 3

1. This northward push had its roots in Allied defensive strategy. Highways of supply are vital to global war. At sea, the protection of the North Atlantic was of the utmost urgency. Ashore, a line of transoceanic air-route "steps" arced from the United States to wartime partner-nations: across Canada and Greenland to Iceland thence Western Europe, and, as well, a Lend-Lease ferry route through northwest Canada to Alaska and the Soviet Far East.
2. See Vladimir O. Pechatnov, "The Big Three After World War Two: New Documents on Soviet Thinking About Post War Relations With the United States and Great Britain," and also Eduard Mark, "Revolution by Degrees: Stalin's National-Front Strategy for Europe, 1941–1947," Cold War International History Project Working Papers No. 13 and No. 31, respectively, for fresh insights into the early Cold War and (particularly) Stalin's postwar policy toward Europe.
3. P. Ostrozhski, "U.S. Polar Strategy," *Trud* 29 (October 1947). U.S. expansion into the far north was deemed "ever-growing." From the same article: "Under the pretext of carrying out the Treaty with Denmark for defence of Greenland, which has

long since lost its power, the USA is retaining hold of its military bases in Greenland. In this way the United States has managed to create all the necessary conditions for basing the American War Department in the Arctic regions. Bases and stations have been set up in territories that do not belong to the United States."
4. Mark, "Revolution by Degrees," 42.
5. "Our adversaries, in the ingrained American way of looking at things, had always to be demonic, monstrous, incalculable, and inscrutable. It was unthinkable that we, by admitting that they sometimes reacted to what we did, should confess to a share in the responsibility for their behavior." (George F. Kennan, *Memoirs, 1925–1950* (Boston: Little, Brown and Co., 1967), 498.)
6. "For forty-five years every domestic initiative was justified, either directly or indirectly, by reference to the long twilight struggle [of the Cold War], from expressways under the National System of Interstate and Defense Highways to college loans (the National Defense Education Act). Even civil rights was argued as essential to American prestige in the international arena." Sidney Blumenthal, "Rendezvousing With Destiny," *The New Yorker*, 8 March 1993, 44.
7. Boris I. Silkin, Ph.D., Academy of Sciences of Russia, Moscow, Geophysical Committee, letter to author, 2 July 1993. "Asked to think about America in the 1950s, any historian will quickly turn to the influence of the Cold War, which weighed heavily on science and technology in particular." (Spenser R. Weart, "Global Warming, Cold War, and the Evolution of Research Plans," *Historical Studies in the Physical and Biological Sciences* 27, pt. 2 (1997), 321.)
8. U.S. Department of the Navy, *Naval Arctic Operations Handbook*, pt. 1, 1949, 6–7.
9. By 1956, missions out of Eielson airfield penetrated daily to latitude 85°–fourteen hours flying. Each took weather observations as well as dropsonde (released-instrument) readings. Except for a small number of ground stations, these "Ptarmigan" missions were the only source of weather data north of latitude 72° N. "Arctic Weather Flights From Alaska, 1947–55," *Polar Record* 8 (January 1956), 24.)
10. Hummocky surfaces and well-defined edges show up well on a radar screen; the rougher the surface the stronger the reflections.
11. Why classified? "Should another world war break out," one account has it, "the Arctic region would most likely be the crossroads of activity. Since the ice island discovery raised many new hopes for its possible role in future U.S. military matters, it was kept secret." (Arctic Air Command, *Ice Islands of the Arctic: Alaskan Air Command's Arctic Experience*, historical monograph, undated (ca. 1969), 3. Courtesy Dana Bell.)
12. "Ice islands" are the most massive features in arctic seas. In austral waters, the world's two largest ice shelves occupy embayments on opposite sides of West Antarctica. Periodically, the downstream ends calve stupendous pieces. In September–October 1987, when Iceberg B-9 broke from the Ross Ice Shelf, the map-outline of the Southern Continent was altered. The drifter measured 96 miles long, 22 miles wide, 750 feet deep.
13. Maxwell E. Britton, Ph.D., interview by author, audiocassette, Arlington, VA, 14 August 1999. Wing Commander Keith R. Greenaway, Royal Canadian Air Force, was navigating the Canadian aircrew that first photographed the slab. "At the time of discovery," he recalls, "we speculated that landings could be made on T-3 without undue risk." (Letter to author, 20 November 1991.)
14. "It is not a very photogenic or easily photographed object in its entirety, and during summer, when it stands out best from the surrounding sea ice, it is blanketed with stratus 90% of the time." (E. F. Roots, letter to David D. Smith, Ph.D., 22 April 1962, courtesy Dr. Smith.)

15. Keith R. Greenaway, interview by author, audiocassette, Ottawa, Ontario, 24 October 1988; "Arctic Ice Islands," *Arctic* 5 (July 1952), 70.
16. Moscow claimed that T-1, T-2, and T-3 were sighted by Soviet airmen before the U.S. Air Force found them.
17. Zalman Gudkovich, Ph.D., interview by author, audiocassette, Arctic and Antarctic Research Institute, St. Petersburg, 21 May 1992.
18. Three primary ridges split the abyssal depths. The Nansen or Arctic Midoceanic Ridge, which bisects the Eurasian Basin, is a continuation of the globe-girdling midoceanic system. Though not readily accessible by air from Alaska, once detected, the nature and origin of these submarine ridges (and the distinct basins they define) rendered the Eurasian of particular geophysical interest. Here also the waters of the Atlantic mix with those of the Arctic.
19. A greater confidence in the data and resulting interpretations is obtainable by exploiting several complementary methods of geophysical and oceanographic exploration.
20. The Soviets unveiled the discovery in a bathymetric map published in 1954. This release caused a sensation among Western governments, which had been unaware of the scope and scale of Soviet programs. Indeed, this newfound mountain complex necessitated a fundamental reevaluation of existing conceptions of the polar seafloor, thereby stimulating and reshaping later investigations.
21. Vodopyanov described these "jumping" sorties. "Generally, the first plane would choose a suitable floe." The ship carries a drill (to determine floe thickness) and explosives, to open the ice for soundings. The second delivers tent, winch, researchers. "It took but a few minutes for the hydraulic station to begin to operate." A sounding then was made, using an automatic winch. Currents at various depths were recorded, seawater samples collected, temperature and salinity measured, a reading taken of the magnetic field. "The pilots and air-mechanics helped the hydrologists and meteorologists to unload, pitch tents and set up the winch. The crew always worked in close contact with the scientific personnel. Their work done, the men would abandon their post and start out for another." (Vodopyanov, *Wings*, 318.)
22. Gudkovich interview, 21 May 1992. "Only continuous observations over a long period of time can capture and portray all the aspects and the laws that govern the complex, contradictory processes and phenomena that are peculiar to the Arctic environment." (G. N. Yakovlev, *Ice Routes of the Arctic,* trans. Ottawa: National Defence, TRS Library No. 1653, 15 August 1977, 7. Originally published by MYSL Publishing House, Moscow, 1975.)
23. Vasily Burkhanov, *New Soviet Discoveries in the Arctic*, Foreign languages Publishing House, 1956, quoted in Max C. Brewer, "The Soviet Drifting Ice Station, NORTH-67," *Polar Record* 13 (October 1967), 266.
24. Yakovlev, *Ice Routes*, 18.
25. SP-2 observational data would be reported under the editorship of Somov in 1954–55. The concern for secrecy was such that letters from the ice were forbidden. Cables also were banned. Every month, the Arctic Institute director informed relatives as to their loved ones, conditions on the ice, and so on. Thus, the station remained confidential even inside the Soviet Union. (Gudkovich interview, 21 May 1992.)
26. J. O. Fletcher, "Origin and Early Utilization of Aircraft-Supported Drifting Stations," in *Arctic Drifting Stations: A Report on Activities Supported by the Office of Naval Research*, proceedings of the symposium, John E. Sater, coordinator, Warrenton, VA, 12–15 April 1966 (published November 1968), 5.
27. Because these stations operate all day every day, and supply problems limit team size, a good deal of versatility is expected of each man.

28. To a large degree, the efficiency with which ocean and atmosphere transfer heat from the tropics to the polar zones determines the global climatic pattern. Sea ice, existing at the interface between ocean and atmosphere, is intimately related to conditions in both the water below and air above.
29. Gudkovich interview, 21 May 1992.
30. The value for gravity is not a constant; instead, it varies according to the mass below the point of measurement–the roots of a mountain range here, an ancient seabed buried there. SP-2 was the only Soviet station for which published scientific data reached the West prior to 1963.
31. Graham W. Rowley, *Cold Comfort: My Love Affair With the Arctic* (Montreal and Kingston: McGill-Queen's University Press, 1996), xii. An archaeologist and recipient of the Northern Science Award for his work in the Arctic, Rowley arrived on Baffin Island in 1936. There he traveled by dogsled and lived as the Inuit did.
32. "The Soviet North/Economic Aspects," paper prepared for the Advisory Committee on Northern Development, Joint Intelligence Bureau, Ottawa, Department of National Defence, Record Group 25, Box 3306, file 9059-D-4, National Archives of Canada.
33. Aerial operations had yet to achieve dull reliability. In 1945 (for example), Canada had decided to conduct frequent long-range penetrations. "After several flights it became apparent that the problems of regular polar flying were more serious than had been realized when the operation was first planned. Mapping was inaccurate, magnetic compasses unreliable over a greater area than had been anticipated, and weather forecasts sketchy and uncertain." (*An Aerial Reconnaissance of Arctic North America*, Defence Research Board, Canada, 1946, 1.)
34. Nathan Reingold, "Choosing the Future: The U.S. Research Community, 1944–1946," *Historical Studies in the Physical and Biological Sciences* 27, pt. 2 (1997), 302. For a full account, see Harvey Sapolsky, *Science and the Navy: The History of the Office of Naval Research* (Princeton: Princeton University Press, 1990). With the creation of the Office of Naval Research, "the navy established the most liberal, effective, and flexible federal patronage agency in history." Gary E. Weir, *An Ocean in Common: American Naval Officers, Scientists, and the Ocean Environment* (College Station: Texas A&M University Press, 2001), xiv.
35. Maxwell E. Britton, "The Role of the Office of Naval Research and the International Geophysical Year (1957–58) in the Growth of the Naval Arctic Research Laboratory," in *Fifty More Years Below Zero* (Fairbanks: Univ. Alaska Press, 2001), 65. Scientific officer and later director of the Arctic Research Program, ONR, Britton reported for duty in September 1955.
36. "NARL [Naval Arctic Research Laboratory] sprang from the fertile minds of ONR [Office of Naval Research] scientists for the sole use of its contractors, grantees, or any other research the Chief of Naval Research found meritorious of his approval. That said it all." (Britton, "The Role," ix.)
37. "The ONR's work is a well-known chapter in the history of American science. As the war effort wound down in 1945 and scientists worried about where they would find support, the United States Navy took the lead in funding basic research. This support, later imitated by other military services, reflected a recognition among some officers that they would need scientists for many purposes." Weart, 332; Britton interview, 14 August 1999. In 1950, the National Science Foundation was created "with much the same philosophical base as ONR, and with a much broader mandate to serve the public scientific interests of the United States." Britton, "The Role," 65.

38. Maxwell E. Britton, "U.S. Office of Naval Research Arctic Research Laboratory, Point Barrow, Alaska," *Polar Record* 13 (January 1967), 422.
39. John F. Holmes and L. V. Worthington, Woods Hole Oceanographic Institution Technical Report, Reference No. 51-67, 1. "The [ONR] organization," Dr. Britton adds, "was in the position of sponsoring research in most any scientific field," and "was very generous in funds, very much like the National Science Foundation is today in the breadth of the sciences [it] could cover."
40. Holmes and Worthington, Woods Hole Report, 1.
41. That March, Ward took 12417 to Kodiak to brief the commander of the Alaskan Sea Frontier. "At first the Admiral was reluctant to approve of our proposed ice landings, referring to it as 'a hair-brained scheme.' Fortunately we were able to convince him otherwise and begrudgingly he approved only if a PBY (navy amphibian) aircraft observed our landings from the air." Cdr. E. M. Ward, U.S. Navy (Ret.), letter to author, 22 June 2001.
42. Among its revelations, SKIJUMP showed that deep Beaufort water was about $0.4°$ warmer than that in the European sector. This led to suspicions (subsequently confirmed) of a bisecting submarine ridge–a natural sill over which the densest water could not pass. The hypothesis was strengthened by soundings from T-3, which revealed depths of less than 2000 meters (6560 feet) over much of its track.
43. Cdr. E. M. Ward, U.S. Navy (Ret.), "Project Ski-Jump 1 & 2, Arctic Ocean Oceanographic Research Expedition, 1951–1952," memoir, 8, courtesy Commander Ward.
44. Lt. Cdr. William V. Kielhorn, U.S. Coast Guard Reserve, "SkiJump II," Naval Institute *Proceedings* 30 (October 1954), 1128; Ward, "Project Ski-Jump," 10. "There is some evidence," *Polar Record* records, reporting on SKI-JUMP, "of a submarine ridge running across the central polar basin." "U.S. Navy Operation 'Skijump II,' 1952," *Polar Record* 7 (July 1954), 149.
45. Ward, "Project Ski-Jump," 11.
46. L. V. Worthington, "Oceanographic Results of Project SKIJUMP I and II in the Polar Sea, 1951–1952," *Transactions American Geophysical Union* 34 (August 1953), 543.
47. Potential military applications: a reporting weather station; an air-sea rescue base; a radar platform for our early-warning network or for tracking guided missiles; a field base for submarines. Scientific interest included investigations of the ocean floor as well as water and atmospheric circulation.
48. Maynard M. Miller, "Floating Islands," *Natural History Magazine* 65 (May 1956), 235.
49. Yakovlev, *Ice Routes*, 93.
50. Sea ice accounts for approximately two-thirds of Earth's frozen sea-skin. This variable lid interacts with the atmosphere above and sea beneath. Its presence or absence is therefore a primary factor in the overall heat exchange and, in turn, global climate.
51. Yakovlev, *Ice Routes*, 105–106.
52. The Air Weather Service of the U.S. Air Force had been formed in 1949 to supplement the work of the U.S. Weather Bureau in Alaska and to act as a coordinating center for meteorological reports received from or required by ground stations in North America and Greenland.
53. Miller, "Floating Islands," 233–39, 274, 276. A glaciologist, Miller was aboard at Fletcher's request.
54. ACC, *Ice Islands*, 19.
55. The request was made on 21 March–in a note from the American ambassador to the Secretary of State for External Affairs. This asked in part "permission of the Canadian Government for use of the joint Canadian-United States Weather Station

at Alert in connection with a project to investigate ice islands in the Arctic Ocean or aground on the northern coast of Greenland. The object of the project is to learn the nature of the ice islands, their sources and composition, and to establish on one of these islands a temporary weather station which will investigate Arctic weather for a period of six months." The T-3 landing was logged on the 19th; hence, the note was a formality, after the fact.

56. Kaare Rodahl, *North: The Nature and Drama of the Polar World* (New York: Harper and Bros., 1953), 172. All Rodahl quotes are from this work.
57. The C-54 navigator was known for a laconic manner. His monotone delivered the word: "I may report that T-3 has been located. The island is underneath us just now—and by the way, there is nobody there." Rodahl, *North,* 173.
58. George Silk, "North Pole 103 Miles; Temperature 60 Below," *Life* 32 (31 March 1952), 13–17. "The two men," *Time* reported, "stood within a foot of each other, their hands over their faces, mumbling against the cold that numbs men's minds. Every now and then they would drop their hands and jump about violently for warmth. Eventually Fletcher won the argument." *Time,* 31 March 1952. Accounts have differed slightly as to what transpired between the two officers.
59. "[General Old]," Fletcher writes, "not only accepted operational responsibility in the midst of dire predictions from above but backed his judgment by unstinting support and personal participation in the more hazardous phases of the project." ("Scientific Studies at Fletcher's Ice Island, T-3, 1952–1955," ed. Vivian Bushnell, Geophysical Research Papers no. 63, Air Force Cambridge Research Center, September 1959, iv.)
60. "The kerosene," Rodahl records, "poured like thick syrup, and when it was spilled on the metal of the primus stove it solidified like stearine dripping from a candle. The fuel oil had the consistency of slush ice." (*North,* 188–189.)
61. Joseph O. Fletcher, "Three Months on an Arctic Ice Island," *National Geographic* 103 (April 1953), 499.
62. Rodahl, *North,* 195.
63. Fletcher, "Three Months," 500.
64. Rodahl, *North,* 201–202.
65. Fletcher, "Three Months," 500.
66. Why sponsor research in the geophysical sciences? See for example Helmut E. Landsberg, Ph.D., *Geophysics and Warfare,* Research and Development Coordinating Committee on General Sciences, Office of the Assistant Secretary of Defense, Research and Development, March 1954. Basin geology is not exposed at sea. Seismic reflection can define, indirectly, the lateral and vertical extent of sedimentary basins and layer sequences. Seismic refraction methods are useful in identifying deep crustal structures of the continental margins and transitions to oceanic crust.
67. Britton interview, 14 August 1999.
68. Norman Goldstein, Ph.D., interview by author (notes), New York City, 5 October 1992. Both Crary and Cotell had participated in the air force's early geophysical studies, in the Beaufort.
69. Rodahl, *North,* 208.
70. Robert D. Cotell, letter to author, 26 September 1991.
71. Britton interview, 14 August 1999.
72. Kenneth L. Hunkins, Ph. D., interview by author, audiocassette, Lamont-Doherty Geological Observatory, Palisades, NY, 21 May 1991. Hunkins, a distinguished geophysicist, is a seasoned arctic researcher.
73. A gravity reading involves measuring the acceleration due to earth's gravitational

field. Small variations in gravity depend upon lateral changes in density of the subsurface in the vicinity of the measuring point. Exceedingly sensitive, gravity meters or gravimeters are mechanical balances in which a mass is supported by a fine spring. The weight responds to subtle changes in the acceleration due to gravity tugging downward.

74. Seismic work requires an energy source to impart an impulse into the subsurface and also receiving/recording instruments. Chemical explosives such as TNT were used (fired in ice or water) as well as mechanical impulses, such as swinging weights. A seismograph records the arrival of reflected or refracted seismic waves with respect to time. These waves are detected by small receivers that transform mechanical energy into electrical voltages relayed by cables to the seismograph, which records voltage output versus time. Instruments for picking up, amplifying, and recording the signal were adapted from prospecting equipment. Moving-coil geophones on the ice as well as hydrophones were used to detect signals.

75. Letter, Commander 7th Air Weather Group to Chief Air Weather Service, 14 April 1952, quoted in ACC, *Ice Islands*, 53.

76. The visit became a visitation. "The accident to the Navy plane," Rodahl explains, "had occurred at a most inopportune moment, when thousands of pounds of vital equipment were scattered in the snow, and when the T-3 party urgently needed to use all its time and effort to consolidate the station and to establish routine operations. Now, however, all this was being held up. To fly a bulky thirty-five-hundred-pound engine in to T-3 was a major operation, for it was too big and heavy to be taken in by a C-47. It now became imperative, therefore, to build a runway to enable a C-54 to land with the engine and the necessary material for an engine change out on the ice." Rodahl, *North*, 215. By way of repayment, the airmen assisted in the building of additional shelters and "generally improving the living conditions" on-island. Kielhorn, "Ski-Jump II," 1130.

77. ACC, *Ice Islands*, 44.

78. Cotell letter, 26 September 1991.

79. Cotell letter, 26 September 1991. Cotell's further remarks are from this source. Joe Fletcher—first off the C-47—was probably the first person to set foot at the geographic pole. Why? Interpretation of the expedition papers of both Robert E. Peary and Frederick A. Cook cast doubt on their 1909 attempts by dogsled. Each party likely fell short of his goal.

80. That September, a veritable crowd gained the pole—thirty-six persons. The group included May Craig, Washington correspondent for the *Portland Press Herald* (Maine)—the first woman to reach the place. Having made the first public inspection of the air base at Thule, and with T-3 shrouded in overcast, the Military Air Transport Service shuttled its passengers (mostly press and radio reporters) to the pole instead.

81. Wrapped in luminescent red silk, the watertight jars held a note asking finders to record the date and position of discovery and report to the air force. "Such information," *New York Times* said, "may supply data on the speed and direction of little-known Arctic Ocean currents."

82. Based on this and follow-up work, on carbon-14 dating of dirt layers on T-3, and on other evidence, the probable history of the shelves was reconstructed. Thick growths of ice had probably begun to build in Yelverton Bay about 5500 years ago; the shelf near Ward Hunt somewhat later. The last accumulation period continued until about four hundred years ago, when the current ablation period began. A remnant of these buildups, T-3 had broken free in the 1940s.

83. *Christian Science Monitor*, 25 July 1952.

84. During the summer season, sea ice melts from the bottom as well as laterally and from the top surface.
85. The major had "a great tolerance for cold. During re-supply missions after a paradrop when all hands were scrambling to recover our goodies Maj. Dorsey would work very hard. He would wear only a field jacket (as contrasted with a heavy parka) and never complained about the cold. He was a hard-working man due for reassignment back to the civilization of the United States." Cotell letter, 26 September 1991.
86. Gerry H. Cabaniss, Ph.D., letter to author, 27 February 1990. A geophysicist, Cabaniss served four tours aboard T-3–531 days total. His last departure: October 1960.
87. John L. Mohr, "Marine Biological Work," in *Scientific Studies at Fletcher's Ice Island, T-3, 1952–1955*, 1, Geophysical Research Papers no. 63, Air Force Cambridge Center, 83.
88. A. Chilingarov, E. Sarukhanyan, and M. Yevseyev, *Pod Nogami Ostrov Ledianoi* (Moscow, 1972), *Life on an Ice Island*, Cold Regions Research and Engineering Laboratory, U.S. Army Corps of Engineers, Hanover, NH, draft translation 502, December 1975, 38–40. The dark time can inspire. "The Arctic winter," an airman mused, "cut us off from the world of every-day life, but it brought us more intimately in touch with Nature, the vast forces of the wind, the biting frost, the silence and the almost limitless spaces about us. As the sun slipped off the edge of the world, the beauty and pageantry of the night sky usurped the sunlit world. Man seemed insignificant and his efforts futile amid this vast drama of the Arctic night." (Guy Blanchet, *Search in the North* (New York: Macmillan Co. of Canada Ltd., 1960), 67.)
89. Guide for Expeditions to the Canadian Arctic Islands, Indian and Northern Affairs, Canada, 32.
90. "One of the few negatives of basing facilities on an ice island, rather than a floe, is that access to the ocean (for soundings, instrument deployments, biological sampling, etc.) requires that the hydro station be on the pack near the edge of the island because an ice island is too thick. It takes a fair amount of work to keep the access open in the winter." Gerry H. Cabaniss, Ph.D., e-mail to author, 10 March 2003.
91. The *New York Times* Index holds no entries regarding Soviet activities from 1950 to 1954. "Soviet post-war activity was unknown to the West," Fletcher (then with the Rand Corporation) wrote in 1966. "It was conducted clandestinely and no announcements were made until mid-1954."
92. The Arctic and Antarctic Research Institute views Alaska as a "local" area–a short coastline compared to that of Siberia. The latter's far-flung littoral is influenced by processes in the central Arctic.
93. Fedorov, *Polar Diaries*, 309, 316.
94. A. F. Treshnikov, "The Soviet Drifting Station NP-3, 1954-55," *Polar Record* 8 (September 1956), 224. Having been kept under observation, the site of SP-2 was revisited (by helo) during the expedition. "Soviet Drifting Expedition in the Arctic Ocean, 1950–51," *Polar Record* 8 (May 1955), 398.
95. Terence Armstrong, *The Russians in the Arctic* (Fair Lawn, NJ: Essential Books, 1958), 73.
96. Fedorov, *Polar Diaries*, 325–326; Fletcher, "Origin and Early Utilization," 7; Vodopyanov, *Wings*, 327.
97. Treshnikov, "Soviet Drifting Station," 227.
98. Gordienko, "Arctic Ocean," 98. A veteran and also station leader (SP-4, 1955–56), Gordienko worked in Antarctica as well. He served as deputy director of the Arctic and Antarctic Research Institute from 1956 to 1964.
99. Charles L. Smith, "A Comparison of Soviet and American Drifting Ice Stations," *Polar Record* 15 (May 1971), 882–883. A submarine eruption in progress might produce seismo-acoustic "noise," temperature anomalies, and, perhaps, discolored

water, bubbles, steam, pumice or dead fish. What if any warning these drifters might have met from sites of active volcanism is not known.
100. When the staff plane bearing Yevgeny Fedorov taxied in for a visit (2 May), "I noticed that the T marking at the beginning of the landing strip was laid out with champagne bottles emptied in the camp the previous night [May Day]." Fedorov, *Polar Diaries*, 319.
101. Fedorov, *Polar Diaries*, 320.
102. "Soviet Investigations in the Arctic Ocean, 1955," *Polar Record* 8 (January 1956), 27–28.
103. "The Soviet High Latitude Arctic Air Expedition and Drifting Stations in the Arctic Ocean, 1954," *Polar Record* 7 (May 1955), 396; Fletcher, "Origin and Early Utilization," in Sater, *Arctic Drifting Stations*, 7.
104. Dated 28 April 1958, Benedict's Travel Orders read in part: "You WP o/a 29 April 55 from Bedford, Mass. to Westover AFB, Mass. rept thereat 30 Apr 55 to proceed to o/s dest under AMD US-80-2D-4430-RF4 to Thule AFB, Greenland thence to Fletcher's Ice Island T-3 on TDY for aprx one hundred and sixty (160) days to conduct fld exper in connection with ARCRC Proj 7606 at Fletcher's Ice Island T-3, and upon compl thereof will rtn to Bedford, Mass." Benedict papers.
105. Goldstein interview, 5 October 1992.
106. Norman Goldstein, "Scientific Activities of GRD on Fletcher's Ice Island, T-3," Progress Report no. 8, Geophysics Research Directorate, Air Force Cambridge Research Center, Air Research and Development Command, 2. Courtesy Mr. Goldstein.
107. In the hut used for seismic shots circa 1958–60, "The procedure included the following: Start the stove, chisel through the ice in the shot hole, prepare the explosive, drop it in the water, attach the shot line to shot box, warm up the developer, start the seismic recorder, fire the shot, stop the recorder, develop the record, and log date, time, and location in the record book. The record was analyzed later." Gerry H. Cabaniss, Ph.D., e-mail to author, 10 March 2003.
108. *Washington Post*, May 1955.
109. Goldstein interview, 5 October 1992.
110. That May, Admiral Burkhanov (in command of the Soviet program) overflew the Ellesmere party while inspecting his operations. As a gesture, vegetables were dropped. The ensuing press speculation about where, precisely, the short-ranged plane had come from might have influenced the announcement as to the existence of SP-3 and SP-4.
111. In 1966, the following could still be written: "Accurate navigational positions, in ordinary latitudes a fairly simple determination, in the Arctic region become extraordinarily difficult to establish. There are neither LORAN nor reliable radio beacons, and skies clear enough for even simple sun lines are rare. Even radar fixes from charted land may be inaccurate, due to inadequate charting. Charted soundings are unreliable, and dead reckoning, normally fairly reliable for modest accuracy in navigation, becomes a highly estimative business in an ice-covered water area with unrecorded currents." Lt. R. D. Wells, U.S. Navy, "Surveying the Eurasian Arctic," Naval Institute *Proceedings* 92 (October 1966), 83.
112. Gerry H. Cabaniss, Ph.D., letter to author, 10 February 1992.
113. *Christian Science Monitor*, 30 September 1955.
114. In 1954, a navy expedition departed Boston for Antarctica. Its mission was to carry out scientific investigations and to examine possible sites for American bases for the IGY. In preparation for IGY, M. M. Somov led a Soviet advance expedition to the Southern Continent in 1955, including (in 1957) an inland reconnoiter of a route to

the projected "Vostok" station, at the South Geomagnetic Pole.
115. Nikolai Dimitrievich Vinogradov, interview by author, audiocassette, Arctic and Antarctic Research Institute, St. Petersburg, 21 May 1992.
116. "Soviet High Latitude Air Expeditions, 1948–1956," *Polar Record* 8 (January 1957), 346, 351.
117. J. O. Fletcher, "Origin and Early Utilization," 9; "U.S.S.R." *Polar Record* 8 (May 1957), 446; and "Soviet Arctic Expeditions, 1957," *Polar Record* 9 (January 1959), 337.
118. Britton interview, 14 August 1990.
119. Cdr. Ronald McGregor interview by author, audiocassette, Washington, DC, 10 August 2002.

CHAPTER 4

1. Krenkel, *RAEM*, 142.
2. Van Allen had taken part in the first rocket launches (from ships) in both the Arctic and Antarctic. That Maryland evening has passed into legend. Among the guests were Sidney Chapman, a geophysicist from Europe and Dr. L. V. Berkner. "The dinner conversation," Van Allen recalled, "ranged widely over geophysics and especially geomagnetism and ionospheric physics. Following dinner, as we were all sipping brandy in the living room, Berkner turned to Chapman and said, 'Sydney, don't you think that it is about time for another polar year?' Chapman immediately embraced the suggestion, remarking that he had been thinking along the same lines himself. The conversation was then directed to the scope of the enterprise and to practical considerations of how to contact leading individuals in a wide range of international organizations in order to enlist their support." (Van Allen, "Genesis of the International Geophysical Year," *Eos Transactions American Geophysical Union* 64 (13 December 1983), 977.)
3. Van Allen, "Genesis."
4. Robert C. Fleagle, "From the International Geophysical Year to Global Change," *Reviews of Geophysics* 30 (November 1992), 306. The IGY, indeed, established a permanent scientific presence in Antarctica.
5. Spencer Weart, "International Cooperation," in Weart, The Discovery of Global Warming, http://www.air.org/history/climate, August 2001, p. 12 of record copy at American Institute of Physics, College Park, MD.
6. Walter Sullivan, *Assault on the Unknown: The International Geophysical Year* (New York: McGraw-Hill, 1961), 30. A layman's account. Science writer for the *New York Times*, Sullivan later was recognized by the American Geophysical Union for his contributions to the understanding of science.
7. Waldo K. Lyon, Ph.D., interview by author, audiocassette, San Diego, CA, 13 October 1996; Robert E. Francois, letter to author, 21 January 2003.
8. Waldo K. Lyon, Personal Journal, 29 October 1956, courtesy Dr. Lyon. He had submitted a request at that meeting "to obtain temp./salinity/current profiles across entire Bering Strait, particularly to gain permission to observe on Russian side." Lyon, Journal. "There has never been a national or international scientific plan to systematically survey the Arctic Ocean; thus, what data exist mostly reflect military operation of submarines in areas of tactical or strategic interest." Lloyd D. Keigwin and G. Leonard Johnson, "Using a Nuclear Submarine for Arctic Research," *Eos Transactions American Geophysical Union* 73, (12 May 1992), 209.
9. Cdr. Ronald Denk, U.S. Air Force (Ret.), e-mail to author, 17 June 2001.

10. Herbert O. Johansen, "Seven Weeks on an Ice Island," *Popular Science Monthly*, August 1957, 76.
11. T-3's original programs had been almost entirely military. The IGY work on T-3 and ALPHA, however, was supported by a number of federal agencies, and the work on CHARLIE in the post-IGY period supported principally by ONR with some assistance from the Air Force Cambridge Research Laboratories.
12. "Soviet Drifting Stations in the Arctic Ocean, 1956–57," *Polar Record* 8 (September 1957), 520; and "Soviet Drifting Stations in the Arctic Ocean, 1957–58," *Polar Record* 9 (September 1958), 242.
13. ARL's logistic support was little-exploited for both ALPHA and BRAVO. "Facilities were pretty meager, after all. But when called upon, the laboratory did things for them. We had our own planes. We had the capability with *small* planes to get out the distances involved. So the laboratory performed many services, kindnesses—mail drop services, things like that." Maxwell Britton, Ph.D., interview by author, audiocassette, Arlington, VA, 14 August 1999.
14. "Father Tom," an old-timer Jesuit missionary, was an acknowledged Arctic expert. The U.S. Air Force had turned to him as an advisor for its floating stations. A dynamic personality, he was to minister to the Eskimos for many years before joining Fletcher's group. He died a major in the Air Force Reserve and was buried with full military honors.
15. ACC, *Ice Islands*, 65.
16. ACC, *Ice Islands*, 67–68. During construction, an experiment was held: flooding two thousand feet of runway with seawater, to gain a level surface. The freezing point of ocean water is depressed by the presence of brine in solution. As the weather warmed, a huge lake formed. The technique would work, it was concluded, if the runway were flooded early in the fall, allowing salt to leach out during the winter months, thus furnishing near indestructible ice.
17. Gerry H. Cabaniss, Ph.D., Kenneth L. Hunkins, Ph.D., Norbert Untersteiner, Ph.D, "US-IGY Drifting Station Alpha Arctic Ocean 1957–1958," Air Force Cambridge Research Laboratories, special report no. 38, Terrestrial Sciences Laboratory, Office of Aerospace Research, U.S. Air Force, Bedford, MA, November 1965, 2.
18. Arnold M. Hanson, "Critique of Ice Station Support," in Sater, *Arctic Drifting Stations*, 108.
19. As part of the U.S. IGY program, a network of scientific stations was installed throughout Antarctica. The Amundsen-Scott South Pole Station, established in November 1956, served as the southern high-latitude anchor for the IGY's international pole-to-pole network of meteorological observation stations.
20. Among IGY legacies for global-change studies: 1) the impetus given to satellite instrumentation and observation; 2) the discovery that ice and ocean sediment cores can reveal records of past climates; and 3) the monitoring of atmospheric carbon dioxide, which, in turn, led to recognition of a startling industrial-age trend in concentration. See Fleagle, "From the International Geophysical Year to Global Change."
21. Most of the scientific effort at the ALPHA, BRAVO, and (later) CHARLIE stations were supported by IGY funds, with both the air force and the navy conducting some specialized research. The navy, in particular, was eager to continue the effort *after* the IGY.
22. Robert Strauss, personal communication, August 1991. Shortly after the orbiting, Caltech aerodynamicist Theodore von Karman found himself on "Face the Nation." His inquisitors were interested mainly in whether the Russians were ahead in space and, if so, how long it would take to catch up. "This type of question," von

Karman wrote, "... unfortunately demonstrates in my view the sin of short-range thinking that besets many Americans. I asked the reporters to define what they meant by the term 'ahead'. For me the term arises out of the American competitive spirit, but means little in the complex field of science and technology where 'aheadness' in one area is often neutralized by 'behindness' in another area. In this sense I said that the Russians were not truly ahead of us in space." Theodore von Karman, *The Wind and Beyond: Theodore von Karman, Pioneer in Aviation and Pathfinder in Space,* with Lee Edson, (Boston: Little, Brown and Co., 1967), 13–14.

23. Boris I. Silken, Ph.D., letter to author, 2 July 1993.
24. The struggle with communism pervaded every sphere. "Even to fund the essential underpinnings of civil society it became necessary to invoke the magic words 'national security.' It was panic at the thought of a Soviet dog orbiting in space [Laika, in *Sputnik II*] rather than a commitment to American children that sparked the concern of the federal government with improving primary and secondary education in the late nineteen-fifties. The bill to create a national highway system was passed as a civil-defense measure." (Richard J. Barnet, "The Disorders of Peace," *The New Yorker,* January 20, 1992, 68.)
25. Such myopia now seems astonishing. That alarming October, *Life* summarized the harsh military implications: "Sputnik proved that there were great military, as well as scientific, advances in the USSR. Getting their heavy satellite up meant that Russia had developed a more powerful rocket than any the U.S. has yet fired and substantiated Soviet claims of success with an intercontinental missile. Putting Sputnik into a precise orbit meant Russia had solved important problems of guidance necessary to aim its missiles at U.S. targets. The satellite could also be a forerunner of a system of observation posts which could watch the U.S. unhindered and with deadly accuracy." *Life,* 21 October 1957, 24.
26. Throughout the 1957–58 ALPHA drift, daily fixes (using an engineering transit) were taken whenever possible——star shots during complete or partial polar darkness, solar observations during the perpetual daylight of late spring and summer, during which weather conditions were generally poor.
27. *Seattle Times,* undated clipping. At the Leningrad institute, a group of "diving oceanologists" was available to join a drift party in its work, penetrating below the ice cover to see and record things that even modern instruments could not reveal.
28. ALPHA's postal address: Drift Station "A" Project Ice Skate, APO 731, Seattle, WA. That of "B" (BRAVO/T-3): Project Ice Skate, APO 23, NY, NY.
29. Chilingarov et al., *Life on an Ice Island,* 36, 47.
30. As Cory Dean, a *New York Times* reporter, shrewdly points out, scientists are uncomfortable with journalists. "For scientists," she says, "an interview with a journalist offers many dangers and few rewards. Even a well-done story in the lay press doesn't necessarily do a scientist any good. Scientists communicate their findings through professional journals and meetings. A newspaper article or television interview does not produce tenure or grant awards. It may only get the scientist a reputation as a publicity hound." Serious questions of public policy depend of course upon scientific findings; the communication of science to the lay public is therefore of inestimable worth.
31. David D. Smith, Ph.D., "Sequential Development of Surface Morphology on Fletcher's Ice Island T-3," Geophysics Research Directorate, U.S. Air Force Cambridge Research Center, scientific report no. 4, 1960, 897–898, courtesy Dr. Smith.
32. ACC, *Ice Islands,* 69. This problem had confronted the Papaninites as well. "A piece of paper or the tiniest chip falling on the ice," Krenkel records, "would thaw it and

sink down to a depth of as much as twenty centimetres. A chocolate wrapping had been lying near the tent; now a deep hole marked the spot, into which one could even put one's foot. Pieces of string placed on the surface of the ice produced particularly attractive loop-like effects as they melted into it." (Krenkel, *RAEM*, 329.)

33. Carl B. Johnston Jr. "Problems of Working on a Drifting Station," in Sater, *Arctic Drifting Stations*, 2; Chilingarov et al., *Life on an Ice Island*, 67.

34. Tatiana Kaliazina, letter to author, 20 September 1992. Tatiana was my host and translator in May 1992, while I was in St. Petersburg.

35. Nikolai Vinogradov, interview by author, audiocassette, Arctic and Antarctic Research Institute, St. Petersburg, Russia, 21 May 1992. Gerry Cabaniss, Ph.D., logged four tours on T-3. "The worst weather was in the spring, especially March. Although the sun was up for a reasonable time, the temperature was 40 below and the wind blew at 40 knots (about 45 mph.) The 'nice' thing about 40 below is that the F&C scales are identical there.... The coldest weather we had on T-3 was −54°F and it was dead calm." Cabaniss, e-mail to author, 10 March 2003.

36. "We at T-3 found that the poor summer radio communications in the IGY were difficult. Recall that the date [1957] was selected because the sun was supposed to be especially active in that year and still high in the next two." (Gerry Cabaniss, e-mail to author, 4 March 2003.)

37. "It is not pleasant," Hanson observed, "to hear a faint reply to your call, 'ARLIS II, ARLIS II [chapter five], I cannot read you, I cannot read you, see you tomorrow, see you tomorrow,' when one of your companions is seriously ill. Ham radio gear would be very helpful at such a time." ALPHA, in contrast, had a facility capable of reaching any point on the globe. "I cannot underscore sufficiently the benefit to the total communications picture," Hanson continues, "that resulted when the ham radios were being used on the stations." Hanson, "Critique," 110.

38. Chilingarov et al., *Life on an Ice Island*, 34.

39. Chilingarov et al., *Life on an Ice Island*, 40. "The situation is very difficult for small babies," Tatiana Kaliazina, *polyarnik* wife and mother, explains. "When their father is absent for twelve months they simply forget him and it will take a long time to adapt to 'an unknown man' in their home." Given the day-to-day domestic grind in Russia, older children don't ease the strain for working mothers. "She has to rush home from the office to her children," Tatiana continues, "find time to buy food (usually during working hours standing in long lines), to cook, to care for her children, etc. She will never have time to go somewhere since it is a problem with children. If they fall ill, the situation becomes simply horrible." (Kaliazina, letter to author, 20 September 1992.)

40. Leonid Vasilyev, chief radioman at SP-19, explored the psychology of choosing to return to pallid, lifeless cold. There's no rational explanation. "I love the woods, fishing, meadows and wild flowers," he writes. "I go on leave. I have a good time. But I know ahead of time that the winter cold will replace the summer heat. Such is the spirit of every polar explorer who, after thawing out, returns again and again to the familiar world of the high latitudes, where the ice is blue and you know every star. What draws me there? I do not know. I cannot explain it. It is a secret, a mystery, and I hope to God that nobody ever figures it out. It's a good life when there is room in the heart for secrets." Chilingarov et al., *Life on an Ice Island*, 45.

41. "There were very few serious injuries on the U.S. ice stations. The military support groups included expert corpsmen for dealing with illnesses or injuries." Still, casualties were inevitable. The use of explosives and electrical pulses—for detonation and seismic-signal recording—required stringent precautions. In the summer of 1960,

on T-3, a shot package was detonated by a test pulse before being pushed down the shot hole. The man kneeling over it was blinded and suffered significant abdominal injuries. "Fortunately, radio contact was good, and a team stabilized him quickly. He was picked up by one of the Cessnas and soon reached the Air Force hospital in Fairbanks." Gerry Cabaniss, e-mail to author, 6 March 2003.

42. Chilingarov et al., *Life on an Ice Island*, 55–56.
43. William S. Carlson, *Lifelines Through the Arctic* (New York: Duell, Sloan and Pearce, 1962), 259.
44. Johnston, "Problems," 98. Leaders of the *Chelyuskin* expedition and the Communist Party had devoted much attention to problems of morale. "Maintaining confidence and good heart," Krenkel writes, "in the circumstances we found ourselves in was no less important–it may have been more important–than keeping us in good health; but, marooned as we were, it was no easy matter." Krenkel, *REAM*, 220.
45. One "ham" was Jules Madey, an industrious New Jersey teenager operating a powerful amateur station. He transmitted thousands of messages and phone dispatches, even baby pictures to new fathers (by radio facsimile) and sports scores for military personnel at both poles. For his public service contributions, he was presented the Edison Radio Amateur Award from the General Electric Company and, in 1959, visited the Antarctic as a guest of the navy's Operation DEEP FREEZE.
46. The conclusion of a September 1958 *Thule Times* editorial by Col. Bryson R. Bailey, Strategic Wing Commander, is expressive of the oratory of the era. "Thule Air base, here on the top of the world, is living proof that WE WILL REMAIN FREE. We are the furthermost outpost of peace. We at Thule truly are–the Guardians of Freedom." Courtesy Charles C. Plummer, Ph.D.
47. "Thule Air Base, Greenland: 1950–1963," U.S. Air Force history, undated.
48. Gerry Cabaniss, letter to author, 27 February 1990.
49. A writer for the *New York Times* has labeled modern gear "the animating force of exploration in our age." "Today, a person dressed in the right set of clothes, and carrying enough food, could live almost indefinitely in below-freezing temperatures without injury.... The impetus to apply new technology to jackets, pants, and shoes came from the military and the burgeoning outdoor recreation industry, both of which were hampered by the inefficiency of natural fibers such as cotton and wool.... Most of the breakthroughs in warm clothes have come from textile makers, whose so-called performance fabrics–blends of synthetic fibers such as polyester, acrylic and nylon–are so efficient and light they make wool and cotton seem as unwieldy as armor." "Braving the Elements," *The Wall Street Journal*, 1 March 2002.
50. Charles C. Plummer, Ph.D., letter to author, 10 February 1990. The incentive for air force personnel was a six-month tour of active duty rather than a year at Thule as well as choice (with restrictions) for next assignment. All were volunteers. Typically, camp commanders were lieutenant colonels. "Without exception," Gerry Cabaniss recalls, "the commander was the only officer. The poor colonels," he continues, "who were used to having at least one major and a brace of captains to command, were lucky to have a master sergeant as the next in line; often they were tech or, at least in one case, a staff." Cabaniss letter, 27 February 1990.
51. Here Collin flew the Canadian flag. When he asked permission, "The CO of T-3 told me he had no difficulty with that a'toll, as long as I flew it lower than his." Always, overflights by Canadian aircraft elicited comment on his flag; once, Collin was even congratulated. Arthur Collin, Ph.D., interview by author, audiocassette, Ottawa, Canada, 25 April 1993.
52. Gerry Cabaniss, letter to author, 10 February 1992.

53. Letters to author: Donald Plouff, 22 June 1989, and Charles C. Plummer, 10 February 1990.
54. Hanson, "Critique," 112.
55. Hanson, "Critique." One T-3 commander, however, resented having to support science. "Sometimes the military contingent lost sight of its primary mission, scientific support," Dr. Cabaniss recalls, "and turned their attention to more military pursuits. For example, our commander rationalized his primary mission to be establishing and maintaining a 10,000-foot runway that any SAC [Strategic Air Command] aircraft could land on, either in emergencies or even for mission support! Since he had a limited number of working over-ice vehicles and drivers, transport to the oceanographic station suffered in the face of round-the-clock runway work." (Cabaniss letter, 27 February 1990. Dr. Carlos Plummer confirms this phenomenon in a letter dated 10 February 1990.)
56. Cabaniss letter, 27 February 1990. Shmidt Camp had experienced this same unspoken division. Like had attracted like—such that tent communities of scientists, sailors, engineers, and so on had formed quite spontaneously.
57. Kenneth L. Hunkins, Ph.D., interview by author, audiocassette, Lamont-Doherty Geological Observatory, Palisades, NY, 3 February 1992.
58. Herman Medwin and Clarence S. Clay, *Acoustical Oceanography* (San Diego: Academic Press), 1998, 2.
59. Robert Francois, e-mail to author, 6 February 2003.
60. Weir, *Ocean in Common*, 292–93.
61. Beaumont M. Buck, e-mail to author, 9 May 2001. The inaugural experiment had been conducted in September 1952, a test of the deep sound SOFAR (Sound Fixing and Ranging) channel. A SOFAR bomb had been detonated at T-3, whereupon a "very strong" signal was received by *Redfish* (SS 395) lying to in the Beaufort, 900 miles (1450 kilometers) off. See William M. Leary, *Under Ice: Waldo Lyon and Development of the Arctic Submarine* (College Station: Texas A&M University Press, 1999), 59–63.
62. Thaddeus G. Bell, *Sonar and Submarine Detection*, U.S. Navy Underwater Sound laboratory, New London, 8 May 1962, 3–6.
63. Beaumont M. Buck, review comments to author, 2002. Antisubmarine warfare (ASW) is based on acoustics, so Underwater Sound Laboratory's major concern in the basin was underwater acoustics, particularly the under-ice transmission of sound and how well and how far sound could be transmitted. Speed was assessed by conducting long-range measurements between the drift-ice camps, one emitting a high-level sound signal and the other measuring what was received. ASW represented approximately four-fifths of the USL program in 1959. (USL *Echo* 6, no. 10, 1, and 14, 4.)
64. Bert Crary, in 1954, had conducted preliminary acoustic experiments aboard Fletcher's Ice Island. The 1955 scientific party on T-3 made tape recordings of ambient noise levels under the ice. Seismic experiments conducted by Ken Hunkins aboard ALPHA were among the first conducted on pack ice in the central Arctic Ocean. Early measurements of ambient noise from ice camps were encumbered by cable flutter during periods of high wind and rapid drift. Background noise in deep Arctic water is due largely to stress events in the canopy, such as, floe collisions or sudden failures.
65. The Polar Research Laboratory (see chapter six) developed a digital profiling system (DIPS) used on all U.S. under-ice submarines. The device recorded "the output of their up-looking fathometer to continuously log ice thickness, providing the only extensive database of that parameter (now of such interest to those scientists

concerned with earth-warming)." Beaumont M. Buck, note to author, November 2002.
66. Waldo K. Lyon, "Submarine Combat in the Ice," Naval Institute *Proceedings*, February 1992, 118/2/1068, 37. A submarine cannot transmit radio signals while submerged; Calvert therefore had to make stationary ascent through suitably thin ice or, instead, locate open water through which an antenna could be raised. The physics of the breakthrough process was but one of many field and laboratory unknowns absorbing Dr. Lyon and associates.
67. Hunkins interview, 3 February 1992.
68. Cdr. James Calvert, U.S. Navy, *Surface at the Pole* (New York: McGraw-Hill, 1960), 107–108.
69. "For the ALPHA station, I think everybody felt that this was something new, important, and they were gearing everything they could," Hunkins explains. After ALPHA, each man continued with polar-related research. "It was a formative period–and a formative platform–for the careers of all those people." Hunkins interview, 3 February 1992.
70. Calvert, *Surface at the Pole*, 112; Malloy, Daily Log, U.S.S. *Skate*, July 29 to August 24, 1958, Lyon Papers, quoted in Leary, *Under Ice*, 138.
71. "Considering the nice break this gave to station routine, the misbalance between giving and receiving by the men of the *Skate* can be overlooked." Hanson, "Critique," 107.
72. Eight other civilian scientists had boarded in Groton: two oceanographers from the Navy Electronics Laboratory (NEL), a gravity-meter specialist, an ice forecaster, two engineers from North American Aviation, one engineer from the Sperry Gyroscope Company, and Malloy. Kelley's on-ice assignment: to make underwater acoustic observations.
73. USL *Echo* 5 (12 September 1958), 2.
74. Malloy, quoted in Leary, *Under Ice*, 139.
75. ACC, *Ice Islands*, 83–85.
76. *Seattle Times*, undated clipping.
77. Cabaniss et al., "U.S.-IGY Drifting Station."
78. Seafloor sediments constitute one climate record against which human modifications might be teased out, that is, human impact as distinguished from natural variation in global climate.
79. Hunkins interview, 3 February 1992.
80. Britton interview, 14 August 1999.
81. ACC, *Ice Islands*, 88.
82. Kenneth Hunkins, Thomas Herron, Henry Kutschale, and George Peter, "Geophysical Studies of the Chukchi Cap, Arctic Ocean," *Journal of Geophysical Research* 67 (January 1962), 235.
83. Hunkins et al., "Geophysical Studies," 235. Seafloor sampling allows direct study. From these the type of sediment and its age can be determined, which can help answer questions pertaining to the evolution of the basin itself. And specimens are natural paleoclimate indicators; given the influence of boreal ice on climate, knowledge of its origin and development–as recorded in bottom oozes–remains vital to understanding long-term climate trends and cycles. See for example *New York Times*, June 2006.
84. During summer, sea ice melts from the bottom and laterally, as well as from the top.
85. "Logistic Problems in the Canadian Arctic," *Polar Record* 10 (January 1961), 385–86.
86. USL *Echo* 6 (8 January 1960), 1. Two P2V missions would be flown that August,

dropping hydrostatically fused depth charges into leads and polynyas in the pack at intervals, to serve as sound sources for listening gear on T-3 and at CHARLIE. To be successful, the flight plan and the beginning-end of the drop period had to be known to both camps. Plainly, good communications were vital.

87. "Soviet Drifting Stations in the Arctic Ocean, 1959–60," *Polar Record* 10 (September 1960), 278.
88. In 1961, Canadian airmen found the floe-rafted remnants of SP-7 grounded on the eastern shore of Baffin Island.
89. Today, runway construction by consecutive flooding is considered preferable (easier) to mechanical means.
90. USL *Echo* 6 (18 September 1959), 1. "The bear had been riddled by a fusillade of pistol and rifle fire," Gerry Cabaniss recalls. "Some felt the crossfire was more dangerous than the bear." The USL team from New London was "especially trigger happy," he continues, "with their .357 Magnums. At one point, the bear kill was getting so large that the ARL [Arctic Research Laboratory] prohibited kills without express permission, except under real emergency conditions." Cabaniss letter, 10 February 1992.
91. USL *Echo* 6 (25 September 1959), 2.
92. En route, *Sargo* logged the first *winter* transit of Bering Strait.
93. Carl J. Milner, "Cold Journey," USL *Echo* 7 (15 April 1960), 6; L. O. Quam, "Station Charlie," in Sater, *Arctic Drifting Stations*, 19 (quote).
94. "Maneuverability is crucial under ice; transits in very shallow water are extremely difficult and require superb shiphandling. It often is necessary to accept 25 feet beneath the keel and 25-foot clearances overhead, while avoiding deeper ice ridges. Under heavy ice, maneuvering decisions may be required as often as every two minutes." (Richard Boyle and Waldo Lyon, "Arctic ASW: Have We Lost?" Naval Institute *Proceedings*, 124/6/1, 144 (June 1998), 33). To transit successfully, "we would have to rely on a new iceberg detector (IBD) under development by Dr. Lyon's people at NEL [Navy Electronics Laboratory]." R. Adm. John Nicholson, U.S. Navy (Ret.), e-mail to author, 9 October 2001.
95. "Arctic Terrain Research 1959," Geophysics Research Directorate, Air Force Cambridge Center, Air Research and Development Command, Bedford, MA (1960), 3, courtesy Dr. Cabaniss.
96. Hunkins et al., "Geophysical Studies," 237.
97. Hunkins interview, 21 May 1991.
98. Carlson, *Lifelines*, 250. "For some reason extensive screening, testing, or training of the men assigned for duty at Station Charlie was ignored. Selections were made from a casual review of personnel records, and no attempt was made to determine if any men would be unsuitable for life on an ice floe." (249)
99. Milner, "Cold Journey."
100. Beaumont M. Buck, e-mail to author, 3 January 2002. Despite problems that "couldn't possibly" be solved in three months, "The complete DIMUS, all analog electronics, and a host of car batteries were packaged into a "Helicop" hut, loaded aboard a Special Air Mission C-123 and flown directly to the arctic logistics center at Ladd Air Force Base in Fairbanks, and from there to the ice island in early January 1960, at a package cost of $20,000, which seemed exorbitant at the time." (Beaumont M. Buck, U.S. Navy, "Accomplishments in Arctic Underwater Surveillance Acoustics, 1960–1988," unpublished manuscript, 3–4.) Acoustic experiments feasible in this ocean were not so in any other. "We envisioned that in the Arctic we would have the opportunity to build and test very large aperture hydrophone arrays at a small

fraction of the cost of trying the same in open water. Advanced processing techniques could be tested on these arrays years in advance of anything comparable to the open ocean." (Beaumont M. Buck, "Arctic Acoustic Transmission Loss and Ambient Noise," in Sater, *Arctic Drifting Stations*, 427.)

101. Milner, "Cold Journey," USL *Echo* 12, no. 13 (22 April 1960), 1; and no. 14 (29 April 1960,) 3. Temperature, though, was but one factor. MONA LISA was the (cursed) project that shipped fuel to the ice camps and the various weather and early warning stations along Arctic Alaska. "These fuels must have been the dregs from the refinery," Gerry Cabaniss recalls. Heavy with particulates, it clogged filters, fuel injectors, and controls. "Worse though was that diesel thickened at temperatures below minus 25 F or so. Between that and the water in it, the lines to our heaters often clogged. That meant that all of us had our favorite blowtorch.... The crudy diesel was the cause of several bad fires on T-3. For example, a stove with a float fuel-control would be lit in a cold shelter (the fuel tank was outside with gravity feed). After the shelter warmed, the fuel became less viscous, the particulates jammed the float valve, and the stove would overflow, spreading fire across the floor." The gasoline also was poor-quality. (Gerry Cabaniss, e-mail to author, 10 March 2003.)
102. *Sargo* patrol report, 15 February 1960 (excerpt), courtesy of Adm. John H. Nicholson, U.S. Navy (Ret.). Eight scientists accompanied *Sargo*, headed by NEL's Waldo Lyon, Ph.D. Ice thickness, water temperature, salinity, and gravity measurements were made throughout the patrol. (*Polar Record* 10 (September 1960), 279.)
103. *Sargo* patrol report, 17 February 1960 (excerpt).
104. Adm. John H. Nicholson, U.S. Navy, personal journal, 17 February 1960; *Sargo* patrol report, 17 February 1960. *Sargo* emerged from beneath ice in the south Bering Sea on 24 February. In August, *Seadragon* (SSN 584), *Skate*'s sister, would log the first submerged transit of the Northwest Passage (the 21st) then the pole (the 25th) before making rendezvous with T-3 off Point Barrow, on 3 September. See George P. Steele, *Seadragon: Northwest Under the Ice* (New York: D. P. Dutton, 1962) for an account of the boat's deep and roving passage. Steele writes of wanting to "tip our hat" to the ice-island base. Sound transmission work, the reason for rendezvous, is not mentioned.
105. This one-of-a kind, CIA-sponsored mission to SP-8 was to assess Soviet progress in meteorology, oceanography, and (especially) the state of acoustical surveillance. Parachuted onto the ice on 28 May 1962, two American officers—one air force, one navy—investigated the abandoned base, making notes, snapping photographs, and collecting documents and hardware before their dramatic (and delayed) skyhook retrieval. See William M. Leary and Leonard A. LeShack, *Project COLDFEET: Secret Mission to a Soviet Ice Station* (Annapolis: Naval Institute Press, 1996) for a full account.
106. "Under-Ice Winter Cruise in the Arctic Basin: U.S.S. *Sargo*," *Polar Record* 10 (September 1960), 279.
107. An iconic figure, Gordienko wore the Orders of Lenin and October Revolution. He accompanied more than fifty expeditions to the circumpolar zones of both hemispheres. In 1956–64, Gordienko was the Arctic and Antarctic Research Institute's deputy director. Inside Russia, he is revered; his initials are "like a password: it opens doors and hearts. [He] evokes great respect and love on both the earth's poles." Chilingarov et al., *Life on an Ice Island*, 3.
108. Station staff continued to use propane gas for cooking and for heating the tents used for field shelters and as temporary accommodation for personnel of the *Sever* air expeditions.

109. P. A. Gordienko, "The Arctic Ocean," *Scientific American* 204 (May 1961), 97–98.
110. Maxwell E. Britton, "Administrative Background of the Developing Program," in Sater, *Arctic Drifting Stations*, 31.
111. Buck, "Accomplishments," 6.

CHAPTER 5

1. The evidence linking greenhouse gases to global warming is derived from coupled ocean-atmosphere general circulation models as well as from empirical data. Although models now are "getting quite remarkable agreement," some caveats are yet necessary. Rather than face up to the threat, it is these climate-forecasting uncertainties that have been seized upon as excuses for inaction. *Science* 292 (13 April 2001), 270, 194.
2. The National Science Foundation is the federal agency that directs the U.S. Antarctic program (transferred from the Defense Department). The U.S. Navy logged its last flight in Antarctica on 17 February 1999; VXE-6 was decommissioned on 31 March, thus ending an era of naval aviation that had begun with Richard E. Byrd in 1929 at Little America.
3. Chilingarov et al., *Life on an Ice Island*, 81.
4. *Polar Record* 10 (September 1961), 617.
5. ACC, *Ice Islands*, 74.
6. A decision to use commercial aircraft to establish the Arctic Research Laboratory Ice Station in the spring of 1960 was scuttled by a last-minute insurance problem. This delayed matters until the melt was too far advanced. Rather than await autumn and the return of conditions suitable for air operations, an icebreaker was enlisted to reach the desired position.
7. Tim Weeks and Ramona Maher, *Ice Island: Polar Science and the Arctic Research Laboratory* (New York: John Day Co., 1965), 132, 134.
8. R. J. Fischer, "Air Support of Drifting Stations–A Decade of Experience," in Sater, *Arctic Drifting Stations*, 82; Maxwell Britton, "The Role of ONR and the IGY," in *Fifty More Years Below Zero: Tributes and Meditations for the Naval Arctic Research Laboratory's First Half Century at Barrow, Alaska* (Winnipeg, Manitoba: Arctic Institute of North America, 2001), 68.
9. John Cadwalader, "Arctic Drift Stations," Naval Institute *Proceedings* 89 (April 1963), 74.
10. Britton interview, 14 August 1999.
11. John F. Schindler, "The Impact of Ice Islands," in Sater, *Arctic Drifting Stations*, 49.
12. John F. Schindler, letter to author, 16 January 2002.
13. Weeks and Maher, *Ice Island*, 78.
14. R. Verrall and D. Baade, "Design and Construction of Prefabricated Plywood Buildings for Use in the High Arctic," report, Research and Development Branch, Department of National Defence, Canada, September 1983, 1; courtesy of George D. Hobson, former director, Polar Continental Shelf Project.
15. Maxwell C. Brewer, "New Applications of Old Concepts," in Sater, *Arctic Drifting Stations*, 27.
16. Seismic soundings would be made between T-3 and ARLIS-II and also by *Burton Island*. "Activities on Ice Island T-3, 1961," *Polar Record* 11 (May 1962), 183.
17. Schindler, "The Impact," 53.
18. David D. Smith, "Ice Lithologies and Structure of Ice Island ARLIS II," *Journal of Glaciology* 5 (February 1964), 19, 25. Max Britton had suggested this particular study.

19. "Until recently, people who worked or played outdoors in inclement weather could be warm or they could be dry, but they could rarely be both. Even sitting, humans sweat, and most people outdoors in the winter are exerting themselves. If their sweat is trapped close to their bodies, they quickly feel cold." ("Braving the Elements," *Wall Street Journal*, 1 March 2002.)
20. Schindler, "The Impact," 57. For insulation between the sleeping bags and the tent bottom, the men used dirty laundry. The navy thereafter forbade Brewer and Schindler to fly together except on commercial flights. "We could not fly together in our own airplanes anymore," Schindler recalls. (Interview by Laura Kissel (transcript), Polar Oral History Program, Ohio State University, 29 May 2001.)
21. Schindler, "The Impact," 57.
22. "The Voyage of the *Lenin*, 1961," *Polar Record* 11 (September 1963), 719.
23. "Soviet Drifting Stations in the Arctic Ocean, 1961–62," *Polar Record* 11 (September 1962), 279.
24. "The Soviet North: Economic Aspects," paper prepared for the Advisory Committee on Northern Development, Joint Intelligence Bureau, Department of National Defence, Ottawa, 30 June 1962, (RG 25, National Archives of Canada, vol. 3306), 22, 25.
25. In May or June 1960, thanks to Louis DeGoes, Irene Brown, and Vivian Bushnell–longtime employees of the Air Force Cambridge Research Center in Massachusetts–were invited to tour ARL and the T-3 facility. "I had the feeling," Bushnell writes, "that I was visiting a place I had known for years because of my contacts with the men who had been there and because of my own part in the data reduction." (Vivian Bushnell, letter to author, 8 April 1991.) "The living spaces were trailers by that time," Robert Cotell says, "but they visited the old Jamesways dating from the initial occupation of the island. They visited the personnel and inspected the equipment that they had written about for so many years. They knew the scientific personnel already." (Robert Cotell, letter to author, 26 September 1991.)
26. Carlson, *Lifelines*, 266. "Yes, America is in grave danger," warned the head of the Strategic Air Command, Gen. Curtis E. LeMay. "If we have not already lost our military superiority we are well on the way to it." (Gen. Curtis LeMay, *America Is in Danger*, with Maj. Gen. Dale O. Smith (New York: Funk and Wagnalls, 1968), viii.)
27. Steven B. Young, *To the Arctic: An Introduction to the Far Northern World* (New York: John Wiley and Sons, 1994), 18; Sergey V. Karpekin, deputy director, Arctic and Antarctic Research Institute, interview by author, St. Petersburg, Russia, 18 May 1992.
28. "Soviet Drifting Stations in the Arctic Ocean, 1937–61," *Polar Record* 11 (September 1962), 292 (table).
29. *Daily-News Miner*, undated clipping. "We used to be in the Fairbanks *News-Miner* all the time," Schindler says. "Ice stations were big news, you know."
30. Robert Francois, e-mail to author, 15 March 2003.
31. In addition to its difficult relief operation, "the [*Sever*] expedition had its customary programme of meteorological, oceanographic, and glaciological studies, and deposited automatic weather stations on the ice. The aircraft used included MI-4 helicopters and AN-2, LI-2, IL-14 and AN-10 fixed-wing aircraft. About 1,000 ice landings were made. The expedition finished by 21 May. Because it was a continuation of the spring expedition, an airlift to SP-10 and 11 that October also was dubbed *Sever-14*." "Soviet Drifting Stations in the Arctic Ocean, 1961–62," 278–279, and "Soviet Drifting Stations, 1962–63," *Polar Record* 11 (September 1963), 722–723.

32. Schindler, "The Impact," 58, 60. Genuine emotions were on display.
33. Weeks and Maher, *Ice Island*, 130.
34. Weeks and Maher, *Ice Island*, 150–152.
35. Barrow-based aircraft twice visited the short-term *Sever-67* camp, in April 1967. Serviced by six flights a day from the mainland, the Americans found busy ice. One hundred pilots were attached to the operation, which somehow accommodated three hundred or more individuals, including an aircraft controller for five types of aircraft. See Max C. Brewer, "The Soviet Drifting Ice Station, NORTH-67," *Arctic* 20 (December 1967), 263-266. "The pilots decided to land . . . in part to encourage friendly relations between two groups engaged in comparable scientific research and combating the same austere environment and in part to satisfy a very human curiosity." (263)
36. Buck e-mail, 9 May 2001; Buck, "Accomplishments," 7; Buck, undated [ca. 1997] audiocassette prepared for author. "We gave the old college try for three years to make T-3 useable by moving our QUIET STATION about 3 miles away from the main camp in 1967. We were largely unsuccessful in this and had to shut most things down in the main camp when we were working with a sub. However, this brought on other problems. . . ." (Buck e-mail, 9 May 2001).
37. Buck e-mail, 9 May 2001. This slushy layer at the bottom of the ice is known to acousticians as the "skeletal transition zone," in which a gradual change of physical and acoustic properties occur. (Cdr. Ronald McGregor, U.S. Navy (Ret.), interview by author, Washington, DC, 10 August 2002.)
38. "Warfare in the ice is like jungle warfare. The sea-ice canopy becomes the jungle in which the submarine must live, work, and fight–not just transit." (Waldo K. Lyon, "Submarine Combat in the Ice," Naval Institute *Proceedings*, 118/2/1068 (February 1992), 38.)
39. Buck, "Accomplishments," 7.
40. Buck, quoted in Weeks and Maher, *Ice Island*, 130.
41. Buck e-mail, 9 May 2001.
42. "Soviet Drifting Stations, 1962–63," 722–723.
43. Capt. Charles Blair, a pilot for Pan American, logged the first solo flight *over* the pole in a single-engine aircraft in 1950. Departing from Norway, Blair landed his F-51 in Fairbanks (Ladd Field) ten hours and twenty-nine minutes later. In November 1965, two DHC Beavers of the Argentine Air Force reached the South Pole and are believed to be the first single-engine fixed-wing aircraft to have landed at that extremity.
44. Fischer, "Air Support," 87.
45. Schindler, "The Impact," 65.
46. Schindler, "The Impact," 66.
47. "Soviet Drifting Stations, 1963–64," *Polar Record* 12 (January 1965), 413–414. Both stations would be visited that November, when the air resupply expedition was reactivated. SP-13 was used as a staging post for aircraft en route to "12," 1000 kilometers beyond. ("Soviet Drifting Stations, 1964–65," *Polar Record* 12 (September 1965), 743.)
48. All ice that is not melted again but is exported from the Arctic represents an actual heat gain for the ocean. The Norwegian Atlantic Current and its northern extension, the East Spitsbergen Current, carry relatively warm and saline upper-ocean waters northward on the eastern side.
49. Schindler, "The Impact," 68.
50. Schindler, "The Impact," 68–69.

51. The pair had flown from Barrow to T-3, from T-3 to Alert, from Alert onto ARLIS II, thence to Iceland.
52. Britton interview, 14 August 1999; Sater, *Arctic Drifting Stations*, 73. Two radio beacons were left behind, to assist tracking. As ARLIS (in pieces) moved down coastal Greenland and through Denmark Strait, its position was noted ashore. Later, ARLIS buildings would be sighted off the west coast of Greenland, drifting north. A door from one of the huts, Schindler says, is in Gothab, Greenland. It probably was salvaged before the station sank.
53. V. P. Hessler, "Critique and Evaluation of Drifting Station Operation–ARLIS III and ARLIS IV," in Sater, *Arctic Drifting Stations*, 102. "No doubt, some clever improvements may yet be made in the design to decrease weight and perhaps facilitate erection, but I fail to see how the hut could be made more useable or comfortable.... They can be heated with a minimum of fuel and when banked properly with snow even a Southern Californian could walk barefoot in the quarters without discomfort."
54. Fischer, "Air Support," 87.
55. "Soviet Drifting Stations, 1965–66," *Polar Record* 13 (September 1967), 779.
56. An April 1966 symposium brought together seasoned hands of U.S. arctic science. Bert Crary was one. "However insignificant the U.S. Air Force scientific discoveries on the drifting station may appear in light of the present-day research efforts with better equipment and facilities," Crary noted, "and the relatively poor showing made in comparison with the massive quantities of data that come from more southerly areas, there was a value in the operations themselves which is far from insignificant, the value in pioneering the trail for others to follow." (A. P. Crary, "Air Force Research Achievements on Drifting Stations," in Sater, *Arctic Drifting Stations*, 119.)
57. "Soviet Drifting Stations, 1966–68," *Polar Record* 14 (January 1969), 489. Both positions lie within multiyear ice poleward of the East Siberian Sea.
58. Weir, *Ocean in Common*.
59. Canadian Forces Station Alert sits 4000 kilometers (nearly 2500 miles) north of Canada's southern cities and 817 kilometers off the geographic pole. Set up in 1950 as a joint Canadian-American weather station, it is the northernmost permanent community on earth. Its nearest neighbor is Thule Air Base. The military arrived in 1958, making the habitation an important listening post for Cold War eavesdropping. Alert is still devoted to "signals intelligence" and accommodates various scientific programs. See David Lanken, "On Alert," *Canadian Geographic* 120 (Nov.–Dec. 2000); also *Toronto Star*, 16 February 1998; and *New York Times*, 10 October 2000.
60. Michael Foster and Carol Marino, *The Polar Shelf: The Saga of Canada's Arctic Scientists* (Toronto: NC Press Ltd., 1986), 64.
61. ARL (Arctic Research Laboratory) became NARL in 1965 to emphasize the navy connection. Labeled a drain on scarce research funds, by 1979 the lab was under review for closure. "It's a national asset," John Kelley, Ph.D., then technical director said in interview. "It's also a white elephant that nobody wants to take on." "This is," he continued, "the only high latitude arctic laboratory in the United States. The beauty of it is it is set up in the region itself." (*New York Times*, 9 December 1979.) Despite protests, 1980 saw the end of navy operation of the facility.
62. Quoted in "Islands in the Midnight Sun: The Story of the Polar Continental Shelf Project," Public Relations and Information Services Branch, 1974, 12.
63. "Achievements of the Program on Drifting Stations of the Office of Naval Research and a Look at the Future: A Summary of Discussions at the Arlie Symposium," in Sater, *Arctic Drifting Stations*, 473.

64. That year, by one estimate, the U.S. government expended about $16 million on basic, unclassified research in the Arctic.
65. *New Scientist*, 20 June 1968, 614.
66. Chilingarov et al., *Life on an Ice Island*, 17.
67. Whereas SP-1 had had to make do with nine tons, "19" received deliveries exceeding two hundred tons during autumn 1969 alone. This included household and office supplies, electrical and power equipment, communications systems, scientific instruments and equipment, coal, diesel fuel, and airplanes—the latter for oceanographic observations off-station. (Chilingarov, *Life on an Ice Island*, 139.)
68. Chilingarov et al., *Life on an Ice Island*, 27 and 33, 36 (cable message).
69. Chilingarov et al., *Life on an Ice Island*, [Valeriy Krivoshein and Mikhail Sudakov], 63–67. "It was frightening to look upon the black water [in a widening crack], but we had to work in order to survive." [Sudakov], 66.
70. The Soviet Hydrometeorological Service requested that Ottawa monitor the deserted camp. Believed to have disintegrated in the Greenland Sea, no further sightings were made. (*Soviet Research From Drifting Ice Stations*, National Defence Headquarters, Ottawa, 1978, 11.)
71. C. R. Greene and Beaumont M. Buck, "Arctic Noise Measurement Experiment Using NIMBUS 6 Data Buoys," *U.S. Navy Journal of Underwater Acoustics* 27 (October 1977), 827. In 1970, no data existed for the summer, fall, and winter seasons. (The all-season measurements taken at T-3 during 1967–70 were not typical of pack ice.) Later measurements found that the levels tend to be lowest in summer, when ice interactions are minimal because of open water.
72. Foster and Marino, *The Polar Shelf*, 66. On 3–5 April 1969, USS *Whale* (SSN 638) provided a sound source for the hydrophone array at T-3–part of SUBICEX 1-69–after which the boat steamed for the pole, arriving on 6 April.
73. "Achievements," 475.
74. Charles L. Smith, "A Comparison of Soviet and American Drifting Ice Stations," *Polar Record* 15 (1971), 885. When asked about data exchange, Ilya Romanov, Ph.D., a distinguished *polyarnik*, remarked, "We are very grateful for American scientists and specialists because all of the data they obtained during their expeditions they have published. They [the data] are not confidential." He and his colleagues, moreover, "badly need" Western results, to compare with their own. Was the reverse true; wasn't Soviet data a state secret? "It is always impossible to get out data." (Ilya P. Romanov, Ph.D., interview by author, audiocassette, St. Petersburg, 19 May 1992.)
75. A less elaborate experiment by Soviet researchers had been conducted in 1961, using SP-8 and SP-9. That March through April, five temporary stations were set up–the array forming a square 75–100 kilometers on a side. Though direct measurements of stress were not attempted and the (brief) test made in a dissimilar drift pattern, "it can be thought of as a pilot experiment for AIDJEX and the results should be useful to AIDJEX participants." (A. V. Bushuyev et al., "Results of a Soviet Experiment Investigating the Drift and Dynamics of the Arctic Basin Ice Cover," translation from the Russian by S. M. Olenicoff and H. Solomon, Rand Corp., May 1970.)
76. Robert E. Francois, letter to author, 21 January 2003; Francois, e-mail to author, 30 January 2003. In summer, surface water is generally diluted by sea-ice melt and, in some areas, by freshwater inflow. Salinity is less than the deep ocean. "A major objective was to investigate the horizontal density gradient variation and rate of change during the melt season. This work was aimed at answering the question of whether or not some special depth/trim/ballast control system might be required of

submarines in transiting a shallow sea (like 130-foot depth), to avoid inadvertent contact with the sea floor or any overhead ice as the density changed." (Francois letter, 21 January 2003.)

77. The buoys were of two types: one employed HF (high-frequency) radio ground-wave telemetry, the other a TIROS satellite. The former held NavSat receivers for precise location and highly accurate barometers and air-temperature sensors. The latter (more numerous) also contained in-buoy processed TIROS-transmitted ambient noise level data from a one hundred-foot deep hydrophone. (Buck e-mail, 9 May 2001.)

78. Buck comments, November 2002. The Soviets were masters at all-season occupancy; typically, two manned stations were continuously operational. "Our [PRL] view was that for many parameters, synoptic measurements could best be handled by those unmanned stations." PRL developed the only air-droppable *Nimbus* 6 data buoy: one version rested atop the ice, the other suspended a hydrophone below the canopy.

79. Andreas G. Ronhovde, "Jurisdiction Over Ice Islands: The Escamilla Case in Retrospect," booklet published by the Arctic Development and Environmental Program of the Arctic Institute of North America, November 1972.

80. T-3's most creative and productive years were over. Routine observations had passed into the hands of contractors: employees of federal agencies, civilian institutions, and private companies under contract to ONR. Alaskan natives were also aboard, as support personnel.

81. Gordon W. Smith, "Ice Islands in Arctic Waters," Department of Indian Affairs and Northern Development, Ottawa, 15 May 1980, 1; R. Adm. C. O. Holmquist, U.S. Navy, "The T-3 Incident," Naval Institute *Proceedings* 98 (September 1972), 42–53. Canada, the only foreign government that could have claimed jurisdiction, yielded. "The reason given for waiving jurisdiction over the T-3 case was a desire not to interfere with the course of justice for the sake of a clarifying a very complex point of international law." (52)

82. Holmquist, "The T-3 Incident," 49–50.

83. Canada has maintained a claim of sovereignty fully to the pole. In this instance, jurisdiction was waived without prejudice to a possible future claim of jurisdiction over the island. In 1972, then–prime minister Pierre Trudeau expressed the opinion that his country's claim does not apply to water and ice.

84. Donat Pharand, "Canada's Jurisdiction in the Arctic," in *The Law of the Sea of the Arctic, With Special Reference to Canada* (Ottawa: University of Ottawa Press, 1973), 121.

85. Holmquist, "The T-3 Incident," 46.

86. See Robert E. Francois, "The Unmanned Arctic Research Submersible System," *Marine Technology Society Journal* 7:1, 46–48. "The elements of UARS formed the basis for weapons system development, testing and training in the Arctic." Francois letter, 21 January 2003. Autonomous vehicles were exploited for under-ice measurements during AIDJEX and, in 1997–98, for the SHEBA drift. (See page 248.) The Arctic Research Commission is studying long-range autonomous underwater vehicles as one means of collecting data.

87. Capt. Brian Shoemaker, U.S. Navy (Ret.), e-mail to author, 1 April 2002. "It was taken out of commission," he continues, "with the sketchy plan that it would be re-manned at some future date if the drift . . . proved conducive to future U.S. national interests." Shoemaker, e-mail to author, 3 June 2002.

88. "I have been studying ice for more than 20 years," *polyarnik* Mikhail Serikov wrote at this time. "Sometimes I ask myself: what am I doing? Hell, ice is ice. And ice, I

tell you, is the whole problem. Mathematicians have devised dozens of complex equations to calculate its strength, to no avail. An equation gives one answer and an experiment gives another." (Chilingarov, et al., *Life on an Ice Island*, 72; *Arctic Research of the United States* (Fall/Winter 2001), 42.)

89. Norbert Untersteiner, Kenneth L. Hunkins, Beaumont M. Buck, "Arctic Science: Current Knowledge and Future Thrusts," in *Science, Technology, and the Modern Navy: Thirtieth Anniversary 1946–1976* (Arlington, VA: Department of the Navy, Office of Naval Research, 1976), 284.

90. Brian Shoemaker, e-mail to author, 2 December 1998; Kenneth L. Hunkins, Ph.D., "Subsurface Eddies in the Arctic Ocean and Baroclinic Instability," in *Climate of the Arctic*, n.p., n.d., 398–406.

91. Polar Continental Shelf Project (George D. Hobson, director), annual report, April 1973. The National Aeronautics and Space Administration's contribution was remote-sensing aircraft overflights during AIDJEX field experiments.

92. Foster and Marino, *The Polar Shelf*, 69. Japanese interest was rooted in that country's need for access to additional oil resources.

93. McGregor interview, 10 August 2002; F. Kenneth Hare, "Natural Environment" (section 2), discussion paper in *Science and the North: A Seminar on Guidelines for Scientific Activities in Northern Canada 1972*, Subcommittee on Science and Technology Advisory Committee on Northern Development publication no. QS-1330-000-EE-A-1, Ottawa, 1973, 73.

94. University of Alaska News Service, 22 March 1972 news release, courtesy John F. Schindler. Acoustics work for naval application persisted. MIZPAC 71 (Marginal Ice Zone-Pacific) had deployed also into the Chukchi. In 1972, its waters hosted the second MIZPAC experiment, exploiting *Burton Island*. That April, the AIDJEX pilot experiment used Polar Research Laboratory–designed SHRAMS (Short-Range Automatic Measuring Station), after which PRL operated the then-abandoned main camp as an acoustically quiet station. MIZEX 73 would comprise several short-term ice camps in the North Bering Sea, helo-supported out of St. Lawrence Island. "The ice was so dynamic, we could only occupy the camps for a few days each." Beaumont M. Buck, e-mails to author, 24 and 26 June 2003.

95. Ernest Frederick Roots, "International and Regional Cooperation in Arctic Science: A Changing Situation," proceedings of the Nordic Conference on Scientific Research, 2–8 August 1984, in *Rapport fra Nordisk Vitenskapelig Konferanse om Arktisk Forskning* (Trondheim, Denmark: Universitet i Trondheim Press, 127–156. Remote sensing is central to providing information on environmental conditions. Nonetheless, interpretation and effective use of the profusion of satellite and airborne remote-imagery demands "ground truth" (verification)–field parties on the ground.

96. The first satellite sensor to provide global information on ice extent was the electrically scanning microwave radiometer (ESMR) launched by the National Aeronautics and Space Administration aboard *Nimbus 5*, in 1972. See Robert H. Thomas, "Satellite Remote Sensing Over Ice," *Journal of Geophysical Research* 91 (15 February 1986), 2493–2502. All submarine deployments to the basin had been classified until AIDJEX. In 1976, a data release was granted after USS *Gurnard* (SSN 662) logged a winter crossing of the Bering and Chukchi shallows as part of SCICEX 1-76, thus duplicating the feat of *Sargo*. The boat collected ice-profile data in support of the experiment. (George B. Newton, "The Science Ice Exercise Program: History, Achievements, and Future of SCICEX," *Arctic Research of the United States* 14 (Fall/Winter 2000), 2; Leary, *Under Ice*, 251.)

97. Fedorov, *Polar Diaries*, 335–336. Compared with that fielded by their U.S. counter-

parts, Soviet equipment was rudimentary, its technology cumbersome. "Remarks were made concerning instruments: everyone realized that many of our instruments were outdated." (337)

98. "The name of Vinogradov is familiar to me," George Hobson recalls, "probably about as familiar as my name would have been to him. When we visited NP 25, the station manager knew quite a bit about me when we arrived and had a chat over some vodka." Hobson, letter to author, 11 November 1993.

99. "Arctic Ice Dynamics Joint Experiment," Norbert Untersteiner, *Arctic Bulletin,* (Spring 1974, 150; *New York Times,* 13 March 1975; *University Record* (Columbia University) 3 (20 March 1975), courtesy Dr. Hunkins.

100. Richard Trowbridge, "The Arctic Ice Dynamics Joint Experiment (AIDJEX)," n.d., 13.

101. "AIDJEX Camp Evacuated," *This Week at NARL,* Office of the Director, Naval Arctic Research Laboratory, October 6–12, 1975.

102. Greene and Buck, "Arctic Noise," 828.

103. During polar summer, in contrast, ice stresses are negligible and drift is essentially wind driven. Hence, underside studies of drag coefficient are particularly useful during that season.

104. Norbert Untersteiner, "AIDJEX Review," in *Sea Ice Processes and Models,* proceedings of the Arctic Ice Dynamics Joint Experiment, International Commission on Snow and Ice Symposium, ed. Robert S. Pritchard (Seattle: University of Washington Press, 1977, 8, 3–4). On the same page, Untersteiner writes, "Despite the risks and difficulties inherent to all field work in the harsh environment of drifting pack ice, the project was completed without loss of life or serious injury to any of the more than two hundred participants." A distinction here deserves emphasis: the Soviets were *operating* whereas, comparatively, the Americans were *experimenting* in the Arctic Ocean on its cover.

105. "Arktika," *Soviet Military Encyclopedia,* trans. CIS Multilingual Section, National Defence Headquarters, Ottawa (Moscow: Voyennoye Izdatel'stuo, 1976), 1A:171. "The determination in both Washington and Moscow to prevent the logic of confrontation from leading to a catastrophic conclusion meant that the bipolar system depended on tacit cooperation and understandings as much as pure antagonism." Lawrence Freedman, "Order and Disorder in the New World," *Foreign Affairs* 71, no. 1, 23.

106. "Walt Whitman, the Navy's foremost ice expert in those days, said that the ice in the area of the Fram Strait was so dynamic, we were going to be all killed! Somewhat of an exaggeration." Beaumont M. Buck, e-mail to author, 24 August 1997.

107. "We are now faced," the Scientific Plan for *Fram* laments, "with the curious situation of knowing that the Eurasian Basin is critical to certain major scientific problems, while at the same time having less available information concerning the physical state of its water and the associated processes, than is probably the case for any other part of the world's oceans. We have only primitive mappings of the large-scale temperature and salinity fields, essentially no information about currents, and can only hypothesize the physically important processes."

108. Hunkins interview, 3 February 1992.

109. National Academy of Sciences, "Scientific Plan for the Proposed Nansen Drift Station," report, Polar Research Board, Washington, DC, 1976. The disciplines were long since standard fare but featured in varying ratios, depending on the expedition: marine geophysics and tectonics, marine geology and paleoclimatology, physical oceanography, biology and biochemistry, marine acoustics, heat and mass balance of the sea-ice cover, troposphere, ice physics and engineering, atmosphere-ionosphere-magnetosphere, and remote sensing.

110. Hunkins interview, 3 February 1992. In 1996, a lone icebreaker tried to drill a deep core on the Lomonosov Ridge. Only a few, small cores were retrieved. "The ice does terrible things up there," a team geologist reported. "When I came out, I said I would never do that again without the proper vessels." With three ice-breaking escort ships and an oil-industry drilling ship, a $12 million return expedition was planned for the summer of 2004, under the sponsorship of the new Integrated Ocean Drilling Program. See "Digging In," *Nature* 426 (4 December 2003), 492–494. For results, see (front page) *New York Times*, 1 June 2006; *Nature*, 441 (1 June 2006).
111. Kenneth L. Hunkins, Yngve Kristoffersen, G. Leonard Johnson, Andreas Heiberg, "The Fram I Expedition," *Eos Transactions American Geophysical Union* 60 (12 December 1979), 1043. "They were called [*Fram*] to get the last word in about the (now sunk) *Fram II* ship fiasco." Buck e-mail, 24 August 1997.
112. Francois letter, 21 January 2003.
113. Starting in 1953, Leningrad obtained information about the movement of near-shore ice by placing radio beacons atop the pack and plotting their movement from shore stations.
114. "An attempt to reach the North Pole is certainly to be expected, for this was the dream of Admiral S. O. Makarov, who first realized the idea of a polar icebreaker when he had *Yermak* built in 1898." *Polar Record* 10 (January 1960), 77.
115. Though less spectacular than *Arktika*'s 1977 cruise, *Sibir* logged a significant sortie early in 1978 by escorting a freighter through the full length of the Northern Sea Route two months before the normal icebreaking season had begun.
116. In 1990, the U.S. research fleet had no dedicated Arctic research vessels, either ice-reinforced, ice-strengthened, or ice-breaking. In addition to its traditional northern missions, the U.S. Coast Guard provides icebreaker services to other agencies and supports research. Its missions can conflict, however; for example, search and rescue takes precedence over science. Icebreaker support is dear; operating costs in the Beaufort Sea in 1990, for example, totaled about $13,000 per day. In 2000, the guard's ice-breaking assets were increased with the delivery of *Healy* (WAGB-20), the first icebreaker designed from the keel up primarily as a research vessel.
117. In July 1978, the Polar Science Center had submitted a proposal to take on major responsibilities in planning and coordination as well as management and field services. After securing ONR approval, the Polar Science Center began purchasing equipment and negotiating contracts for aircraft and personnel support services. The operations manager was Andreas "Andy" Heiberg. (Hunkins et al., "The Fram I," 1043.) RUBY and PEARL were planned by Buck as quiet stations in conjunction with SUBICEX-1-77. Submarines working ice camps—serving as a target for experimental ice-borne sonars—were only part of a total SUBICEX. "And, similarly, our AREA [camp] work was more involved than just tracking subs." Buck e-mail, 24 August 1997.
118. Buck comments, November 2002.
119. Francois letter, 21 January 2003; Beaumont M. Buck, e-mail to author, 1 December 2002. "Our role," Francois adds, "was primarily to provide the ground truth for these [P-3] observations such as location and air-photo reference points, environmental attributes (snow and ice properties, measured albedo, etc.)."
120. Several methods of establishing an ice camp were exploited in the eastern Arctic. If the Twin Otter landed on floe ice, the Tri Turbo-3 followed, landing in the ski tracks. Another used two Cessna 185s on skis to make the first landing on a refrozen lead, to test thickness. If three feet or more, the TT-3 came in. "Another method em-

ployed two parajumpers who set up a temporary camp and scouted and marked a suitable landing area on the floe for the TT-3 to return to the next day." Beaufort M. Buck, summary written at author's request, 4 August 1997.
121. Hunkins et al., "The Fram I, 1043." Cracks can open at any time, but if a station is located well into multiyear pack, particularly north of 81°, the pieces cannot separate readily because of volume restrictions, that is, tight ice. The Tri-Turbo 3 was a specially modified, long-range DC-3 purchased and operated by the Polar Research Laboratory for the installation and support of ice camps well removed from shore-support bases.
122. Monitoring for microearthquakes in the vicinity of the Midoceanic Ridge, seismic events near the ridge axis were recorded at the rate of two per hour.
123. Hunkins et al., "The Fram I, 1043."
124. Robin Falconer, Ph.D., "Fram: Second Drift Research Station," *Geos*, Summer 1980, 1.
125. The device was an Applied Physics Laboratory product. "In the early '70s, we developed a lightweight CTD profiler . . . that reached over 300 meters [984 feet] depth and weighed less than 100 pounds. This allowed extensive oceanographic surveys to be made using unusual platforms, such as an SEV, helicopter, and landings on refrozen leads with light aircraft." Francois letter, 21 January 2003.
126. The helicopters usually were on contract from Greenland Air. The pilot and his co-pilot-mechanic, Dr. Hunkins recalls, were well-seasoned. "The mechanic, you know, he could do anything–take that thing completely apart if he had to." Good ground reference is needed for such flying; blowing snow or fog might necessitate a set-down, to await better conditions. One time, grounded atop vacant ice, Hunkins queried his pilot as to what was the longest he had had to sit, powered down, waiting. "Three days," he was told. (Hunkins interview, 21 May 1991.) In Antarctica also, helicopter pilots often are obliged to camp out.
127. Hunkins interview, 21 May 1991.

CHAPTER 6

1. Richard Monastersky, "Global Change: The Scientific Challenge," *Science News* 135 (14 April 1989), 235. See also Monastersky's "The $1.5 Billion Question: Can the U.S. Global Change Program Deliver on its Promise?" *Science News* 144 (4 September 1993). Several issues fall under the rubric of global change, among them ozone depletion, deforestation, and desertification.
2. *Winnipeg Free Press*, 6 June 1979. Substantial austral camps remote from the National Science Foundation's McMurdo Station, the largest in Antarctica, can require several years for planning and coordination.
3. J. Tuzo Wilson, "Room at the Top of the World," *Nature* 316 (29 August 1985), 768.
4. Falconer, "Fram," 9.
5. Francois letter, 21 January 2003. "We didn't have much use for the four *Fram* stations," Beau Buck says. "Too noisy and too many uncleared people, both foreign and U.S. We used them mainly for logistic jumping-off places for installations of our quiet camps and unmanned, long-term autonomous acoustic systems. (Beaumont M. Buck, e-mail to author, 25 June 2003.)
6. T. O. Manley, L. A. Codispoti, K. L. Hunkins, H. R. Jackson, E. P. Jones, V. Lee, S. Moore, J. Morison, T. T. Packard, and P. Wadhams, "The Fram 3 Expedition," *Eos Transactions American Geophysical Union* 63 (31 August 1982), 629.
7. Manley et al., "Fram 3," 627. The *Fram III* evacuees left behind an oceanographic buoy. The device measured temperature to two hundred meters and transmitted

through the ARGOS system as its ice tracked south off Greenland then through Denmark Strait.
8. J. R. Weber, "Maps of the Arctic Basin Sea Floor; Part II: Bathymetry and Gravity of the Alpha Ridge: The 1983 CESAR Expedition," *Arctic* 40 (March 1987), 6.
9. A fifty-five-gallon drum of fuel weighs approximately 450 pounds.
10. George D. Hobson, interview by author, audiocassette, 25 April 1993, Manotik, Ontario, Canada.
11. Every third spring since 1983, aircraft and surface measurements of gases, aerosols, and radiation have been collected Arctic-wide with international coordination. In 1986, for instance, a half-dozen aircraft from five countries participated with surface readings conducted at eighteen locations. "The program has shown that the Arctic basin is the airshed of northeastern Europe and air-pollution levels regularly equal or exceed those observed in urban North America." ("Arctic Gas and Aerosol Sampling Program (AGASP)," in Arctic Sciences Program Summary FY89, Office of the Chief of Naval Research (report OCNR 1125AR90-13), 46.) Soviet industrial pollution accounted for half of the air pollution in the Arctic. (*Science* 241 (26 August 1988), 1035.)
12. Juan E. Roederer, "Understanding the Arctic: Research Policies and Responsibilities," in *Pollution of the Arctic Atmosphere* (London and New York: Elsevier Science Publishers, 1991), 3–4.
13. The first ozone measurement for the Arctic was recorded on 27 January 1953, on a flight over the geographic pole by a Royal Canadian Air Force aircraft. The pilot was 2nd Lt. C. Torontow, the navigator was 2nd Lt. K. R. Greenaway who had planned the flight so that the moon was exactly where it was wanted at latitudes between 85° and 90° N. D.C. Rose, "Ozone Over the North Pole," *Arctic Circular* 6, no. 5, 57.
14. The disintegration may be related to the pronounced warming documented for the twentieth century. An increase in the calving rate of *Antarctic* ice shelves has spawned gargantuan bergs. In October 1987, for instance, the Ross Ice Shelf calved a chunk the size of Long Island. In February–March 2002, a Rhode-Island sized piece of the Larsen B. Ice Shelf separated from the Antarctic Peninsula, splintering into a plume of icebergs.
15. J. R. Weber, D. A. Forsyth, A. S. Judge, and Ruth Jackson, "A Geoscience Program for the Canadian Ice Island Research Project," internal report (Department of Energy, Mines and Resources/Geological Survey of Canada, May 1984), 4.
16. "A Unique Opportunity: An Oceanographic Program for the Canadian Arctic Ice Island," internal report, Marine Sciences and Information Directorate, Department of Fisheries and Oceans (Ottawa, 2 April 1985), 1.
17. Andreas Heiberg, letter (with attachment) to author, 14 November 1989.
18. Capt. Brian Shoemaker, U.S. Navy (Ret.), e-mail to author, 31 March 2002.
19. Terence Armstrong, "Soviet Drifting Stations in the Arctic Ocean, 1984–87," *Polar Notes*, 249.
20. Maintenance proved to be a huge headache. Opening up in February 1990 after September's shutdown proved to be a job in itself: coaxing the generator and other equipment into reluctant life, digging out and warming the cabins and kitchen so that staff could have shelter and sustenance—after which came the annual "major battle": the runway. (David Maloney, interview by author, Ice Island, 12 May 1990. Maloney was camp manager that field season.)
21. Robert H. Thomas, "Satellite Remote Sensing Over Ice," *Journal of Geophysical Research* 91 (15 February 1986), 2494.

22. Konstantin Y. Vinnikov, Alan Robock, Ronald J. Stouffer, John E. Walsh, Claire L. Parkinson, Donald J. Cavalieri, John F. B. Mitchell, Donald Garrett, Victor F. Zakharov, "Global Warming and Northern Hemisphere Sea Ice Extent," *Science* 286 (3 December 1999), 1934. Among the sources for the authors: Leningrad's Arctic and Antarctic Research Institute. The Navy/NOAA (National Oceanic and Atmospheric Administration) Joint Ice center in Suitland, Maryland, is the national center for ice analysis and forecasting. One example of its product is the finding that the position of the multiyear ice-edge in spring is an important predictor of the earliest date on which barges will be able to resupply the North Slope. See "Remote Sensing Aids in Sea-Ice Analysis," *Eos Transactions American Geophysical Union* 74 (15 June 1993).
23. Thomas, "Satellite," 2493.
24. "NASA Earth Science and Applications Division: The Programs and Plans for FY 1987–1988–1989," May 1987, 41, 45.
25. All tenants were required to have "proper" arctic clothing, including parkas and boots as well as their own sleeping bags and personal kit plus towels.
26. George Hobson, memorandum to Messrs. Schmidt and Gorveatt, "Some Guidelines and Information for Ice Island Camp Managers," 5 March 1986. What was the island's legal status? "In brief, the legal character of ice islands used for the purpose of marine scientific research is not at all clear in international law. There is very little case law, and what exists is far from definitive, and jurists are divided." Letter, Director Legal Operations Division, Department of External Affairs, to Hobson, 10 December 1986, files Polar Continental Shelf Project, courtesy Director Bonnie Hrycyk.
27. Commander Peter Y. Haydon, "The Strategic Importance of the Arctic: Understanding the Military Issues," *Canadian Defence Quarterly*, Spring 1988, 27.
28. The first weather report broadcast to the mainland was considered a kind of "calling card"–official notification that a new floating center of polar science had begun to operate.
29. *Vodnyi Transport* (27 May) and newspaper *Trud* ["Labor"] (21 May 1986), in "Extracts From the Soviet Press on the Soviet North and the Antarctic," Department of External Affairs, Ottawa, May 1986. The post-Soviet economic crisis deepened the problem. "Doing science is a challenge for researchers in the former Soviet Union (FSU), many of whom can barely afford to eat, let alone scrape up money for supplies." (*Science* 265 (1 July 1994), 19.)
30. Armstrong, "Soviet Drifting," 249.
31. Ice Island 1986 status report, November 1986, files of the Polar Continental Shelf Project, courtesy Director Hrycyk. Had Ice Island gained Soviet waters, could it have continued to be serviced? "We weren't too sure" if the problems were surmountable, Hobson replied, when asked this by the author.
32. Michael Schmidt, "Ice Island 1987, Status Report," files Polar Continental Shelf Project, courtesy Director Hrycyk.
33. Armstrong, "Soviet Drifting," 249.
34. Vladimir Sokolov and Valery S. Ippolitov, interview by author (notes), Arctic and Antarctic Research Institute, St. Petersburg, 22 May 1992. Each had led the much-cracked SP-30 (1987–91) camp. Ippolitov was station head the second shift, Sokolov its last.
35. Sokolov and Ippolitov interview, 22 May 1992.
36. Ivan Frolov, "The 1987 Expedition of the Icebreaker *Sibir* to the North Pole," in *The Soviet Maritime Arctic* (Annapolis, MD: Naval Institute Press, 1991), 36, 44.
37. T. Wen et al., "Environmental Measurements in the Beaufort Sea, Spring 1988,"

Applied Physics Laboratory, Seattle, WA, Technical Report, APL-UW 8822 (1989), 1.
38. George Hobson, letter to author, 20 November 1990.
39. "CEAREX Drift Experiment," *Eos Transactions American Geophysical Union* 71 (2 October 1990), 1115–1118.
40. "CEAREX Drift Experiment," 1116–1117.
41. *Arctic Sciences Summary FY 89*, Office of Naval Research, April 1990, 204–205.
42. Spenser Weart, "International Cooperation" in Weart, The Discovery of Global Warming, http://www.aip.org/history/climate, p. 12 of August 2001 text (record copy at American Institute of Physics, College Park, MD). For a critical discussion of the greenhouse effect and global warming as the eighties closed, see S. F. Singer, ed., *Global Climate Change* (New York: Paragon House Publishers, 1989).
43. Kevin E. Trenberth and Bette L. Otto-Bleisner, "Toward Integrated Reconstruction of Past Climates," *Science* 300 (25 April 2003), 589.
44. "The Arctic: A Key to World Climate," *Science* 243 (17 February 1989), 873. Evidence of ancient climates can be read, for instance, from analyses of interstitial air held within glacial ice. Glaciers and ice sheets grow annually, layer by layer, "capturing" information about climate. Dust and ash reveal can wind patterns and volcanic eruptions.
45. See, for instance, William K. Stevens, "Global Warming: The Contrarian View," *New York Times*, 29 February 2000.
46. Beaumont M. Buck, e-mail to author, 10 March 2003. The initial visit was "quite a surprise, but probably shouldn't have been since they had overflown our camp several times on their way to SP-28, rolling "hellos" and in one case dropped us a present (a roll of magazines I believe it was). We waved to them to land their twin-engine fixed-wing DC-3-like plane, but they preferred to make it with a helo. They brought us a cake." (Buck, e–mail to author, 12 March 2003.)
47. Lawrence Jackson, "Ice Island: An Ideal Platform," *Canadian Geographic*, December 1988/January 1989, unpaginated reprint.
48. Jerome Groopman, "Medicine on Mars," *The New Yorker*, 14 February 2000, 40.
49. Antarctic Research: Program Announcement and Proposal Preparation Guide, National Science Foundation 91-41, 10.
50. Thomas Krochak, interview by author, audiocassette, Resolute, Northwest Territories, 10 May 1990. Twenty-five percent of logistics plans include bad weather and breakdowns; timelines must be flexible. One expert put it this way: The Arctic's not a good place to visit if one has a scheduled meeting in Washington. (Imnats J. Versnieks, interview by author, Thule AB, Greenland, 15 April 1994.)
51. Krochak interview, 10 May 1990; Jackson, "Ice Island."
52. Krochak interview, 10 May 1990.
53. Franklyn Griffiths, "The Arctic in the Russian Identity," in *The Soviet Maritime Arctic* (Annapolis: Naval Institute Press, 1991), 83–84.
54. *Polarnie Novosti* ("Polar News") magazine, 1992, 46. The resupply and refueling of SP-30 and SP-31, as conducted, "increased considerably" fuel consumption for aerial operations.
55. Sokolov and Ippolitov interview, 22 May 1992.
56. Valery S. Ippolitov, *Polarnie Novosti* [1991], 50. Valery Ippolitov was then head of research expeditions' department at Arctic and Antarctic Research Institute. During May 1992, the author had the pleasure of meeting him and fellow polyarniks at the St. Petersburg institute.
57. Valery S. Ippolitov, *Polarnie Novosti* [1991?], 50.
58. Jude Wanniski, "The Future of Russian Capitalism," *Foreign Affairs*, 71, Spring 1992, 22.

59. Juan G. Roederer, "From East-West Confidence Building to West-East Assistance: A Critical Review of the Effects of Recent Changes in the Former Soviet Union on its Scientific Establishment," report, Geophysical Institute and Department of Physics, University of Alaska, Fairbanks, May 1992.
60. Sergey V. Karpekin, interview with author, Deputy Director, Arctic and Antarctic Research Institute, St. Petersburg, Russia, 18 May 1992.
61. *Arctic Research of the United States* 7 (Spring 1993), 14.
62. Courtesy of then-director George Hobson, Polar Continental Shelf Project, the author visited Ice Island during this span–the camp's sixth full field season. ("Polar Continental Shelf Project is pleased to provide the support as set out below for the 1990 field season. Attached are the Conditions of Support.") Ferried from Resolute–the base camp for logistics–the author's Twin Otter reached the ice-island skiway on 11 May. The cost to Polar Shelf to sustain one man on-island: $150/day (Canadian). "And you get the scenery for free."
63. Martin O. Jeffries and M. Amanda Shaw, "The Drift of Ice Islands From the Arctic Ocean Into the Channels of the Canadian Arctic Archipelago: The History of Hobson's Choice Ice Island," *Polar Record* 29 (October 1993), 307, 310.
64. George Hobson, letter to author, 21 November 1991.
65. "Research Base Floats Away," *The Globe and Mail*, 21 March 1992; letter to author, 17 November 1992; "When a Floating Island Makes a Wrong Turn," *Washington Post*, 22 March 1992.
66. Sarah T. Gille, "Warming of the Southern Ocean Since the 1950s," *Science* 295 (15 February 2002), 1275.
67. News release, Office of Public Information, Columbia University, February 1992.
68. *New York Times*, 1 January 1991.
69. News release, Office of Public Information, Columbia University, 5 June 1992. See also, "In Antarctica, Scientists Go With the Floe," *Science News* 14 (22 February 1992), 119.
70. The author is reminded of a remark pertaining to *Sever-67*, a temporary Soviet ice station: "Sanitary facilities were designed solely for necessity, not for comfort; they consisted of a simple trench, unheated, with canvas protection from the wind." Brewer, "Drifting," 1967.
71. News release, 5 June 1992.
72. Arnold Gordon, personal communication, 28 July 1992; Sergey Priamikov, fax to author, 16 November 1992.
73. Lloyd D. Keigwin and G. Leonard Johnson, Using a Nuclear Submarine for Arctic Research," *Eos Transactions American Geophysical Union*, 73 (12 May 1992), 209; "For Attack Subs, New Roles in a Changing World," *Washington Post*, 5 March 2001; interview by author, audiocassette, Arlington, VA, 14 August 1999. In the eighties, conversion of an outdated SSBN-type (fleet ballistic missile submarine) submarine into an Arctic surface research ship had garnered an analysis. Altered to civilian standards in a noncombat, nonnuclear (diesel-electric retained), nondiving configuration, the boat could have been operated by a crew of fifty-one supporting thirty or more research personnel. The notion came to nothing. (Norman Polmar, e-mail to author, 17 August 2002. See his "Submarines in the Ice," Naval Institute *Proceedings*, 117/8/1062, August 1991), 105–106.)
74. *Arctic Research of the United States* 7 (Spring 1993), 5. Under-ice combat is the province of submarines alone. SCICEX did nothing for warfare under the ice. "Science cruises, carried out in the deep central Arctic Ocean, benefit only the scientists. They do not support warfighting capabilities." (Richard Boyle and Waldo Lyon, "Arctic ASW: Have We Lost?" Naval Institute *Proceedings*, 124/6/1/144, 34.)

75. *Arctic Research of the United States* 7 (Fall 1993), 44. A two-year drift had been planned for 1993–95, by Norwegian scientists. The platform was to have been "a small robust ice-breaker." "The overall goal is to improve our understanding of climatically significant air-snow-ice-ocean processes, and collect basic data on atmosphere, ice and ocean parameters for use in large scale climate models." (Nansen Centennial Arctic Programme, Plan for a Scientific Expedition to the Arctic Ocean, 1993–95 (preliminary draft), Norwegian Research Council for Science and the Humanities, Oslo, May 1991, courtesy Tore O. Vorren.)
76. *Orange County (CA) Register*, 1997. In 1997, the U.S. Navy released formerly classified data collected by submarines between 1957 and 1982. Garrett W. Brass was then head of the U.S. Arctic Research Commission. "This is at least a factor of two, and maybe three or four, times more data for the central Arctic than was available before," he said. *Science News* 152 (13 September 1997), 172.
77. *Eos Transactions American Geophysical Union* 74 (12 October 1993), 466. This issue announced three sessions for its 1993 meeting on the climate record profiled by the Greenland Ice Sheet Project 2 (GISP2) ice core–"the longest continuous high-resolution record of climate response available in the Northern Hemisphere, covering approximately 250,000 years."
78. R. G. Barry, M. C. Serreze, and J. A. Maslanik, "The Arctic Ice-Climate System: Observations and Modeling," *Reviews of Geophysics* 31 (November 1993), 416–417.
79. *Science Times*, 29 February 2000; *Science* 293 (13 July 2001), 1999.
80. Donald K. Perovich et al., "Year on Ice Gives Climate Insights," *Eos Transactions American Geophysical Union* 80 (12 October 1999), 481.
81. Perovich et al., "Year on Ice," 481.
82. Holly Menino, "The Long, Cold Journey of Ice Station SHEBA," *Smithsonian* 29 (September 1998), 42–44. The year 1998 saw launch of the first Earth Observing System, a long-term series of earth-orbiting satellites that are part of the U.S. Global Change Research Program.
83. D. K. Perovich, T. C. Grenfel, B. Light, and P. V. Hobbs, "The Seasonal Evolution of Multiyear Arctic Sea Ice Albedo," *Journal of Geophysical Research*, 10.1029/2000 JC000438, 2002.
84. *New York Times*, 3 October 2000. Russia is nearly bankrupt (but improving, because of the price of crude oil), so the symptoms are endless. See for example "Fuel-Short Region in Russia Is Now Freezing in the Dark," *New York Times*, 12 February 2001.
85. John Cassidy, "Master of Disaster," *The New Yorker*, 15 July 2002, 86.
86. Perovich et al., "Year on Ice," 481.
87. Imnats Versnieks, e-mail to author, 23 April 1998. Courtesy of Versnieks, a Space and Naval Warfare Systems Global Change Research Program employee, this author visited the navy's latest (1994) AREA [Arctic Research in Environmental Acoustics] camp. Deployed in March, these seasonal bivouacs are manned for forty-five to sixty days. Further on frozen-in ships: "[Shipboard] studies are much inferior to the studies from a natural site–the drifting ice. The ship hull itself influences the instruments. That is why the drifting station located on sea ice remains the most efficient method of investigating the processes in the 'atmosphere-ice-ocean' system in the polar regions of the World Ocean." Sergey M. Priamikov, Ph.D., e-mail to author (attachment), 26 August 2003. Dr. Priamikov was the head of the international science cooperation department, AARI. "Even if the ship they froze in survived, the ice around it for a large radius would probably become jumbled up and not sustain a good fixed-wing strip for long. This would necessitate basing a helicopter permanently at the ship (actually two would be required to be sure one is

NOTES   331

flyable. Very expensive) . . . (Buck e-mail, 24 August 1997.)
88. Donald K. Perovich, e-mail to author, 22 February 2003.
89. Donald K. Perovich, personal communication, 17 August 2001.
90. "Ottawa Cuts Leave Arctic Scientists out in the Cold," *Toronto Star*, 11 October 1998; "Ottawa Warms up to Arctic Research," *Toronto Star*, 13 April 1999.
91. "Ottawa Warms." Prime Minister John Diefenbaker had ordered Polar Shelf's creation in 1958, when the government realized how little it knew about a region Ottawa insisted was its own—and to help counter an American scientific invasion of Arctic Canada following launch of the first artificial satellite. 'If Polar Shelf was to be eliminated today [1993], there would be *no* university science in the Arctic. . . . You've got to get out [of Resolute]. And if you don't have flying support, you don't *get* out." Hobson interview, 25 April 1993.
92. Don Walsh, "Being There: The Case for the Oceanographic Submarine," Naval Institute *Proceedings*, December 2000, 87. A one-of-a-kind fuel core would be very expensive. "Coupled with a 130+ Navy crew, it was just out of reach of post-Cold War budgeting. . . . Remember, when the Cold War ended the Navy had about 95 SSNs [nuclear attack submarine] in service; they are now down to just under 50. Obviously the older boats would go first, and the current SSN force is too heavily committed for SCICEX operations." Polmar e-mail, 17 August 2002. See also "The Science Ice Exercise Program," *Arctic Research of the United States*, Fall/Winter 2000, 2–7.
93. W. Krabill, E. Frederick, S. Manzade, C. Martin, J. Sonntag, R. Swift, R. Thomas, W. Wright, and J. Yungel, "Rapid Thinning of Parts of the Southern Greenland Ice Sheet, *Science* 283, 5407 (5 March 1999), 1522; Richard E. Kerr, "Will the Arctic Ocean Lose All Its Ice?" *Science* 286 (3 December 1999), 1828, and "A New Force in High-latitude Science," *Science* 284 (9 April 1999), 242; Dennis M. Conlon, Ph.D., Office of Naval Research, personal communication, 23 May 2001.
94. Kerr, "Will the Arctic Ocean." See also "Thinning Sea Ice Stokes Debate on Climate," *New York Times*, 17 November 1999. This high reflectance or albedo delays spring melt and limits summertime ice decay.
95. Peter Lemke, "Open Windows to the Polar Oceans," *Science* 292 (1 June 2001), 1670. This feedback mechanism is noted also in professional journals.
96. An interpretation of the new warming range in probabilistic terms is presented in T. M. L. Wigley and S. C. B. Raper, "Interpretation of High Projections for Global-Mean Warming," *Science* 293 (20 July 2001), 451–454.
97. John Reilly, Peter H. Stone, Chris E. Forest, Mort D. Webster, Henry D. Jacoby, and Ronald G. Prinn, "Uncertainty and Climate Change Assessments," *Science* 293 (20 July 2001), 430.

EPILOGUE

1. Modern-day ocean drilling will open up the basin. "You could lay out all the existing Arctic cores in my office," a U.S. oceanographer remarks. A successful mission to the deep Arctic will provide "an entire history from fifty million years ago to the present." See "A Sea Change in Ocean Drilling," *Science* (18 April 2003), 410–412; and "Arctic Is First Call for New Global Program," *Science* 302 (12 December 2003), 1878; *Nature*, 441 (1 June 2006).
2. An understanding of global systems will *not* come from either observations or models alone, but from a marriage of the two, that is, measuring plus modeling. Current climate models suggest that the amount of greenhouse warming may be significantly

greater in northern high-latitude regions than in lower latitudes, but the models do not agree well on the relative amount of warming to be expected.
3. Jack Oliver, "Solid Earth Science During the 21st Century," *Eos Transactions American Geophysical Union* 72 (12 March 1991), 124.
4. *Science* 301 (22 August 2003), 1021.
5. The climate has probably always been changing, at whatever rate. And change does not always equate with "warming." "Parts of the eastern Arctic are cooling both as predicted by general circulation models and as measured empirically." (David Norton, Ph.D., University of Alaska, e-mail to author, 2 July 2003.)
6. Dr. Kenneth L. Hunkins, e-mail to author, 31 January 2002; R. A. Kerr, "Whither Arctic Ice? Less of it, for Sure," *Science* 297 (30 August 2002), 1491; Mauri S. Pelto and Maynard M. Miller, "The Impact of Global Warming on Glacier Mass Balance in the North Cascades, Washington and the Juneau Icefield, Alaska," *World Resource Review* 13 (2001), 91, 104.
7. Research on climate change and its effects on natural systems has exploded. "There are funds, programmes, academic offices . . . all built around the premise that global warming is underway, and the implications from that 'given.' In other words, doomsaying has become 'big business.'" (Norton e-mail, 2 July 2003.)
8. James D. Johnston letters, *Science* 292 (11 May 2001), 1063–1064. "And often [this issue editorializes] the questions are outside the envelope of known science, and the risks can only be guessed at. This is especially awkward for a public that experiences science—in school, in university, and on quiz shows—as the certainties of established knowledge, not the unknown terrain at or beyond the frontiers." (1021)
9. Editorial, *Science* 292 (15 June 2001), 1965. In a report from the National Academy of Sciences, eleven leading atmospheric scientists, including previous skeptics, reaffirmed the mainstream scientific view that the earth's atmosphere was getting warmer and that in fact human activity was largely responsible.
10. *Science* 295 (4 January 2002), 29.
11. Editorial, *Science* 297 (2 August 2002), 737. "Communicating the fact of climate change [the editorial concludes] is a complex process involving political leadership, science, public pressure, and even perhaps a useful catastrophe or two to illuminate the issues. We should not forget the moral dimension: a sense of responsibility to future human generations and a respect for the totality of ecosystems."
12. Editorial, *Science* 299 (17 January 2003), 309.
13. "Human Impacts on Climate," statement voted on and adopted by American Geophysical Union, adopted by American Geophysical Union Council December 2003.
14. Editorial and "A Year to Remember at the Ends of the Earth," *Science* 303 (5 March 2004), 1437, 1459.
15. Fourteen drifting buoys were deployed onto Arctic pack ice during 1990, for instance; twelve in 1991. They are used to derive ice-drift velocity vectors and to report air temperature and barometric pressure. Satellite methods cannot do everything, however, such as observing processes within the water column.
16. "Drifting Ice Stations," *Encyclopedia of Oceanography*, ed., Rhodes Whitmore Fairbridge, (Stroudsberg, PA: Dowden, Hutchinson & Ross, 1966), 232–233.
17. *Arctic Research of the United States* 13 (Fall/Winter 1999), 72.
18. Norbert Untersteiner, Ph.D., personal communication, 18 May 1999.
19. *Arctic Research of the United States* 7 (Fall 1993), 29–32.
20. Like the Mediterranean, the Arctic Ocean is a basin mostly surrounded by land. Its riverine inflow, for instance, is largely from major Russian rivers. So, evaluating the impacts of climate change on the Arctic without Russian data and collaborative

studies would be difficult. Eurasian river discharge into the basin has increased significantly—with possible implications for Northern Europe.
21. Keith R. Greenaway, memorandum to Aurora Institute, 21 March 2000, courtesy Brig. Gen. Greenaway. See for example *Toronto Star*, 4 May 2003.
22. Until its demise, the Soviet Union had a longstanding policy and clearly defined goals for its Arctic regions. Approximately 25,000 scientists and engineers were at work there, along with 170 scientific institutes linked to offshore oil and gas exploration alone, thirty-seven polar research vessels, and nineteen icebreakers, several of them nuclear-powered. (Juan G. Roederer, "Neglected Arctic Gains Attention," Forum for Applied Research and Public Policy, Spring 1991, 62–63.)
23. Dimitri K. Simes, "America and the Post-Soviet Republics," *Foreign Affairs* 712, no.3 (Summer 1992), 78.
24. Irina Dezhina and Loren Graham, "Science and Higher Education in Russia," *Science* 286 (12 November 1999), 1303. "Scientific research is much less popular in today's Russia than it was in the Soviet Union. . . . Being a scientist in Russia today requires great dedication, and the situation is not likely to improve quickly."
25. Priamikov e-mail attachment, 26 August 2003.
26. Untersteiner, personal communication, 18 May 1999; Sergey M. Priamikov, Arctic and Antarctic Research Institute, e-mail to author, 25 January 2001. A June 1997 visitor to the institute—the sixtieth anniversary of the Shmidt-Papanin drift—found mood and comment "perfectly clear that there would be no more SPs." (Christopher Pala, e-mail to author, 27 October 1997. Pala is a freelance journalist.)
27. *The New Yorker*, 21 May 2001, 37.
28. Arkadiy S. Safray, e-mail to author, 10 February 2006.
29. Christopher Pala, *The Oddest Place on Earth: Rediscovering the North Pole* (San José: Writer's Showcase, 2002), 62.
30. Deanna Swaney, *The Arctic* (Lonely Planet Publications: Melbourne-Oakland-London-Paris, November 1999), 407.
31. Following the 21 August 1991 coup attempt, pressure became irresistible. "The [name] change symbolized a larger development: everything communism had built since 1917 was now either repudiated or destroyed. It was as if all of the twentieth century had suddenly been repealed. The economic and political systems that had bound together the three hundred million people of what had once been the Soviet Union lay in ruins, leaving uncoordinated fragments." (Michael Mandelbaum, "Coup de Grace: The End of the Soviet Union," *Foreign Affairs* 71 (1992), 173.)

GLOSSARY

ablation:	the disappearance of an ice or snow surface by melting and/or evaporation
albedo:	reflectivity
AUV:	autonomous underwater vehicle
calving:	the breaking away of a mass of ice from a glacier or iceberg
close pack ice:	pack ice composed of floes that are mostly in contact; ice cover of seven-tenths to nine-tenths of the sea surface
consolidated ice:	ice that covers ten-tenths of the sea surface; the floes are frozen together
dropsonde:	instrument package dropped from aircraft
floe:	any piece of floating sea ice, large or small, other than fast ice (ice attached to the shore)
geophysics:	the application of physics to the study of the Earth (geology) e.g., gravity, magnetics, and seismic methods
Glavsevmorput:	see GUSMP
GUSMP:	*Glavnoe upravlenie Severnogo morskogo puti,* the Central Administration of the Northern Sea Route (in the West, Northeast Passage)
halocline:	the layer of rapid salinity increase; a region in the water column containing a marked salinity gradient, probably implying a marked density gradient
heavy ice:	any sea ice more than ten feet thick
hummocking:	pressure ridge formation by which level ice becomes broken up into hummocks and ridges

ice island:	fragment calved from an ice shelf–nonsaline ice of freshwater origin and the most massive ice features in the Arctic Ocean
ice shelf:	a floating ice sheet of considerable thickness, and usually of great horizontal extent, with a level or undulating surface
komsomol:	Communist Party youth league
KSMP:	*Komitet Severnogo morskogo puti,* the Committee for the Northern Sea Route
lead:	a long narrow passage through pack ice
multiyear ice:	ice that is more than one year old; multiyear ice has undergone deformation and cracking
NP:	(North Pole) semi-permanent Arctic ice stations (SP in Russian)
open pack ice:	pack ice composed by floes seldom in contact and with many leads and pools; ice cover four-tenths to six-tenths of the sea surface
open water:	water that is less than one-tenth covered with floating ice
pack ice:	any area of sea ice composed of a heterogeneous mixture of size and age types; pack ice (unlike fast ice) is usually in motion
paleoclimate:	fossil or ancient climate, analyzed using proxy records such as ice and marine-sediment cores, lake and sea-level changes, pollen, and tree rings
phytoplankton:	microscopic, photosynthetic organisms that float passively in surface waters
polyarniks:	polar researchers
polynya:	natural runways of new-formed flat ice. Any enclosed sea water area in pack ice, other than a lead
pressure ridge:	a ridge or wall of hummocked ice formed where one floe has been pressed against another
sea ice:	any form of ice at sea which has originated from the freezing of sea water
self-noise:	man-made acoustic interference
SP:	(*Severnyy Polyus*) semi-permanent Arctic ice stations (NP in English)
SSN:	nuclear attack submarine
UARS:	Unmanned Arctic Research Submersible

SELECTED BIBLIOGRAPHY

PRIMARY UNPUBLISHED SOURCES

In his research the author has relied, in significant part, on repositories in the United States, Canada, and the former Soviet Union (Russia). In St. Petersburg (former Leningrad), the Arktika Musee and the Arctic and Antarctic Research Institute made available materials pertaining to Soviet expeditions. A number of AARI *polyarniks* met with the author for interview. The National Library of Canada and the National Archives of Canada hold a wealth of materials relating to the circumpolar North and to sorties for science off the Canadian Polar Margin. Private papers from investigators and also military personnel were no less important. Among these, the personal files/libraries and reminiscences of Beaumont M. Buck, George D. Hobson, Brig. Gen. Keith R. Greenaway, Royal Canadian Air Force (Ret.), Gerry H. Cabaniss, Ph.D., Kenneth L. Hunkins, Ph.D., and Robert Francois were especially informative. In addition, interviews recorded in three countries along with an extensive correspondence were rich sources (see below).

As well, primary and selected secondary sources were essential for background, for corroboration of certain information, and for helping round out the narrative.

CONVERSATIONS AND(OR) ONE-ON-ONE INTERVIEWS

Maxwell E. Britton, Ph.D., ecologist, Office of Naval Research*
Gerry H. Cabaniss, Ph.D., geologist field assistant, T-3 (1958–60)
Norman Goldstein, geophysicist, U.S. Air Force Cambridge Research Laboratory
Keith R. Greenaway, Brig. Gen., Royal Canadian Air Force (Ret.)*
Zalman Gudkovich, Ph.D., hydrologist, Arctic and Antarctic Research Institute (AARI), St. Petersburg, Russia*
George D. Hobson, geophysicist and director (1972–88), Polar Continental Shelf Project, Canada ("Polar Shelf")*
Bonni A. Hrycyk, director, Polar Shelf (1994)
Kenneth L. Hunkins, Ph.D., geophysicist, Lamont-Doherty Geological Observatory*

Valery S. Ippolitov, AARI
Sergey V. Karpekin, deputy director, AARI
Thomas Krochak, chef contracted to Polar Shelf*
Waldo K. Lyon, Ph.D., physicist, Naval Electronics Laboratory/Arctic Submarine Laboratory*
David A. Malony, Ice Island Base Manager (1990), Polar Shelf *
Ronald McGregor, Cdr., U.S. Navy (Ret.), naval aviator, Office of Naval Research*
Wayne Parton, senior pilot, Canadian Air*
Albert A. Romanov, Ph.D., AARI*
Ilya P. Romanov, Ph.D., AARI*
John L. Schindler, assistant director, Naval Arctic Research Laboratory (1960–71), director (1971–73)
Vladimir Sokolov, AARI
Igor Tsigelnitsky, AARI*
Imants J. Versnieks, Arctic Systems Undersea Warfare Systems Command*
Nikolai D. Vinogradov, AARI*
Nikolai Yagodnitsin, director, Arktika Musee, St. Petersburg, Russia*

* = audio-recorded interview

CORRESPONDENCE (LETTERS, E-MAILS, DOCUMENTS)

Terence Armstrong, Ph.D., Scott Polar Research Institute, United Kingdom
Hazard C. Benedict, Staff Sgt., U.S. Air Force (T-3, 1955)
James Bitterman, Canadian Broadcasting Corp., 1977 (now ABC News, Paris)
Beaumont M. Buck, Special Projects Officer, Undersea Branch, Office of Naval Research/Polar Research Lab
Vivian C. Bushnell, Air Force Cambridge Research Center
Gerry H. Cabaniss, Ph.D., geologist (T-3, 1958–60)
Robert Cotell, Air Force Cambridge Research Center (T-3, 1952, 1955)
Honorable Barnett J. Danson, Minister of Defence, Canada (1977)
Ronald Denk, Cdr., U.S. Air Force (Ret.), 18th Air Force
Robert E. Francois, Applied Physics Laboratory, University of Washington
Brig. Gen. Keith R. Greenaway, Royal Canadian Air Force (Ret.)
Andreas Heiberg, Polar Science Center, University of Washington, Seattle
George D. Hobson, director, Polar Continental Shelf Project (1972–88)
Sergey Karpekin, deputy director of foreign relations, AARI
Thomas B. Kilpatrick, chief, Ice Reconnaissance Division, Ice Branch, Atmospheric Environment Service, Environment Canada
Jeffrey Lord, Woods Hole Oceanographic Institution (AREA 85 through SHEBA)
Waldo K. Lyon, Ph.D., Naval Electronics Laboratory/Arctic Submarine Laboratory
Charles E. Myers, Ph.D., National Science Foundation
V. Adm. John H. Nicholson, U.S. Navy (Ret.) (commanding officer, USS *Sargo*)
David Norton, Ph.D., University of Alaska, Fairbanks
D. K. Perovich, Ph.D., Cold Regions Research and Engineering Laboratory (Ice Station SHEBA)
Donald Plouff, Ph.D., U.S. Geological Survey (T-3, 1958)
Carlos Plummer, Ph.D., California State Univ., Sacramento, geologist (T-3, 1958)
Sergey M. Priamikov, Ph.D., AARI
Juan G. Roederer, Ph.D., Geophysical Institute, University of Alaska

John L. Schindler, Naval Arctic Research Laboratory, Office of Naval Research
Michael Schmidt, Ice Island Base Manager (1988–89), Polar Continental Shelf Project
Capt. Brian Shoemaker, U.S. Navy (Ret.) (final occupation-years, T-3)
Boris I. Silken, Ph.D., Russian [former Soviet] Academy of Sciences
Norbert Untersteiner, Ph.D., University of Washington (ALPHA, AIDJEX)
James A. Van Allen, Ph.D., University of Iowa (genesis IGY)
Imants J. Versnieks, Arctic Systems Undersea Warfare Systems Command
Cdr. Edward M. Ward, U.S. Navy (Ret.), officer-in-chief, Project SKI-JUMP I and II (1951, 1952)

PUBLISHED SOURCES

Books

Althoff, William F., *Arctic Mission: By Airship and Submarine to the Far North*, 2nd ed., Auckland, New Zealand: Lighter-Than-Air Institute, 2000.
American-Russian Chamber of Commerce, *Handbook of the Soviet Union*, New York: John Day Company, 1936.
Armstrong, Terence, *The Russians in the Arctic*, Fairlawn, NJ: Essential Books, 1958.
———, George Rogers, and Graham Rowley, *The Circumpolar North*, London: Methuen and Co. LTD, 1978.
Blanchet, Guy, *Search in the North*, Toronto: Macmillan Co., 1960.
Britton, Maxwell E., *Foreword* and "The Role of the Office of Naval Research and the International Geophysical Year (1957–58) in the Growth of the Naval Arctic Research Laboratory," in *Fifty More Years Below Zero: Tributes and Meditations for the Naval Arctic Research Laboratory's First Half Century at Barrow, Alaska*, Winnipeg, Manitoba: Arctic Institute of North America, 2001.
Calvert, Cdr. James, U.S. Navy, *Surface at the Pole: The Extraordinary Voyages of the USS Skate*, New York: McGraw-Hill Book Company, Inc., 1960.
Carlson, William S., *Lifelines Through the Arctic*, New York: Duell, Sloan and Pearce, 1962.
Chilingarov, A., Sarukhanyan, E., and M. Yevseyev, *Pod Nogami Ostrov Ledianoi*, Moscow, 1972, Trans. no. 502, *Life on an Ice Island*, Cold Regions Research and Engineering Laboratory, U.S. Army Corps of Engineers, Hanover, NH, December 1975.
Davies, R. E. G., *Aeroflot: An Airline and Its Aircraft: An Illustrated History of the World's Largest Airline;* Rockville, MD: Paladwr Press, 1992.
———, with Yuri Salnikov, *The Chelyuskin Adventure: Exploration, Tragedy, Heroism*, McLean, VA: Paladwr Press, 2005.
Dolan, Edward F., Jr., *White Battleground: The Conquest of the Arctic*, New York: Dodd, Mead & Co., 1961.
Dunbar, Moira, and Wing Cdr. Keith R. Greenaway, Royal Canadian Air Force, *Arctic Canada From the Air*, Ottawa: Canada Defence Research Board, 1956.
Fedorov, Yevgeny, *Polar Diaries*, Moscow: Progress Publishers, 1983.
Foster, Michael, and Carol Marino, *The Polar Shelf: The Saga of Canada's Arctic Scientists*, Toronto: NC Press Limited, 1986.
Frolov, Ivan, "The 1987 Expedition of the Icebreaker Sibir' to the North Pole," in *The Soviet Maritime Arctic*, edited by Lawson W. Brigham, Annapolis: Naval Institute Press, 1991.
Grierson, John, *Heroes of the Polar Skies*, New York: Meredith Press, 1967.
———, *Challenge to the Poles*, London: G. T. Foulis & Co. LTD, 1964.
Griffiths, Franklyn, "The Arctic in the Russian Identity," in *The Soviet Maritime Arctic*, edited by Lawson W. Brigham, Annapolis: Naval Institute Press, 1991.

Hattersley-Smith, G., *North of Latitude Eighty*, Ottawa: Defence Research Board, 1974.

Kerber, L. L., *Stalin's Aviation Gulag: A Memoir of Andrei Tupolev and the Purge Era*, Washington, DC: Smithsonian Institution Press, 1996.

Krenkel, Ernst, *RAEM Is My Call-Sign*, Moscow: Progress Publishers, 1978.

Leary, William M., *Under Ice: Waldo Lyon and the Development of the Arctic Submarine*, College Station: Texas A&M University Press, 1999.

———, and Leonard A. LeShack, *Project COLDFEET: Secret Mission to a Soviet Ice Station*, Annapolis: Naval Institute Press, 1996.

Medwin, Herman, and Clarence S. Clay, *Fundamentals of Acoustical Oceanography*, San Diego: Academic Press, 1998.

Members of the Expedition, *The Voyage of the Chelyuskin*, trans. Alec Brown, New York: Macmillan Company, 1935.

Nansen, Fridtjof, Ph.D., *Farthest North: Being the Record of a Voyage of Exploration of the Ship "Fram" 1893-96 and of a Fifteen Months' Sleigh Journey by Dr. Nansen and Lieut. Johansen*, 2 vols., New York: Harper & Brothers Publishers, 1897.

Pala, Christopher, *The Oddest Place on Earth: Rediscovering the North Pole*, San Jose: Writer's Showcase, 2002.

Papanin, Ivan, *Life on an Ice Floe*, New York: Julian Messner, Inc., 1939.

Petrov, Richard, *Across the Top of Russia*, New York: David McKay Company Inc., 1967.

Rodahl, Kaare, *North: The Nature and Drama of the Polar World*, New York: Harper and Brothers, Publishers, 1953.

Rowley, Graham W., *Cold Comfort: My Love Affair With the Arctic*, Montreal and Kingston: McGill-Queens University Press, 1996.

Smolka, H. P., *40,000 Against the Arctic*, New York: William Morrow and Company, 1937.

Sullivan, Walter, *Assault on the Unknown*, New York: McGraw-Hill Book Company, 1961.

Taracouzio, T. A., *Soviets in the Arctic*, New York: Macmillan Company, 1938.

Vodopyanov, M. V., *Wings Over the Arctic*, Moscow: Foreign Languages Publishing House, 1957.

Weart, Spencer R., *The Discovery of Global Warming*, Cambridge, MA: Harvard University Press, 2003.

Weeks, Tim, and Ramona Maher, *Ice Island: Polar Science and the Arctic Research Laboratory*, New York: John Day Company, 1965.

Weir, Gary E., Ph.D., *An Ocean in Common: American Naval Officers, Scientists and the Ocean Environment*, College Station: Texas A&M University Press, 2001.

Yakovlev, G. N., *Ice Routes of the Arctic (Ledovyye puti Arktiki)*, [trans.] Ottawa: Natioanl Defence, 1977, courtesy Keith R. Greenaway. Originally published by MYSL Publishing House, Moscow, 1975.

TECHNICAL PAPERS, JOURNALS, DOCUMENTS

Alley, R. B., J. Marotzke, W. D. Nordhaus, J. T. Overpeck, D. M. Peteet, R. A. Pielke Jr., R. T. Pierrehumbert, P. B. Rhines, T. F. Stocker, L. D. Talley, and J. M. Wallace, "Abrupt Climate Change," *Science* 299 (28 March 2003), 2005–2010.

Armstrong, Terence, "The Northern Sea Route Today," *Cold Regions Science and Technology* 7 (1983), 251–57.

———, "The Northeast Passage as a Commercial Waterway, 1879–1979," *Ymer* 86, (1980), 87–130.

———, "The Recording and Reporting of Floating Ice," *The Polar Record* 9 (September 1958), 184–90.

Barry, R. G., M. C. Serreze, J. A. Maslanik, and R. H. Preller, "The Arctic Sea Ice-Climate

System: Observations and Modeling," *Reviews of Geophysics* 31 (November 1993), 397–422.
Beal, M. Allan, "The Use of Light Aircraft for Oceanographic Observations in the Arctic Ice Pack," in *Arctic Drifting Stations: A Report on Activities Supported by the Office of Naval Research*. Proceedings of the Symposium, Warrenton, VA, 12–15 April 1966 under the auspices of the Arctic Institute of North America and the Office of Naval Research, John E. Sater, Coordinator, November 1968, 37–48.
Bell, Thaddeus G., *Sonar and Submarine Detection*, U.S. Navy Underwater Sound Laboratory, in *Papers on History of Sonar*, New London, 8 May 1962.
Brewer, Max C., "New Applications of Old Concepts of Drifting Station Operations," in Sater, *Arctic Drifting Stations*, 23–28.
Britton, M. E., "Administrative Background of the Developing Program," in Sater, *Arctic Drifting Stations*, 29–36.
Broecker, W. S., "Does the Trigger for Abrupt Climate Change Reside in the Ocean or in the Atmosphere?" *Science* 300 (6 June 2003), 1519–22.
Buck, Beaumont M., "Arctic Acoustic Transmission Loss and Ambient Noise," in Sater, *Arctic Drifting Stations*, 427–38.
Bushnell, Vivian, editor, *Scientific Studies at Fletcher's Ice Island, T-3, 1952-1955*, Geophysics Research Directorate, Air Force Cambridge Research Center, Air Research and Development Command, United States Air Force, geophysical research paper no. 63.
Cabaniss, Gerry H., Kenneth L. Hunkins, and Norbert Untersteiner, compilers, U.S.-IGY Drifting Station Alpha Arctic Ocean 1957-1958, Air Force Cambridge Research Laboratories, special report no. 38.
CEAREX Drift Group, "CEAREX Drift Experiment," *Eos Transactions American Geophysical Union* 71 (October 2, 1990), 1115–8.
Cottell, Irene Browne, "United States Research at Drifting Stations in the Arctic Ocean," *The Polar Record* 10 (September 1960), 269–74.
Crary, A. P., "Air Force Research Achievements on Drifting Stations," in Sater, *Arctic Drifting Stations*, 113–25.
———, R. D. Cotell, and T. F. Sexton, "Preliminary Report on Scientific Work on 'Fletcher's Ice Island,' T-3," *Arctic* 5 (December 1952), 211–23.
———, "Seismic Studies on Fletcher's Ice Island, T-3," *Transactions American Geophysical Union* 35, no. 2 (April 1954), 293–300.
———, and Norman Goldstein, "Geophysical Studies in the Arctic Ocean," *Deep-Sea Research* 4 (1957), 185–201.
Crickard, F. W., Rear Admiral (CF, Ret.), "An Anti-Submarine Warfare Capability in the Arctic a National Requirement," *Canadian Defence Quarterly*, April 1987, 24–30.
Curry, Ruth, Bob Dickson, and Igor Yashayaev, "A Change in the Freshwater Balance of the Atlantic Ocean Over the Past Four Decades," *Nature* 426 (18/25 December 2003), 826–29.
Dezhina, Irina, and Loren Graham, "Science and Higher Education in Russia," *Science* 286 (12 November 1999), 1303–4.
Director General Intelligence and Security, National Defence Headquarters, Department of National Defence, Ottawa, "Soviet Research from Drifting Ice Stations," 8 September 1978.
Dunbar, Moira, "Arctic Ice Islands: Historical References to Ice Islands," *Arctic* 5 (July 1952), 83–95.
Ewing, Maurice, Kenneth Hunkins, and E. M. Thorndike, "Some Unusual Photographs in the Arctic Ocean," *Marine Technology Society Journal* 3 (January 1969).
Fischer, R. J., "Air Support of Drifting Stations–A Decade of Experience," in Sater, *Arctic*

Drifting Stations, 81–90.
Fleagle, Robert C., "From the International Geophysical Year to Global Change," *Reviews of Geophysics* 30 (November 1992), 305–313.
Fletcher, J. O., "Origin and Early Utilization of Aircraft-Supported Drifting Stations," in Sater, *Arctic Drifting Stations*, 1–13.
Forest, Chris E., Peter H. Stone, Andrei P. Sokolov, Myles R. Allen, Mort D. Webster, "Quantifying Uncertainties in Climate System Properties with the Use of Recent Climate Observations," *Science* 295 (4 January 2002), 113–6.
Francois, Robert E. "The Unmanned Arctic Research Submersible System," *Marine Technology Society Journal*, 7, 46–48.
———, K. L. Williams, R. G. Garrison, P. D. Mourad, and T. Wen, "ICE KEELS I: Intrinsic Physical/Acoustic Properties of Sea Ice and Scattering from Ice Surfaces," *U.S. Navy Journal of Underwater Acoustics* 39, part 2 (October 1989), 1203–28.
Greenaway, K. R., "Arctic Ice Islands: Additional Information From Flights and Air Photographs in the Canadian Arctic," *Arctic* 5 (July 1952), 75–82.
Greene, Charles R., and Beaumont M. Buck, "Arctic Noise Measurement Experiment Using NIMBUS 6 Data Buoys," *U.S. Navy Journal of Underwater Acoustics* 27 (October 1977), 827–38.
Hanson, Arnold M., "Critique of Ice Station Support," in Sater, *Arctic Drifting Stations*, 107–12.
Hattersley-Smith, Geoffrey, "Arctic Ice Islands: Comments on the Origin of Ice Islands," *Arctic* 5 (July 1952), 95–103.
Hayton, Peter T., "The Strategic Importance of the Arctic: Understanding the Military Issues," *Canadian Defence Quarterly*, Spring 1988, 27–34.
Hobson, George, "Ice Island Field Station: New Features of Canadian Polar Margin," *Eos Transactions American Geophysical Union* 70 (September 12, 1989), 838–39.
Holmes, John F., and L. V. Worthington, "PROJECT SKIJUMP conducted during the period February 1951–May 1951," reference no. 51–67 (technical report submitted to the Office of Naval Research), Woods Hole Oceanographic Institution, September 1951.
Hunkins, Kenneth, Yngve Kristoffersen, G. Leonard Johnson, and Andreas Heiberg, "The Fram I Expedition," *Eos Transactions American Geophysical Union* 60 (12 December 1979), 1043–4.
———, "Inertial Oscillations of Fletcher's Ice Island (T-3)," *Journal Geophysical Research* 72, no. 4 (February 15, 1967), 1165–1174.
———, Allan W. H. Be', Neil D. Opdyke, and Gary Mathieu, "The Late Cenozoic History of the Arctic Ocean," (chapter 8) *The Late Cenozoic Glacial Ages*, New Haven and London: Yale University Press, 1971.
———, Thomas Herbron, Henry Kutschale, and George Peter, "Geophysical Studies of the Chukchi Cap, Arctic Ocean," *Journal Geophysical Research* 67 (January 1962), 235–47.
Jeffries, Martin O., "Arctic Ice Shelves and Ice Islands: Origin, Growth and Disintegration, Physical Characteristics, Structural-Stratigraphic Variability, and Dynamics," *Reviews of Geophysics* 30, no.3 (August 1992), 245–67.
———, and William M. Sackinger, "Analysis and Interpretation of an Airborne Synthetic Aperture Radar Image of Hobson's Ice Island," draft, submitted to 10th International Symposium on Port and Ocean Engineering Under Arctic Conditions, 12–16 June 1989, Lulea, Sweden.
Johannessen, Ola M., Elena V. Shalina, Martin W. Miles, "Satellite Evidence for an Arctic Sea Ice Cover in Transformation," *Science* 286 (3 December 1999), 1937–39.
Johnston, Carl B., Jr., "Problems of Working on a Drifting Station," in Sater, *Arctic Drifting Stations*, 91–99.

Keigwin, Lloyd D., and G. Leonard Johnson, "Using a Nuclear Submarine for Arctic Research," *Eos Transaction American Geophysical Union* 73 (12 May 1992), 209, 220–21.
Kerr, Richard A., "A Warmer Arctic Means Change for All," *Science* 297 (30 August 2002), 1490–92.
Koenig, L. S., "Arctic Ice Islands: Discovery of Ice Islands on U.S.A.F. Flights Over the Arctic Ocean," *Arctic* 5 (July 1952), 68–75.
Krabill, W., E. Frederick, S. Manijzade, C. Martin, J. Sonntag, R. Swift, R. Thomas, W. Wright, and J. Yungel, "Rapid Thinning of Parts of the Southern Greenland Ice Sheet," *Science* 283 (5 March 1999), 1522–24.
Kutschale, Henry, "Seismic Studies on Ice Island ARLIS II," in Sater, *Arctic Drifting Stations*, 439–58.
Langseth, Marcus, Theodore Delaca, George Newton, Bernard Coakley, Roger Colony, Peter McRoy, James Morison, Jeff Gossett, Walter Tucker, and William Smethie, "SCICEX-93: Arctic Cruise of the U.S. Navy Nuclear Powered Submarine USS Pargo," *Marine Technology Society Journal* 27 (Winter 1993–94), 4–12.
Laskey, Marvin, "Historical Review of Undersea Warfare Planning and Organization 1945–1960 with Emphasis on the Role of the Office of Naval Research," *U.S. Navy Journal of Underwater Acoustics* 26 (April 1976), 327–57.
Lyon, Waldo K., "The Navigation of Arctic Polar Submarines," *The Journal of Navigation* 17 (May 1984), 155–79.
———, "Ocean and Sea-Ice Research in the Arctic Ocean via Submarine," *Trans. New York Academy of Sciences,* series 2, 23, (June 1961), 662–74.
———, "The Polar Submarine and Navigation in the Arctic Ocean," reissue of final report (no. 88), U.S. Navy Electronics Laboratory, San Diego, 21 May 1959.
Macdonald, Robie, "Awakenings in the Arctic," *Nature* 380 (28 March 1996), 286–87.
Manley, T. O., L. A. Codispoti, Kenneth L. Hunkins, H. R. Jackson, E. P. Jones, V. Lee, S. Moore, J. Morison, T. T. Packard, and P. Wadhams, "The Fram 3 Expedition," *Eos Transactions American Geophysical Union* 63 (31 August 1982), 627–36.
Mark, Eduard, "Revolution by Degrees: Stalin's National-Front Strategy for Europe, 1941–1947," Cold War International History Project, Working Papers Series, no. 31 (February 2001), Woodrow Wilson International Center for Scholars, Washington, D.C.
Mellen, R. H., "Underwater Sound in the Arctic Ocean," in Sater, *Arctic Drifting Stations*, 419–26.
Mitchell, John F. B., "The 'Greenhouse' Effect and Climate Change," *Reviews of Geophysics* 27 (February 1989), 115–39.
Moritz, Richard E., Cecilia M. Bitz, and Eric J. Steig, "Dynamics of Recent Climate Change in the Arctic," *Science* 297 (30 August 2002), 1497–1502.
Nature 441 (1 June 2006), 426 (4 December 2003), 412 (16 August 2001), 410 (19 April 2001).
1992 Arctic Summer West Scientific Party, "Cruise to the Chukchi Borderland, Arctic Ocean," *Eos Transactions American Geophysical Union* 74 (1 June 1993), 249, 253–54.
Normile, Dennis, and Richard A. Kerr, "A Sea Change in Ocean Drilling," *Science* 300, no. 5618 (18 April 2003), 410–12.
Norwegian Research Council for Science and the Humanities, Oslo, "Nansen Centennial Arctic Program: Plan for a Scientific Expedition to the Arctic Ocean, 1993–1995," May 1991.
Office of the Chief of Naval Research, "Arctic Sciences Program Summary FY89," report OCNR 1125AR90-13, Office of Naval Research High Latitude Dynamics Research Summaries, 1992, 1995.
Pechatnov, Vladimir O., "The Big Three After World War II: New Documents on Soviet Thinking About Post War Relations With the United States and Great Britain," Cold

War International History Project Working Papers Series, no. 13 (July 1995), Woodrow Wilson International Center for Scholars, Washington, DC.

Pharand, Donat, "State Jurisdiction Over Ice Island T-3: The Escamilla Case," *Arctic* 24 (June 1991), 82–89, 139.

Plouff, Donald, "Gravity Measurements in the Beaufort Sea Area," *Arctic* 17 (September 1964), 151–61.

———, G. V. Keller, F. C. Frischkecht, and R. R. Wahl, "Geophysical Studies on IGY Drifting Station Bravo, T-3, 1958 to 1959," in *Geology of the Arctic,* Toronto: University of Toronto Press, 1961, 709–16.

Quam, L. O., "Station Charlie," in Sater, *Arctic Drifting Stations,* 17–21.

Reed, Richard J., and William J. Campbell, "The Equilibrium Drift of Ice Station Alpha," *Journal Geophysical Research* 67 (Jan. 1962), 281–97.

Reilly, John, Peter H. Stone, Chris E. Forest, Mort D. Webster, Henry D. Jacoby, and Ronald D. Prinn, "Uncertainty and Climate Assessments," *Science* 293 (20 July 2001), 430–33.

Rignot, Eric, and Robert H. Thomas, "Mass Balance of Polar Ice Sheets," *Science* 297 (30 August 2002), 1502–6.

Roederer, Juan G., "After Gorbachev: Science in the Former Soviet Union," *Eos Transactions American Geophysical Union* 73 (1 September 1992), 369, 379.

———, "Effects of Glasnost and Perestroika on the Soviet Scientific Establishment: Relevance to Arctic Research," background report no. 3, U.S. Arctic Research Commission, March 1991.

———, "Understanding the Arctic: Research Policies and Responsibilities," (chapter one) *Pollution of the Arctic Atmosphere,* edited by W. T. Sturges, London and New York: Elsevier Science Publishers, 1991.

Safon, Teresa, D., "Stalinist Myths and the Saga of the Arctic Explorers," paper presented at the Southern Slavic Conference, Jacksonville, FL, 1991.

Sater, John E., coordinator, *Arctic Drifting Stations: A Report on Activities Supported by the Office of Naval Research.* Proceedings of the Symposium, Warrenton, VA, 12-15 April 1966 under the auspices of the Arctic Institute of North America and the Office of Naval Research, November 1968.

Schindler, John F., "The Impact of Ice Islands—The Story of ARLIS II and Fletcher's Ice Island, T-3, Since 1962," in Sater, *Arctic Drifting Stations,* 49–78.

Science 311 (24 March 2006), 309 (16 September 2005), 303 (5 March 2004), 426 (4 December 2003), 292 (11 May 2001), 286 (3 December 1999), 284 (9 April 1999), 281 (10 July 1998).

Severinghaus, Jeffrey P., and Edward J. Brook, "Abrupt Climate Change at the End of the Last Glacial Period Inferred From Trapped Air in Polar Ice," *Science* 286 (29 Oct. 1999), 930–34.

Shindell, Drew, "Whither Arctic Climate?" *Science* 299 (10 January 2003), 215.

Smith, David D., "Ice Lithologies and Structure of Ice Island ARLIS II," *Journal of Glaciology* 5 (February 1964), 17–38.

Smith, Gordon W., "Ice Islands in Arctic Waters," Department of Indian Affairs and Northern Development, Ottawa, Canada, 15 May 1980.

Smith, Walker O. Jr., and H. J. Niebauer, "Interactions Between Biological and Physical Processes in Arctic Seas: Investigations Using Numerical Models," *Reviews of Geophysics* 31 (May 1993), 189–209.

Trenberth, Kevin E., and Bette L. Otto-Bliesner, "Toward Integrated Reconstruction of Past Climates," *Science* 300 (25 April 2003), 589–91.

Treshnikov, A. F., "The Soviet Drifting Station SP-3, 1954–55," *Polar Record* 8 (September 1956), 222–29.

Untersteiner, Norbert, K. L. Hunkins, B. M. Buck, "Arctic Science: Current Knowledge and Future Thrust," *Science Technology and the Modern Navy: Thirtieth Anniversary, 1946–1976*, ed., E. I. Salkovitz (Arlington, VA: Department of the Navy, Office of Naval Research, 1976).

Van Allen, James A. "Early Days of Space Science," *Journal of Interplanetary Society* 41 (1988), 11–15.

———, "Genesis of the International Geophysical Year," *Eos Transactions American Geophysical Union* 64 (13 December 1983), 977–78.

———, "Scientific Value of the Earth Satellite Program," *Proceedings of the I.R.E.* 44 (1956), 764–67.

———, "What Is a Space Scientist? An Autobiographical Example," *Annual Review of Earth Planet Science* 18 (1990), 1–26.

Vaughn, David G., Gareth J. Marshall, William M. Connolley, John C. King, Robert Mulvaney, "Devil in the Detail," *Science* 293 (7 September 2001), 1777–9.

Verrall, R., and D. Bade, "Design and Construction of Prefabricated Plywood Buildings for Use in the High Arctic," Research and Development Branch, Department of National Defence, Canada, September 1983.

Vinnikov, Konstantin Y., Alan Robock, Ronald J. Stouffer, John E. Walsh, Claire L. Parkinson, Donald J. Cavalieri, John F. B. Mitchell, Donald Garrett, Victor F. Zakharov, "Global Warming and Northern Hemisphere Sea Ice Extent." *Science* 286 (3 December 1999), 1934–37.

Weart, Spencer, "International Cooperation," in Weart, The Discovery of Global Warming, http://www.aip.org/history/climate, of August 2001 text (record copy at American Institute of Physics, College Park, MD.

———, "Ocean Currents and Climate," in Weart, The Discovery of Global Warming.

Weber, J. R., "Maps of the Arctic Basin Sea Floor: Bathymetry and Gravity of the Alpha Ridge: The 1983 CESAR Expedition," *Arctic* 40 (March 1987), 1–15.

Wen, T., W. J. Felton, J. C. Luby, W. L. J. Fox, and K. L. Kientz, "Environmental Measurements in the Beaufort Sea, Spring 1988," technical report APL-UW 8822, Applied Physics Laboratory, Seattle, WA, March 1989.

Worthington, L. V., "Oceanographic Results of Project Skijump I and Skijump II in the Polar Sea, 1951–1952," *Transactions American Geophysical Union* 34 (August 1953), 543–51.

Other Periodicals, Technical/Nontechnical

Barr, William, "Imperial Russia's Pioneers in Arctic Aviation," *Arctic* 38 (September 1985), 219–30.

Boyle, Richard, "Arctic Passages of North America," *U.S. Naval Institute Proceedings* 95 (January 1969), 48–55.

Brewer, Max C., "The Soviet Drifting Ice Station NORTH-67," *Arctic* 20 (December 1967), 263–65.

Britton, M. E. "US Office of Naval Research Arctic Research Laboratory, Point Barrow, Alaska," *Polar Record* 13 (January 1967), 421–23.

Broecker, Wallace S., "Glaciers That Speak in Tongues," *Natural History* 110 (October 2001), 60–68.

Cadwalader, John, Captain, U.S. Naval Research, "Arctic Drift Stations," *U.S. Naval Institute Proceedings* 89 (April 1963), 67–75.

Carr, James, "The Ultimate Ice Ship," *Wooden Boat* 85 (November/December 1988), 44–51.

Christie, R. L., "The Polar Continental Shelf Project—A Scientist's Friend in the Arctic," *Geos* 17 (Summer 1988), 13–16.

Descheneau, Timothy E., "The New Central Front," U.S. Naval Institute *Proceedings*, 103/9/895 (September 1977), 38–46.
Everett, A. Long, "Has the Levanevsky Mystery Been Solved?" *Air&Space Smithsonian* 2 (December 1987/January 1988), 52–53.
Falconer, Robin, "Fram: Second Drifting Research Station," *Geos* (Summer 1980), 1–2.
Finnie, Richard, "Flying Beyond Sixty," (parts 1, 2, 3), *Canadian Aviation* 12 (February, March, April 1939).
Fletcher, Joseph O., Lieutenant Colonel, U.S. Air Force, "Three Months on an Arctic Ice Island," *National Geographic* 103 (April 1953), 489–504.
Freedman, Lawrence, "Order and Disorder in the New World," *Foreign Affairs* 71 (1992), 20–37.
Frey, Darcy, "George Divoky's Planet," *New York Times Magazine*, 6 January 2002, 24–33, 47–48, 54–55.
Gordienko, P. A., "The Arctic Ocean," *Scientific American* 204 (May 1961), 88–102.
Hardesty, Von, "Soviets Blaze Sky Trail Over Top of World," *Air&Space Smithsonian* 2 (December 1987/January 1988), 48–54.
Holmquist, C. O., Rear Admiral, U.S. Navy, "The T-3 Incident," *U.S. Naval Institute Proceedings* 98 (September 1972), 44–53.
Iseman, Peter A., "Lifting the Ice Curtain," *New York Times Magazine*, 23 October 1988, 48–51, 59–62.
Jackson, Lawrence, "Ice Island: An Ideal Platform," *Canadian Geographic*, December 1988/January 1989, unpaginated reprint.
Johansen, Herbert O., "Seven Weeks on an Ice Island," *Popular Science Monthly*, 171.
Kielhorn, William V., "Skijump II," U.S. Naval Institute *Proceedings*, 80/10/620 (October 1954), 1122–30.
Klemmer, Harvey, "Lend-Lease and the Russian Victory," *National Geographic* 88 (October 1945), 499–512.
Koby, Victor, "SAC–CONAD," *Canadian Aviation* 29 (February 1956), 36–40.
Kozyrev, Andrei, "Russia: A Chance for Survival," *Foreign Affairs* 71 (Spring 1992), 1–16.
LaForest, T. J., Captain, U.S. Navy, "Strategic Significance of the Northern Sea Route," U.S. Naval Institute *Proceedings* 93 (December 1967), 56–65.
Lore, David, "Polar Dreams, Polar Nightmares: The Quests of Lincoln Ellsworth," *Timeline* 5 (December 1988–January 1989), 14–27.
Lyon, Waldo K., "Submarine Combat in the Ice," U.S. Naval Institute *Proceedings*, 118/2/1068, (February 1992), 33–40.
Mandelbaum, Michael, "Coup de Grace: The End of the Soviet Union," *Foreign Affairs* 71 (1992), 164–183.
Miller, Maynard M., "Beyond The Pole," *Appalachia*, (June 1954), 10–15.
———, "Floating Islands," *Natural History Magazine* 65 (1956), 233–39, 274, 276.
———, "Vanishing Lands of The Polar Sea," *Science World* 2 (3 December 1957), 4–7.
Monastersky, "Sea Change in the Arctic," *Science News*, 155, no. 7 (13 February 1999), 104-106.
Morgan, Lael, "The Aleutians: Alaska's Far-Out Islands," *National Geographic* 164 (September 1983), 336–63.
Neil, Robert, "Ice Islands: Drifting Stations in the Arctic," *All Hands*, (December 1974), 18–23.
Polar Record, volumes 8 through 13, 1950–1970, various notes, reviews, field work (excluding articles).
Polarnie Novosti (Polar News), 1991–1992, courtesy Boris I. Silkin, Ph.D., Russian Academy of Science.

Polmar, Norman, "Submarines in the Ice," U.S. Naval Institute *Proceedings,* 117/8/1062 (August 1991), 105–6.
Roberts, Brian, and Terence Armstrong, "The Arctic Institute, Leningrad," *Polar Record* 8 (January 1957), 306–16.
Ronhovde, Andreas G., "Jurisdiction Over Ice Islands: The Escamilla Case in Retrospect," Arctic Institute of North America, November 1972.
Rose, D.C., "Ozone Over the North Pole," *Arctic Circular* 6, no. 5, 57.
Ryback, Timothy W., "Letters From the Dead," *The New Yorker,* 1 February 1993.
Silk, George, "Life Near the Pole [photographic essay]," *Life* 32, no. 18, (5 May 1952), 60–65.
———, "North Pole 103 Miles: Temperature 60 Below," *Life* 32, no. 13 (31 March 1952), 13–17.
Simes, Dimitri K., "America and the Post-Soviet Republics," *Foreign Affairs* 71, no. 3 (Summer 1992), 73–89.
Smith, Charles L., "A Comparison of Soviet and American Drifting Ice Stations," *Polar Record* 15, no. 99 (1971), 877–885.
"Soviet Aviation Today," *Flight,* 31, (3 June 1937).
Sullivan, Walter, "Adrift on the Bottom of the World," *New York Times Magazine,* 1 November 1992, 34–38, 47–48, 51, 53.
Thomas, Lowell, Jr., "Scientists Ride Ice Islands on Arctic Odysseys," *National Geographic* 128 (November 1965), 670–91.
Tierney, John, "Going Where a Lot of Other Dudes With Really Great Equipment Have Gone Before," *New York Times Magazine,* 26 July 1998, 18–23, 33–34, 46–48.
Treadwell, T. K., Jr., Captain, U.S. Navy, "Soviet Oceanography Today," U.S. Naval Institute *Proceedings* 91 (May 1965), 26–37.
Turnbull, Jim, "Coast Guard Patrols Iceberg Alley," *Naval Aviation News,* January–February 2001.
Wanniski, Jude, "The Future of Russian Capitalism," *Foreign Affairs* 71 (Spring 1992), 17–25.
Weber, J. R., "Exploring the Arctic Seafloor," *Geos,* Summer 1980, 2–7.
———, "Maps of the Arctic Basin Sea Floor: Bathymetry and Gravity of the Alpha Ridge: The 1983 CESAR Expedition," *Arctic* 40 (March 1987), 1–15.
Wells, R. D., Lieutenant Commander, U.S. Navy, "Surveying the Eurasian Arctic," U.S. Naval Institute *Proceedings* 92 (October 1966), 79–85.
———, U.S. Naval Reserves, "The Naval Arctic Research Laboratory," U.S. Naval Institute *Proceedings* 95 (September 1969), 39–45.

OTHER DOCUMENTS AND SOURCES

Alaskan Air Command, *Ice Islands of the Arctic: Alaskan Air Command's Arctic Experience,* historical monograph, n.d.
Benedict, Hazard C., "Arctic Studies 1950–1952 (Barter Island and T-3" and "Arctic Studies 1953–1955 (T-3 and Ellesmere)," bound scrapbooks with photographs, courtesy Mr. Benedict.
Buck, Beaumont M., "Spring to the Ice: A Memoir," privately published, 2004.
Energy, Mines, and Resources Canada, "Islands in the Midnight Sun: The Story of the Polar Continental Shelf Project," 1974.
Files, Polar Continental Shelf Project, Energy, Mines and Resources Canada, Ottawa, courtesy Mr. George Hobson (former director) and Bonni Hrycyk, director.
Goldstein, Norman, "Scientific Activities of GRD On Fletcher's Ice Island, T-3," progress report no. 6 (10 October 1955–14 May 1954), progress report no. 8 (21 June 1955–24

September 1955) and unnumbered report (7 March 1957–6 June 1957), courtesy Mr. Goldstein.

Greenaway, Keith R., memorandum to Aurora Institute, 21 March 2000, courtesy Brig. Gen. Greenaway, RCAF (Ret.).

Hare, F. Kenneth, discussion paper in "Science and the North: A Seminar on Guidelines for Scientific Activities in Northern Canada 1972," publication no. QS-1330-000-EE-A-1, The Subcommittee on Science and Technology/Advisory Committee on Northern Development, Ottawa, Canada, 1973.

Indian and Northern Affairs Canada, "Guide for Expeditions to the Canadian Arctic Islands," 3rd ed. (August 1985).

Miller, Maynard M., Captain, U.S. Navy, "Observations on Selected Problems Relating to Fleet Readiness for Arctic Operations and the Question of Research Relevancy," prepared for OCEANAV and CNR, July 1974.

Monastersky, Richard, "Sea Change in the Arctic," *Science News* 155 (13 February 1999), 104–6.

National Academy of Sciences, Polar Research Board, Committee for the Nansen Drift Station, *Scientific Plan for the Proposed Nansen Drift Station,* Washington, D.C., 1976, courtesy Kenneth L. Hunkins, Ph.D.

National Science Foundation, Arlington, VA, *Arctic Research of the United States,* 1998–2202, various issues.

Tore O. Vorren, Ola M. Johannessen, Egil Sakshaug, Torgny Vinje, eds., Nansen Centennial Arctic Programme, Plan for a Scientific Expedition to the Arctic Ocean, 1993–1995, Norwegian Research Council for Science and the Humanities, Oslo, May 1991.

USL *Echo,* U.S. Navy Underwater Sound Laboratory (now Naval Underwater Systems Center), New London, Connecticut, various issues, 1958–1960, courtesy Kathleen P. O'Beirne.

Ward, Edward M., Commander, U.S. Navy (Ret) "Project Ski-Jump 1 and 2, Arctic Oceanographic Research Expedition, 1951–1952," memoir, 12 pp., courtesy Commander Ward.

Weber, J. R., D. A. Forsyth, A. S. Judge, and Ruth Jackson, "A Geoscience Program for the Canadian Arctic Ice Island Research Project," Department of Energy, Mines and Resources/Geological Survey of Canada, May 1984.

NEWSPAPERS

Air Force Times
Anchorage Daily Times
Boston Globe
Calgary Herald (Alberta)
Chicago Tribune
Daily News-Miner (Fairbanks)
Globe and Mail (Toronto)
International Herald Tribune
New York Times
The Christian Science Monitor
The National Observer
The Ottawa Citizen
The Thule Times (Greenland)
Toronto Star
Winnipeg Free Press

INDEX

acoustic research
   ALPHA and, 132–134
   ARLIS V and, 185–186
   Fram III and, 213–214
   ice island T-3 and, 146–148, 161, 170, 312n64
   list of U. S. ice stations involved in, 271–274
   MIZPAC and, 187
   short-term bivouacs (GMIS) and, 170–172
Admiral Tegetthoff, 290n25
AIDJEX (Arctic Ice Dynamics Joint Experiment), 194, 322n94
   establishment of, 193
   ice fractures at, 197–198
   participants in, 191–192
   research at, 186–188, 191–199
aircraft. *See* aviation
Akademik Federov, 245
Akpik, Frank, 145, 220
Alderfer, Cliff, 173–174
Alexeyev, Anatoly, 44
ALPHA
   airlift support for, 154–155
   established, 114–116, 128
   evacuated, 137
   life at, 120–121, 123–127, 131
   moved, 130–132
   personnel at, 117, 132, 269–270
   research at, *118*, *119*, 132, 137–138, 308n11
   Skate visits, 134–137, *136*
Amudsen, Roald, 13
Anderson, Neil, 180

APLIS, 187, 201, 213, 214, 231–232
Applied Physics Laboratory (APL), 187
Archer, Colin, 8
Arctic and Antarctic Research Institute (AARI), 17, 259, 263
Arctic Ocean/Arctic region. *See also* North Pole
   acoustics and, 171
   debate about feasibility of landing aircraft in, 35–37
   lack of data about, 33–35
   pollution of, 217, 326nn13, 14
   research needed today, 255–256
   seasonal changes in, 1–3
Arctic Research Laboratory (ARL), 74, 319n61
Arktika, 202, 324n114
ARLIS (Arctic Research Laboratory Ice Station). *See also* ARLIS II
   ARLIS I, 155–157
   ARLIS IV, 177–178
   ARLIS V, 185–186
   ARLIS VI, 185–186
ARLIS II
   established, 157–160, 316n6
   evacuation of, 176–177, 319n52
   life at, 126, 160–163, 174–175
   research at, 172–173, 175–176, 275–276
   Russian territory and, 168, 169
Artiki Musee, 263
Aviation. *See also Sever* airlift
   aircraft adapted for polar landing, 36–38, 40, 44

349

best ice for landing, 97
charter rates in 1986, 221, 222, 228
debate about feasibility of landing in Arctic region, 35–37
difficulties of, 123, 310n32
early Russian, 15, 23–28
importance of, 74, 120
long range Arctic probes before World War II, 63–64
smaller planes utilized, 150–151
Twin Otter planes and, 222, 228
U.S Ptarmignan missions, 67–68, 299n9

Babushkin, Mikhail, 40, 43
Baidukov, Georgiy, 27
Beck, John, 158, 162–163
Bedrick, C., 105
Belyakov, Alexander, 27
Belyakov, L., 173
Benedict, Hazard C., 103
Benedict, William "Wild Bill," 92–93
Bennett, James Gordon Jr., 6
Bennington, K., 156
Bering, Vitus, 5
Bilotta, Joseph P., 121, 131, 134–135, 137
Bitters, John, 220
Black, Jan, 173–174
Blair, Charles, 318n43
Blinov, N. I., 149
Boarfish, 134
bouys. *See* AIDJEX
Braun, Don, 180
BRAVO. *See* ice island T-3
Brewer, Max, 150–151, 162–163
 ARLIS I and, 156, 157
 ARLIS II and, 158, 169
 CHARLIE and, 139
Brinegan, Michael, 86
Britton, Maxwell E., 68, 74, 107, 150–151
 ARLIS II and, 176–177
 CHARLIE and, 144
 on Crary, 89
 on Fletcher, 90
 on ice island T-3 reoccupation, 166
 on SCICEX, 246
Browne, Irene, 104, 317n25
Bryazgin, N. N., 169
Buck, Beaumont M., 151, 204
 on acoustic research, 170
 on *Fram* stations, 325n5
 ice island T-3 and, 145
 MIZPAC and, 187
 SP-28 and, 237
Bulatov, L. V., 180

Burroughs, John, 263
Burton Island, 155–156, 200
Bushnell, Vivian, 317n25
Buzuyev, A., 175
Byrd, Robert E., 48

Cabaniss, Gerry, 95, 123, 128–129, 130
Calvert, James, 134–146
Canada, 4, 186
 AIDJEX and, 192
 ice island T-3 shooting incident and, 188–190
 list of ice stations of, 267
 research today, 260–261
CEAREX (Coordinated Eastern Arctic Experiment), 232–235
CESAR 83 (Canadian Expedition to Study the Alpha Ridge), 214–217
CHARLIE
 established, 138–139
 evacuated, 143–145
 life at, 142, 145
 research at, 140, 145, 308n11
Chelyuskin, 20–23, 22, 28–31, 30
Cherevichny (airman), 60, 63, 97
Chernyshev, A., 227
Chilingarov, A., 184–185
Chkalov, Valeriy, 27, 36, 48
Chukhnovsky, B. G., 24
climate change, 153, 209–210, 253–254
 AIDJEX and, 186–188, 191–199
 global warming consensus, 257–258, 332nn2, 5
 Intergovernmental Panel on Climate Change (IPCC), 235–236, 248, 254, 257–258
 ozone, 217, 326n13
 research needed today, 256
 water currents and, 2–3, 33–34, 46, 289n4, 294n3
Coley, Vernon "Jack," 76, 77–78
Collin, Arthur, 128
Conlon, Dennis, 246, 251
Cotell, Robert D., 76, 92–93, 317n25
 ice island T-3 and, 89, 92, 94
Crary, Albert P., 75, 76, 92–93
 ice island T-3 and, 89, 91, 92, 103, 104
CRYSTAL, 237
Cunningham, Thomas, 114, 308n14

Davies, Ron, 38
DeGoes, Louis, 317n25
DeLong, George W., 6–7
Dench, Ron, 112

Derrickson, R., 94
Des Groseillers, 249–251
Dorsey, H. G. Jr., 94, 305n85
Dubovtsev, V. F., 179
Dzerdzeyevsky, Boris, 40

Edisto, 177
Eielson, Carl, 33, 35
English, Thomas, 117, 135
Erhart, Lewis, 84–85, 89
Escamilla, Mario, 188–190
Evans, Griffith C. Jr., 156
Explorer I, 121

Feathers, Edward, 141, 142–143
Fedorov, Yevgeny
    North Pole-1 and, 34, 39, 40–42, 43, 45–46, 49, 56, 57, 61–62
    on *Sever* airlift, 98
    SP-19 and, 195–196
Fendley, M., 138
Fischer, Robert
    ARLIS II and, 158
    CHARLIE and, 139
    North Pole landing and, 173–174
    SP-11 and, 169
Fletcher, Joseph
    AIDJEX and, 193
    ALPHA and, 114
    ice island T-3 and, 75, 76, 84–89, 94
    North Pole landing and, 92–93
    Project Icicle and, 79–80
    on Russian polar operations, 98, 102
    on SP-2, 72
Fram, 9, 200
    design of, 8
    life aboard, 10–11
    nuclear version proposed in 1968, 182
    voyage of, 8–13
Fram series, 199–207, 325n5
    Fram I, 201–207, 324n117, 324nn119, 120, 121
    Fram II, 212–213
    Fram III, 203–204, 213–214
Francois, Robert E., 111, 132, 168, 187
Freeman, Richard E., 114
Frolov, Ivan, 228–229

Garrett, Pat, 114
Germany, 290n25
Giles, James J., 128
GMIS I and II (General Motors Ice Stations), 172
Godden, James, 243

Goldstein, Norman, 89, 103, 104, 105, 113
Golovin, Pavel, 40
Goodall, Clifford E., 103
Gorbachev, Mikhail, 222, 237, 246
Gordienko, P. A., 62, 100, 149, 184
Gordon, Arnold, 245
Greely, Adolphus, 7
Green, Paul L., 89, 92
Greenaway, Keith, 69
Gromov, Mikhail, 28
Gudkovich, Zalman, 69, 70, 73

Hale, Robert D., 92
Hanson, Arnold, 117, 125, 129, 130, 135, 173–174
Hattersley-Smith, Geoffrey, 93
Hawkbill, 252
Healy, 324n116
Heiberg, Andy, 233, 235
Hobson, George D., 218–219, 222, 225–226, 227, 232, 243–244, 323n98
Hobson's Choice, 218–219, 220–228, 232, 238, 243–244
    life at, 237–240
    research at, 223
Holmes, John F., 75, 76
Holmquist, C. O., 188–190
Hopson, Eddie, 145
Horvath, Charles, 95, 103, 105
Hunkins, Kenneth L.
    AIDJEX and, 193
    ALPHA and, 117, 131, 135, 137–138
    CHARLIE and, 145
    on *Fram* series, 200, 203–207
    on Hobson's Choice, 243
    on today's research, 255, 260

ice islands. *See also* Hobson's Choice; ice island T-3
    origins of, 93, 304n82
    T-1, 69, 83, 300n16
    T-2, 79
ice island T-3, 83–96, 103–105, 123, 166, 173
    camp established on, 84–90
    deaths at, 174, 188–190
    drift track of, 287
    evacuated, 149, 190–191
    expanded, 91–92, 304n76
    found and reoccupied, 165–167
    International Geophysical Year and, 112–113, 127–128
    last confirmed position of, 219–220
    life at, 95–96, 104, 105, 121, 126–129, 132, 142–143, 145–146, 174–175, 305n88

programs suspended, 96
research at, 90–91, 93, 103–104, 140–142, 145–148, 161, 167, 170, 181, 304n73, 74, 308n11, 312n55
shooting incident at, 188–190
sortie to Pole from, 92–93
spotted from air, 79
temperature data for, 122
Ice Station Weddell Sea, 244–246
Intergovernmental Panel on Climate Change (IPCC), 235–236, 248, 254, 257–258
International Geophysical Year (IGY)
genesis and objectives of, 109–110
Russia and, 107, 110–111, 116–119, 307n114
United States and, 111–112, 116–117, 119–120, 307n114, 308nn19, 20, 21
International Polar Year (IPY), 18, 109, 258
Ippolitov, Valery, 242
Ivanov (radioman), 41–42

Jackson, Frederick, 12
Jeanette, 6–7
Johansen, F. H., 12
Johns, Robert, 138
Johnson, Carl, 177

Kalinin, M. I., 61
Kamanin, Nikolai, 31
Kelley, Elton, 132, 134–135, 136–137, 145
Kelley, John, 319n61
Kennan, George, 70
Khrushchev, Nikita, 117
Knutsen, Willie, 112
Koenig, Lawrence S., 79–80, 83
Konstantinov, Yu. B., 179
Kornilov, N. A., 164
Koshelev, Vladimir, 262
Kotov, I. S., 97
Krenkel, Ernst, 18, 34
on *Chelyuskin*, 21, 22–23, 29, 31
on International Polar Year, 109
Matochkin Shar and, 15–16
North Pole-1 and, 39, 40, 41–42, 48, 52, 53, 55, 57, 58, 60
on *Sibiryakov*, 19–20
Krochak, Tom, 222, 223, 237, 239–240

LeCloirec, Alain, 220
Lenin, 156, 163–164, 202
LeShack, Leonard A., 159
Levanevsky, Sigismund, 27, 28, 54
Lewis, Roger, 145
Lightsy, Bernie P., 188

Lindbergh, Charles, 292n47
LOREX (Lomonosov Ridge Experiment), 181, 205, 210–211, 212
Lukin, Valery, 230
Lyapidevsky, Anatoly, 29
Lyon, Waldo K., 111, 132, 134, 187, 253

Main, Robert, 139
Makarov, Stan Osipovich, 14
Makhotkin, V. M., 38
Makurin, N. V., 196
Malloy, Arthur D., 136, 137
Maloney, Dave, 239
Matochkin Shar, 15–16
Mazuruk, Ilya, 44–45, 97
McComas, William, 159
McGregor, Ronald, 107, 171, 176–177, 192
Mellen, Robert H., 145
Miller, Maynard, 83, 84
Milner, Carl T., 143, 145–147, 168
MIZPAC (Marginal Ice Zone-Pacific), 187, 322n94
Molokov, Vassily, 31, 44
Moorhead, D. L., 75
Morozov, V., 196
Murman, 59–60
Murmanets, 57–58
Murphy, Bob, 193

Nagursky, Ivan Josifovich, 24
Nansen, Fridjtof, 4
Fram and, 8–13
influence on Arctic oceanography, 13, 290n24
proposes drifting station, 7–8, 33, 35, 294n1
on Russian explorations, 5
Nathan B. Palmer, 245
Nautilus, 111, 134
Newton, George, 247
Nicholson, John, 134–135, 147–148
Nickerson, Norvel E., 177
Nikitin, M. M., 107
North Pole
aircraft landing at, 173–174
Arktika reaches, 202
"race" to, 13, 290n25
sortie from ice island T-3 to, 92–93
North Pole-1 expedition, 33–64
drift of, 49, 51, 55–59
established, 35–45, 295nn19, 22, 27
life at, 50–55, 296nn36, 38
media coverage of, 42–43, 47–48, 53, 55, 56, 60–61

rescue from, 57–60
research at, 45–50, 62–63
Northwind, 181, 187, 204
Norway, list of ice stations of, 267–268

Office of Naval Research (U.S.), 74
Old, William D., 85–86, 303nn58, 59
OPAL 88, 232
Ostenso, Ned, 10
Ostrekin, M. Ye., 106

Panov, V. V., 179
Papanin, Ivan
  described, 38–39
  North Pole-1 and, 39, 42, 43, 45, 47–63
Pargo, 247
Parry, Sir Edward, 3–4
Paul I, emperor of Russia, 5
Peary, Robert, 4, 13, 35, 289n8
Perovich, Donald K., 248–251
Peter the Great, czar of Russia, 5
Petrovsky, T. V., 262
Pharand, Donat, 190
Plouff, Donald, 129
Plummer, Charles "Carlos," 128, 129
Polarbjorn, 232–233
Polar Continental Shelf Project, 180–181. *See also* Hobson's Choice
  charter aircraft rates, 221
  occupancy in 1987, 229
  regulations of, 279–282
Polar Research Laboratory, 187–188, 236–237
Porter, Dick, 237
Priamikov, Sergey, 246
Project Icicle Field, 79–80, 82–83. *See also* ice island T-3
Ptarmigan missions, 67–68, 80, 299n9

Quam, Louis, 138, 144

Reynolds, Bert, 143
Rigby, Bruce, 252
Ritsly, Alexei, 44
Rodahl, Kaare, 85–88, 93, 304n82
Rogachev, V. M., 175
Romanov, Ilya, 154, 320n74
Roots, Fred, 181
Russia
  Arctic lands of, 4, 13–14, 15, 17, 25
  Arktika reaches North Pole, 202
  Artiki Musee in, 263
  Chelyuskin's voyage and crew's rescue, 20–23, 28–31, *30*
  climate research in cooperation with U.S., 186–187
  Cold War politics and exploration, 65–67, 167–169, 199, 299nn5, 6, 7, 11
  collapse of Soviet Union, 222, 237, 241–242, 246
  creates Committee for the Northern Sea Route, 16–18
  current economic situation and research, 260–263
  early Arctic explorations, 4–5, 14–18
  early polar aviation, 15, 23–28
  International Geophysical Year and, 107, 110–111, 116–119, 307n114
  list of ice stations of, 265–266
  releases bathymetric map, 300n20
  Sibiryakov's voyage, 18–20
  space program of, 117, 119

Sargo, 143–144, 146–148
satellite imaging, 223–225
SCAMP (Seafloor Characterization and Mapping Pod), 252
Schindler, John. F.
  ARLIS II and, 158, 159, 176, 177
  in crash landing, 162–163, 317n20
Schmidt, Michael, 227
SCICEX (Science Exercise), 246–247, 253
Scientific Results of Drifting Stations, 167
Sedov, 200, 297n45
Serikov, Mikhail, 322n88
Sever airlifts, 69–71, 96–98, 300n21. *See also Severnyy Polyus* stations
Severnyy Polyus stations, 69, 71, 184
  dangers of ice break up, 99–100
  drift tracks of, 286
  International Geophysical Year and, 113–114
  life at, 124–127, 149
  research of, 99, 102
  SP-2, 71–74, 80–83, 96, *100*
  SP-3, 98, 99, 100, 102
  SP-4, 98, 99, 100–102, 106, 116
  SP-5, 99, 102, 106
  SP-6, 100, 106, 113, 142, 286
  SP-7, 106, 107, 114, 116, 126, 142, 286
  SP-8, 99, 100, 142, 148–149, 154, 164–165, 168, 286, 320n75
  SP-9, 106, 149, 154, 320n75
  SP-10, 107, 154, 156, 163–164, 169, 173, 175
  SP-11, 169, 173
  SP-12, 175
  SP-13, 175, 179–180
  SP-14, 179

SP-15, 179–180, 181
SP-16, 180, 181
SP-17, 121, 180, 181
SP-18, 182, 184–185
SP-19, 121, 182, 184–185, 195–196, 320n67
SP-20, 196
SP-22, 196, 206, 211–212, 277–278
SP-23, 199
SP-24, 204
SP-25, 213–214, 216
SP-26, 217, 220, 226
SP-27, 228
SP-28, 226–227, 237
SP-29, 229–231
SP-30, 231, 240–241
SP-31, 237, 240–243
SP-32, 262
SP-33, 262
SP-34, 262
Sexton, R. F., 94
SHEBA (Surface Heat Budget of the Arctic Ocean), 248–252, 254, 288
Shevchenko, N., 196
Shirshov, Pyotr
  Chelyuskin and, 20
  North Pole-1 and, 39, 45–46, 49, 51, 53, 54, 57, 59
Shmidt, Otto Julievitch, 16–17
  Chelyuskin and, 20–23, 28
  on importance of polar aviation, 26–27
  North Pole-1 and, 34–35, 36, 37, 40, 42, 43, 45, 47, 54, 57, 58, 62
  Sibiryakov and, 18–20
Shoemaker, Brian, 191, 220
Shorey, R. R., 94
Sibir, 202, 228–231, 324n115
Sidorov, V. C., 237
Silk, George, 85–86, 89
Silkin, Boris I., 117–118
Simonov, I. M., 196
Sir John Franklin, 260
Skate, 111, 134–137
SKI-JUMP, 74–79, 84, 302n42
  occupants visit ice island T-3, 91–92
Smith, James F., 139, 144
Somov, M. M., 71
space program
  of Russia, 117, 119
  of U.S., 121, 309n25
Spirin, Ivan, 40–42
Sputnik, 117, 119
Stalin, Josef, 25–26, 28
  North Pole-1 and, 36, 39, 43, 61

Staten Island, 161, 162
Stefansson, Vilhjalmur, 3, 48
Sturgeon, 246
SUBICEX (Submarine Ice Exercise), 170, 324n117
submarines, 111–112, 132–137, 143–144, 146–148. *See also* acoustic research
Superior, 5
Sverdrup, Otto, 10–11, 12–13
Sweeney, Jack, 211

Taimyr, 59–60
Thomas, Lowell, 120–121
Thomas, Lowell Jr., 177
Thompson, Herbert, 93
Tikhonov, Yu. P., 196, 220
Tolstikov, Ye. I., 98
Treshnikov, A. F., 72, 97, 99
Tupolev, Andrei Nikollayevich, 37, 61

UARS, 190, 321n86
Ulyanov (ship's master), 59–60
United States
  climate research in cooperation with Russia, 186–187
  Cold War politics and exploration, 65–67, 167–169, 199, 299nn5, 6, 7, 11, 309n24
  early Arctic explorations, 5–7
  expenditures for Arctic activities (1968), 183
  International Geophysical Year and, 111–112, 116–117, 119–120, 307n114, 308nn19, 20, 21
  list of research budgets, by agency, 283–284
  lists of ice stations of, 266–267, 271–274
  research policies of, 153–154
  space program of, 121, 309n25
Untersteiner, Norbert, 260
  AIDJEX and, 193, 198–199, 323n104
  ALPHA and, 117, 120, 131, 135, 137
  Fram series and, 200

Van Allen, James K., 109, 121, 307n2
Vedernikov, V. A., 114
Vinogradov, Nikolai D., 106, 125, 196, 241, 323n98
Vitus Bering, 231
Vlassov, Gennadi, 60
Vodopyanov, Mikhail Vasilievitch, 23, 31, 98
  North Pole-1 and, 36–37, 38, 40–42, 60, 294n10
Volkov, N. A., 102
Volovich, V. G., 98

von Karman, Theodore, 309n22

Ward, Edward M., 75–77, 79
Ward Hunt Ice Shelf, 218
Weber, Hans, 210, 214
Weddell Sea, ice station at, 244–246
Weigle, Francis, 136–137
Weyprecht, Karl, 290n25
Whitmer, J. R., 76
Wilkins, Sir Hubert, 33, 35

Wilkniss, Peter, 244
Williams, Frederick, 143
Woodward, E. C., 75
Worthington, L. V., 75, 76, 78, 95

Yakovlev, G. N., 71, 81–82, 97
Yeltsin, Boris, 241
Yermak, 14, 59, 60, 231

Zimmerman, Lloyd, 220

ABOUT THE AUTHOR

William F. Althoff enjoys two careers. A geologist by profession, he is also a writer-researcher of military history, particularly U.S. naval aviation, polar aeronautics, and the history of twentieth-century technology. Mr. Althoff has published extensively in both technical and historical journals, and is the recipient of several grants to help support travel and research. Most recently (1999–2000), he was named a Ramsey Fellow in Naval Aviation History at the National Air and Space Museum, Smithsonian Institution. This is his fourth book of history. He resides in west-central New Jersey.